Translation In Eukaryotes

Translation In Eukaryotes

Edited by

Hans Trachsel

Institute für Biochemie und Molekularbiologie
Universität Berne
Berne, Switzerland

CRC Press
Boca Raton   Ann Arbor   Boston   London

Library of Congress Cataloging-in-Publication Data

Cataloging information is on file with the Library of Congress

Developed by Telford Press

Direct all inquiries to CRC Press, Inc., 2000 Corporate Blvd., N.W., Boca Raton, Florida 33431.

© 1991 by CRC Press, Inc.

International Standard Book Number 0-8493-8816-3

Printed in the United States

PREFACE

The synthesis of a protein is a multi-step biochemical pathway in which numerous components such as RNAs, proteins and small molecules cooperate to translate the nucleotide sequence of an RNA into the corresponding amino acid sequence of the protein. Several steps in this pathway are regulated and for a number of genes regulation at the level of translation contributes significantly to overall regulation of their expression.

Translation in eukaryotes has been studied in many laboratories and for more than twenty years biochemical methods were mainly used. These include the preparation of cell-free extracts competent for *in vitro* translation, fractionation of translational components, and reconstitution of translation systems from purified components. More recently, genes encoding translational components were cloned and genetic and molecular genetic methods were introduced to study the mechanism and regulation of eukaryotic translation. This opened the door for studies of translation at the molecular level *in vivo*. Important contributions to our understanding of eukaryotic translation are also more and more often made by investigators studying the expression of their gene of interest and finding themselves studying a translational phenomenon. These developments have led to rapid accumulation of new data and make it increasingly difficult to keep up with the literature.

This multi-author book was written for students, newcomers in the field, and others interested in obtaining overviews of specific aspects of eukaryotic translation in the form of short reviews. (The authors restrict themselves to the description of concepts and main findings and give references for the study of details.) Therefore this book should serve the reader as a guide to the vast amount of literature in this field. In order to keep the chapters short and to limit their content of information, aspects of mechanism of translation (Part I) were separated from aspects of regulation of translation (Part II) and special aspects such as comparison of eukaryotic and prokaryotic translation, mode of action of inhibitors and nomenclature of initiation factors asre treated in Part III.

I would like to thank all the authors for writing their chapters and for making many helpful suggestions. I am specially grateful to my secretary, Mrs. Marianne Berger who did most of the correspondence involved in the preparation of this book.

June 1991
Hans Trachsel

CONTRIBUTORS

J. Bag
*Department of Molecular Biology
   and Genetics
College of Biological Science
University of Guelph
Guelph, Ontario, Canada*

Juan P. G. Ballesta
*Centro de Biología Molecular
Universidad Autonoma de Madrid
Campus de Cantoblanco
Madrid, Spain*

Mike J. Clemens
*Department of Cellular and
   Molecular Sciences
St. George's Hospital
Medical School
London, England*

Richard Giegé
*Laboratoire de Biochimie
Institute de Biologie Moléculaire
   et Cellulaire du CNRS
Strasbourg, France*

Anne-Lise Haenni
*Institute Jacques Monod
Université de Paris VII
Paris, France*

John W. B. Hershey
*Department of Biological
   Chemistry
University of California
Davis, California*

Alan Hinnebusch
*National Institutes of Health
Laboratory of Molecular Genetics
Bethesda, Maryland*

Markus Huembelin
*F. Hoffmann-La Roche, Ltd.
Basel, Switzerland*

Richard J. Jackson
*Department of Biochemistry
University of Cambridge
Cambridge, England*

Jacques Lapointe
*Départment de Biochimie
Faculté des Sciernces et de Génie
Université Laval
Québec, Canada*

Richard D. Klausner
*National Institutes of Health
Cell Biology and Metabolism
   Branch
Bethesda, Maryland*

Karen Meerovitch
*Department of Biochemistry
McGill University
Montreal, Québec, Canada*

Virginia M. Pain
*School of Biological Sciences
University of Sussex
Falmer, Brighton, England*

Jerry Pelletier
*Department of Cancer Research
Massachusetts Institute of
   Technology
Cambridge, Massachusetts*

Robert E. Rhoads
Department of Biochemistry
University of Kentucky
College of Medicine
Lexington, Kentucky

Alexey G. Ryazanov
Institute of Protein Research
Academy of Sciences of the
  U.S.S.R.
Puschino, Moscow Region,
  U.S.S.R.

Brian Safer
Section on Protein Biosynthesis
National Institutes of Health
Bethesda, Maryland

Lawrence I. Slobin
Department of Biochemistry
University of Mississippi
School of Medicine
Jackson, Mississippi

Nahum Sonenberg
Department of Biochemistry and
  McGill Cancer Center
McGill University
Montreal, Québec, Canada

Alexander S. Spirin
Institute of Protein Research
Academy of Sciences of the
  U.S.S.R.
Pushchino, Moscow Region,
  U.S.S.R.

George Thomas
Fredrich Miescher Institut
Basel, Switzerland

Rosaura P. C. Valle
Institut Jacques Monod
Université de Paris VII
Paris, France

Harry O. Voorma
Department of Molecular Cell
  Biology
University of Utrecht
Utrecht, The Netherlands

Ira G. Wool
Department of Biochemistry and
  Molecular Biology
The University of Chicago
Chicago, Illinois

CONTENTS

PART III: APPENDIX

Chapter 16
NOMENCLATURE OF INITIATION, ELONGATION AND TERMINATION FACTORS FOR TRANSLATION IN EUKARYOTES
Brian Safer

Mechanism of Translation

Eukaryotic Ribosomes: Structure, Function, Biogenesis, and Evolution

Ira G. Wool
The Department of Biochemistry and Molecular Biology
The University of Chicago
Chicago, Illinois

1.1. INTRODUCTION

Ribosomes are ribonucleoprotein organelles that link the genotype to the phenotype. They catalyze protein synthesis in all of the organisms in our biosphere; indeed, they are at one and the same time, universal, essential, and complicated. The grand imperative of research on ribosomes is to know their structure so as to be able to account for their function. This has as a prerequisite knowledge of the chemistry of the constituents. A great deal is known about prokaryotic (principally *Escherichia coli*) ribosomes: the primary structures of the nucleic acids[1] and of the proteins,[2] the secondary structures,[1] from comparative sequence analysis,[3, 4] and preliminary proposals for the tertiary structures[5, 6] of the ribosomal RNAs (rRNA); the binding sites for the ribosomal proteins on the rRNAs[5, 7–9] the topography of the proteins from neutron scattering[10] and from immune electron microscopy;[11, 12] and there are even preliminary, low-resolution X-ray crystallographic data for the structure of the ribosomal subunits.[13] Far less is known of eukaryotic ribosomes because their structure is more complicated, because of the difficulty of applying genetic analysis, and because of the lack of a means for reconstituting ribosomal subunits. Nonetheless, progress is being made with ribosomes from yeast and from rat. It is the burden of this review to provide a synthesis of what is known and to demarcate what remains to be determined.

1.2. STRUCTURE OF EUKARYOTIC RIBOSOMES

Importance attaches to obtaining a solution to the structure of ribosomes since knowledge of the structure is believed, with cause, to be essential for a rational, molecular account of the function of the organelle in protein synthesis. For a solution of the structure a requisite is the sequences of nucleotides and of amino acids in the constituent nucleic acids and proteins. An attempt is underway to acquire the chemical data for eukaryotic ribosomes particularly for the particles from yeast[14] and from rat.[15]

The Sequence of Nucleotides and the Secondary Structure of Eukaryotic Ribosomal RNA—The sequences of nucleotides in the four species of RNA—5S,[16] 5.8S,[17] 18S,[18] and 28S[20, 21]—have been determined for rat ribosomes and for a large number of other eukaryotic species.[22] Indeed, the sequence of nucleotides for some 400 16S-like and about 75 23S-like rRNAs have now been determined. The significance is that this library of sequences provides the data for the determination of the secondary structure using comparative sequence analysis.[4] In this procedure, one begins with the assumption that two rRNAs, say *E. coli* 16S and rat 18S, have the same secondary structure. The primary structures are aligned using regions of identity as a guide—this is the critical operation—and putative helices are constructed. Compensated base changes in the nucleotide sequences in the helices is taken as evidence for the structure, whereas, uncompensated changes leading to mismatches are evidence against it. Generally, at least two independent examples of compensated base changes are required to establish the existence of a helix. In principle, the phylogenetic method amounts to having the data from a pre-existing genetic experiment in which all the organisms in the biosphere are considered pseudorevertants of the various mutations in rRNA genes that have occurred during evolution. Most of the structure derived by phylogenetic comparison has now been confirmed by direct experimental test, principally by modification of the rRNA in intact ribosomes with chemical reagents followed by identification of the altered nucleotides by the extension of primers with reverse transcriptase.[23] The conclusion is that all rRNAs can be folded into the same secondary structure. The secondary structures of rat 18S and 5.8S·28S rRNAs[18, 24] are given in Figure 1.

The sequence of amino acids in eukaryotic ribosomal proteins—A commitment has been made to the determination of the sequences of amino acids in all of the proteins in a single mammalian species, the rat. This is an arduous, tedious, time consuming, expensive project. What is more, now that there is in place a feasible plan for accomplishing the chore it lacks excitement and intellectual challenge. How then does one justify this program. The abiding belief is that the data are important, indeed, that all analytical and structural chemistry is laden with theory. Moreover, the data is an absolute requirement for the resolution of the structure. That this information is already available for *E. coli* ribosomes[2] in no way diminishes the importance of the enterprise. Indeed, each of the sets (*E. coli* and rat) is likely to enhance the value of the other just as subsequent sequences of rRNAs (after determination of *E. coli* 16S and 23S rRNAs) were increasingly valuable because they made possible phylogenetic comparisons and, hence, led to the secondary structures of the nucleic acids. One is not so naive as to believe that the determination of the primary sequences will lead inevitably to the structure of the particle or that the structure will tell us inevitably how the organelle functions in protein synthesis. However, it is a near certainty that without the basic chemistry it will be impossible to fully comprehend the structure and to fully account for function. The sequences of amino acids may also help in understanding the evolution of ribosomes, in unraveling the function of the ribosomal proteins, in fathoming the reasons for the discrepancy between the number of ribosomal proteins in *E. coli* which has 52 and the 70 to 80 in eukaryotes, in defining the rules that govern the interaction of the ribosomal proteins and the rRNAs, and in uncovering the amino acid sequences that direct the ribosomal proteins to the nucleolus for assembly on nascent rRNA. Indeed, it is difficult to predict all of the uses for the data and it may be that their greatest value will be for some purpose we cannot envision now.

Eukaryotic ribosomal proteins are relatively small—they range from about 7000 to 37,000 Da; they are nearly all basic—most have pIs in the range of 10 to 12. Eighty-four proteins have been isolated from rat ribosomes.[25] That is more than was expected, indeed, may be more than are in ribosomes; without genetics and without reconstitution it is difficult to be certain that all of them are constituents of the particle. The actual number is thought to be between 70 and 80 and it is unlikely that there are any that have not been characterized.

To date the sequences of amino acids in 35 rat ribosomal proteins have been determined (Table 1-1); 5 were established directly from the proteins, the others were deduced from the sequences of nucleotides in recombinant cDNAs.[15] Since there are 70 to 80 eukaryotic ribosomal proteins the undertaking is about half completed.

For each of the ribosomal proteins and for their cDNAs the following analyses have been carried out:

1. The identification of the ribosomal protein encoded in the open reading frame and the confirmation of the sequence by determining the order of amino acids in a portion of the protein, usually at the NH_2 terminus, often at the carboxyl terminus as well, and sometimes from the sequence in a peptide fragment.
2. The identification of the hexamer, AATAAA, in the 3′ noncoding sequence that directs posttranscriptional cleavage-polyadenylation of pre-mRNA.[26]

A

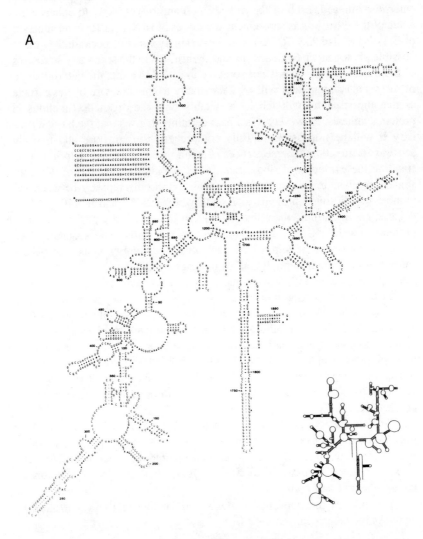

Figure 1A. The secondary structure of rat 18S and 28S rRNAs. In A, the secondary structure of rat 18S rRNA; the inset at lower right is the secondary structure of *E. coli* 16S rRNA for comparison—note how similar is the folding of the two rRNAs.

B

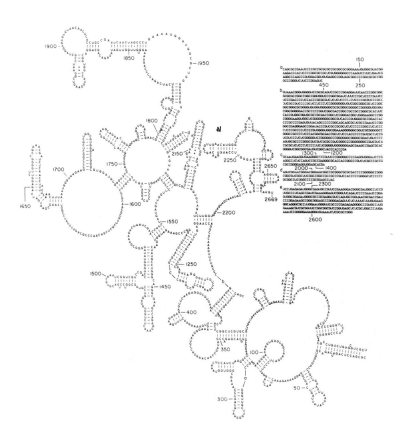

Figure 1B. The secondary structure of the 5′ half of rat 28S rRNA—note the association of 5.8S rRNA with the 5′ end of 28S rRNA at lower right.

C

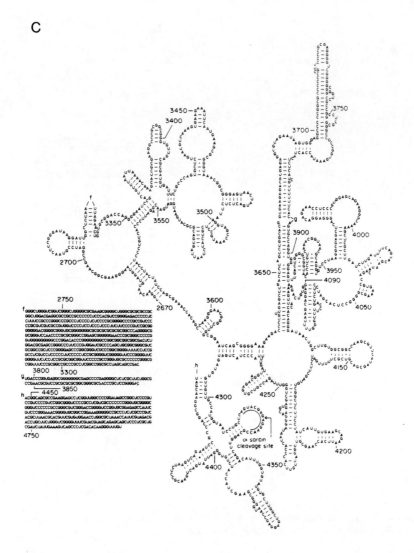

Figure 1C. The secondary structure of the 3′ half of rat 28S rRNA with the α-sarcin domain designated.

3. The identification at the start of the 5′ noncoding region of a pyrimidine sequence. These stretches are present in most if not all eukaryotic ribosomal protein mRNAs and may play a role in the regulation of their translation.[27-30] Indeed, the presence of a polypyrimidine stretch confirms that the cDNA includes the region where transcription starts. Seven rat ribosomal protein mRNAs have in common the hexanucleotide sequence CTTTCC. This may be a *cis* acting translation promoter.
4. The determination, by the hybridization of the cDNA to restriction enzyme digests of genomic DNA, of the number of copies of the ribosomal protein gene.[31] Mammalian ribosomal protein genes are present in multiple copies, generally 7 to 20;[15, 32] moreover, they are dispersed in the genome.[33] However, in no instance has it been shown that more than one of the genes is functional.[27-29] The presumption is that for each ribosomal protein the genome contains only one gene that is expressed, that the other copies are nonfunctional pseudogenes;[34] we note, however, that this presumption derives from a detailed analysis of only a limited number of families.
5. The determination, by the hybridization of the cDNA to poly(A)$^+$ mRNA, of the size of the mRNA encoding the ribosomal protein.[35] In general, the mRNA is about 150 to 300 nucleotides longer than the open reading frame; the poly(A) tail has 100 to 150 nucleotides and there are 25 to 80 residues in the 5′ and 3′ noncoding sequences.
6. The general chemical and physical properties of the ribosomal protein.
7. The presence of internal duplications in the ribosomal proteins.
8. The identification of related proteins by searching a library that has the primary structure of more than 500 ribosomal proteins. These last two are taken up separately.

Internal duplications in rat ribosomal proteins—A number of rat ribosomal proteins have duplications of amino acid sequences. Protein L7 has five repeats of a segment of 12 amino acids arranged in tandem near the NH$_2$ terminus.[36] The repeats are very basic; four to six of the residues are lysyl or arginyl and there are no acidic amino acids. Protein S8 has internal repeats of 12 or 13 residues that are basic and that occur five times in the protein.[37] Protein S10 has a possible tandem repeat of ten residues of which five are basic, and other potential duplications as well.[35] Protein P1 has a duplication of the sequence GGPAP-A-AHPA (Chan YL and Wool IG: unpublished). Protein P2 has two short sequences that are repeated; EEKK(D/E) which contains only charged residues and GS-APAA (Chan YL and Wool IG: unpublished). The sequence VQV-R-KYK occurs three times in L26.[38] Finally, the sequence PRRL—(K/R)(K/R) occurs four times in S6.[31]

The occurrence of multiple, related, generally basic repeats in ribosomal

TABLE 1–1.
Rat Ribosomal Proteins Whose Primary
Sequences Have Been Determined

Protein	Reference	Protein	Reference
S3[a]	d	P0	d
S4	101	P1	d
S6	31	P2*	45
S8	37	L5	42
S10	35	L7	36
S11	102	L7[a]	107
S12*	103	L18	108
S14	d	L18[a]	109
S16	d	L19	30
S17	104	L21	110
S21*	105	L26	38
S23/24[b]	d	L27	111
S26	106	L30	104
URF-S22[c]	d	L31	112
		L32	113
		L34	114
		L35[a]	115
		L36[a]	116
		L37*	117
		L37[a]	118
		L39*	119

[a]Identification is provisional; related to *X. laevis* S1.
[b]Identification is provisional; related to *X. laevis* S19.
[c]Identification has not yet been made; related to *X. laevis* S22.
[d]Chan YL and Wool IG: unpublished.
*Sequence determined directly from the protein.

proteins insinuates that they have functional significance, but there is no indication as yet what this might be. Possibilities that suggest themselves are that they play a role in the interaction with RNA (ribosomal, transfer, or messenger) or that they are involved in directing the proteins to the nucleolus for assembly of ribosomes. There is evidence that information specifying the localization of proteins in the nucleus is encoded in short sequences of amino acids,[39] although, it is not yet possible to derive a consensus sequence nor to formulate general rules for the structure of the peptide. In the best characterized examples, entry into the nucleus is contingent on a consecutive sequence of several basic amino acids preceded by a prolyl residue. For example, the first 21 amino acids of yeast ribosomal protein L3 which is needed for entry into the nucleus contains the sequence -PRKR-.[40] Several of the repeats we have found in rat ribosomal proteins also have these characteristics. Although, there is no experimental evidence available to evaluate the proposal, and we cannot even be sure if the repeats have statistical significance, their structures have sufficient similarity to known nuclear local-ization sequences as to require consideration of the possibility that they serve the same function.

1.3. Evolution of Eukaryotic Ribosomes

It has been apparent for some time that individual ribosomal proteins from different eukaryotic species are derived from common ancestral genes. The homologies are obvious when the sequences of amino acids in rat and yeast ribosomal proteins, the two eukaryotic species for which the largest sets of data have been collected (35 for rat and 25 for yeast; there are also for the latter some partial sequences), are compared. Despite the two species being evolutionarily distant eukaryotes, 17 can be correlated (Table 1-2). All of the comparisons give significant scores; the percentage of identities in the align-ments range from 40 to 80. The data are sufficient to provide confidence that most if not all of the ribosomal proteins from the two species are homologous and that it will be possible to establish a protein to protein correlation. This has at least one practical consequence: The correlation might be used to establish a uniform nomenclature for the two species; this would substitute a measure of order for the chaos that now confounds the designation of indi-vidual yeast ribosomal proteins.

Of the eight *Xenopus laevis* ribosomal protein amino acid sequences that have been determined in full or in part 6 (S1, S19, S22, L5, L14, L32) can be correlated with a rat protein (S3, S23/S24, URF-S22, L5, L18, AND L35a, respectively).[15]

It has been surmised that archebacterial ribosomes are transition particles in the evolution of the organelle and that they combine properties of those from prokaryotes and from eukaryotes.[41] For example, if we consider the two 5S rRNA binding proteins from *Halobacterium cutirubrum* one, HcL19, is related to *E. coli* L5, whereas the other, HcL13, is related to rat L5.[42] Nonetheless, the amino acid sequences of archebacterial ribosomal proteins are in general closer to those of eukaryotes than to the prokaryotic counter-parts.[43] Twelve of the rat ribosomal proteins can be correlated with arche-bacterial proteins.[15]

Until recently, common wisdom has had it that there are no close sequence similarities between eukaryotic and prokaryotic ribosomal proteins with the exception of the acidic or 'A' proteins[43–45] and it was suggested that a rela-tionship might only be found in their three-dimensional structure. The thinking with regard to this issue was strongly influenced by the results of the com-parison of the structure of ribosomal RNAs. The motif for these molecules is conservation of secondary structure rather than primary sequence.[1] This is not to say there are no conserved nucleotide sequences; most assuredly there

TABLE 1–2.
Related Rat and Yeast Ribosomal Proteins

Proteins		Scores[b]		
Rat	Yeast[a]	Relate	Align	Identities[c]
S3	YS33[d]	6.6	6.7	9/20
S4	S7[d]	14.6	12.4	19/28
S6	S10	42.7	83.9	149/236
	S6[e]	41.1	72.2	164/237
S8	YS9[d]	20.3	24.9	30/49
S14	rp59	41.9	58.4	108/137
S17	rp51	32.2	52.5	76/132
S21	S25	17.8	32.6	45/83
	S28[e]	23.5	33.6	47/83
P0	A0	48.9	83.7	170/312
P1	A1	14.8	26.2	48/106
	L44'	14.6	31.6	42/106
P2	A2	20.3	26.4	52/106
	YPA1	17.7	29.8	59/109
	L44	20.0	26.4	52/106
	L45	13.7	28.2	57/109
	L40c[d,e]	13.5	21.9	18/40
L5	YL3 (CN-1)[d]	13.3	12.9	19/30
	YL3 (CN-2)[d]	9.5	11.5	25/82
L18	rp28	39.5	50.1	102/184
L18a	L17[d,e]	14.6	16.1	22/45
L19	L15[d,e]	13.9	9.6	14/29
	YL14[d]	4.3	6.8	15/30
L26	YL33[d]	10.6	11.2	20/40
L30	L32	28.9	38.5	61/105
L31	L34	26.0	38.0	65/112
L36a	rp44	26.3	45.8	74/101
L37	L27[d,e]	11.0	9.6	17/28
	YP55[d]	6.4	9.0	16/30
L39	L46	17.1	21.9	31/50
	L36[d,e]	15.0	21.1	22/42

[a]The species is *Saccharomyces cerevisiae* unless otherwise indicated.
Because of limitations of space we do not provide references to the
yeast sequences; they are available on request.
[b]The scores are in SD units; all are highly significant.
[c]The number of identities in the alignment over the possible matches.
[d]A partial sequence.
[e]*Schizosaccharomyces pombe.*

are. The conserved sequences tend to be (but are not exclusively) in nonhelical regions, and there is evidence that they are important for function. There is a lesson here: It is possible that only the functionally important amino acids in ribosomal proteins, which need not be contiguous in the sequence, will be conserved. This may account for the difficulty in establishing relationships between eukaryotic and prokaryotic ribosomal proteins.

Intuition leads one to surmise that ribosomes arose on a single occasion and, hence, that the proteins as well as the rRNAs of eukaryotic ribosomes are likely to be related to their prokaryotic progenitors. The problem is how to trace the chemical spoor and unravel the mechanism by which the proteins evolved. The analysis is complicated because there are so many ribosomal proteins; because there are still far fewer data than for the nucleic acids (the only complete set is for *E. coli*); and because the algorithms for the comparison of proteins are less satisfactory and less reliable since one must deal with 20 different amino acids rather than four different nucleotides. Nonetheless, we have for some time argued for homology of eukaryotic and prokaryotic ribosomal proteins.[44–46] This seems certain for the acidic proteins.[47] The ribosomes of all the species that have been analyzed have a protein or proteins homologous to *E. coli* L12; the rat equivalent is P2.[45] There is convincing evidence that five other rat ribosomal proteins (S3, S11, S14, S16, and URF-S22) are related to *E. coli* ribosomal proteins (S3, S17, S11, S9, and S10, respectively).[15] It is surprising that more proteins cannot be correlated since the sequences of the entire *E. coli* set has been determined. One can only imagine that the divergence has been so great (the acidic proteins and a few others excepted) as to make it difficult to trace the relationships without improved methods. It is possible that the amino acid sequences of archebacterial ribosomal proteins, since they seem intermediates between eukaryotes and prokaryotes, will provide a connecting link for the correlation; that they will serve as a kind of molecular rosetta stone.

There is another aspect to the problem of the evolution of ribosomes. Eukaryotic ribosomes are larger than prokaryotic ribosomes; they contain a greater number of proteins and an additional molecule of RNA. Albeit, the latter is a difference more apparent than real; prokaryotes lack 5.8S rRNA but have a related sequence at the 3′ end of 23S rRNA—prokaryotes have not gone to the trouble of processing the 5.8S-like RNA out of the primary rRNA transcript, whereas eukaryotes have.[48] The individual protein and RNA molecules in eukaryotic ribosomes are, on the average, larger than those in prokaryotes. There is no convincing explanation for these differences. Moreover, it is a paradox since eukaryotic and prokaryotic ribosomes perform precisely the same function (the catalysis of protein synthesis) and more importantly they do so by appreciably the same biochemical means; albeit, the initiation of protein synthesis is more complicated in animal cells. The critical question is what was the evolutionary pressure for the accretion of the extra proteins and for the increase in the size of the proteins and of the RNAs in eukaryotic ribosomes. Some have interpreted this question to imply that eukaryotic ribosomes have functions lacking in prokaryotic particles, perhaps related to the regulation of translation.

One can, of course, look at the problem the other way around and ask why prokaryotic ribosomes are smaller. One answer is that they had to be streamlined. During log phase growth as much as 35% of cellular protein in a bacterium is ribosomal; the bacterium might have difficulty supporting a particle with as much protein as eukaryotic ribosomes have. The assumptions implicit in this suggestion is that ribosomes evolved once, which seems eminently reasonable, and that prokaryotic ribosomes at one time had a greater number of proteins, a number comparable to that in eukaryotes. Prokaryotic ribosomes then responded to selective pressure by discarding proteins and reducing the size of those that were retained—the advantage being a greater number of smaller ribosomes capable of performing the same function as their larger progenitors. This would have had to have happened after the divergence of primitive eukaryotes. The argument does not change the fundamental nature of the problem; it merely alters the way it is put. What ribosomal proteins could be dispensed with by prokaryotes without loss of function and if the extra proteins could be discarded with impunity why have they been retained by eukaryotes.

One reconciliation would have it that the earliest cells, progenitors of prokaryotes and eukaryotes, had nuclei and eukaryotic-like ribosomes and that a number of the proteins were needed, not for protein synthesis, but to manage the complicated traffic between nucleus and cytoplasm.[49] Recall ribosomal proteins are synthesized in the cytoplasm and then transported to the nucleus where they are assembled on nascent rRNA transcripts; after the processing of the rRNA, which most likely occurs during assembly, the ribosomal subunits must pass through nuclear pores to get to the cytoplasm. Thus, eukaryotic ribosomal proteins are likely to contain amino acid sequences that serve as zip codes for nuclear localization; ribosomes may also have proteins whose sole function is to facilitate transnuclear membrane traffic. Obviously, bacteria no longer require either the nuclear localization amino acid sequences or the transport proteins.

There is still another rationalization of the paradox and it derives from the recent momentous discoveries of Cech.[50] The initial finding was that ribosomal RNA has the capacity to mediate self-splicing and self-ligation, that is to say, the removal of a sequence of nucleotides (an intron) from a rRNA, and the religation of the ends so as to reestablish the integrity of the molecule. Before this observation nucleic acids were considered to be relatively inert chemically—capable of serving in information transfer by providing a template for polymerization reactions (as with messenger and transfer RNAs) and capable of providing the scaffolding for proteins in ribonucleo-

The coordination of the synthesis of the ribosomal proteins is more complicated than that of the rRNAs: the transcription of 70 to 80 unlinked genes must be balanced. In exponentially growing cells the balanced synthesis of the various ribosomal proteins is due primarily to their mRNAs being present in similar amounts and to their being translated with similar efficiencies.[54, 55] Thus, coordination of synthesis would seem to be determined in the first instance, and most importantly, by regulation of transcription,[56, 57] although there is evidence also for regulation of pre-mRNA processing[58] and of translation.[59] What this implies is that one would be well advised to seek the source of the coordination in *cis* acting elements of the promoters of ribosomal protein genes and in *trans* acting factors.

In *E. coli,* the 53 ribosomal protein genes are organized into 20 operons; the largest has 11 genes, the smallest a single one.[60, 61] Almost half of these genes, all of which are present in single copies, are at one locus *(Str)* on the chromosome; the remainder are scattered over the genome from 0 to 94 min. The ribosomal protein operons are complex transcriptional units; some contain nonribosomal protein genes and some have two promoters. In bacteria the main control of ribosomal protein formation is not of transcription of mRNAs but is mediated by autogenous regulation of their translation.[62] This regulation exploits the arrangement of the genes in operons and specific interactions of the ribosomal proteins with RNAs. To wit: Certain ribosomal proteins when synthesized in excess function as repressors of the expression of their own operons by binding to a control region on the polycistronic mRNA rather than to rRNA. This follows from similarities in the structure of the control region of the polycistronic mRNA and the repressor ribosomal protein binding site on rRNA; of course the affinity for the latter has to be greater than for the former. As to the synthesis of rRNA, there probably is regulation (repression) of transcription by nontranslating ribosomes and perhaps by ppGpp.[60]

Information on the organization of ribosomal protein genes in eukaryotes is just beginning to accumulate. What is certain is that both their structure and the regulation of their expression is different than in prokaryotes.[63] The number of copies of the genes for individual ribosomal proteins varies amongst eukaryotes. In yeast there are either one or two copies, most often two, of the genes that have been analyzed (more than 20 of them); where there are two they are both functional. In *Xenopus* there are two to six genes; no nontranscribed ribosomal protein genes have been identified. In mammals there are 7 to 20 copies of each; however, in no instance has it been shown that more than one of the ribosomal protein genes in a family is functional.

In eukaryotes the genes are not clustered, not even for the multiple copies of the same gene no less for genes for different ribosomal proteins. Indeed, different genes for the same ribosomal protein may be on different chromosomes.[33] In yeast only two ribosomal protein genes are physically linked (rp28 and S16a are 600 nucleotides apart) and there is no evidence that the tran

protein (RNP) particles and organelles (as in ribosomes) but not of catalysi:
The latter was assumed to be the sole province of proteins. Cech's exper
ments, which confounded common wisdom and even intuition, has change
forever our ideas about nucleic acids and like all momentous discoveries ha
had far reaching consequences. Perhaps, the single most important of thes
is the bearing his discovery has had on theories of molecular evolution. It i
apparent now that RNA can be both the repository of information that ca
be transmitted and that it can have enzymatic activity. This makes it possible
even likely, that the precellular biological world was dominated by RNA
that RNA preceded DNA and protein. It is of relevance that Cech's wor
lends support to the idea that ur-ribosomes had only RNA as had been propose
early on by Crick and Orgel. The concept is that the basic biochemistry o
protein synthesis (the binding of aminoacyl-tRNA, peptide bond formation
and translocation) are intrinsic properties of ribosomal RNA; that the ribo
somal proteins, a later evolutionary embellishment, facilitate the folding an
the maintenance of an optimal configuration of the ribosomal RNA[51] and ii
this way confer on protein synthesis speed and accuracy. Perhaps then eu
karyotic ribosomes, which have far more RNA than prokaryotes (1300 mor
nucleotides), need more proteins to tune their nucleic acids. There is stil
another scenario: If the ur-ribosome had only RNA and divergence of eu
karyotes and prokaryotes occurred before the emergence of a complete set o
ribosomal proteins then one would not expect in the amino acid sequence th
close resemblance that is apparent in the structure of the rRNA; rather one
would expect homology only of those ribosomal proteins (like the acidic
proteins and a few others) that had evolved before divergence.

1.4. BIOGENESIS OF RIBOSOMES: THE REGULATION OF THE SYNTHESIS OF THE MOLECULAR COMPONENTS AND THE ASSEMBLY OF THE PARTICLES

The assembly of ribosomes is an extraordinarily complex process that
must test the cell's capacity for regulation and coordination. Biogenesis of
eukaryotic ribosomes requires the synthesis of equimolar amounts of the four
rRNAs and of 70 to 80 proteins. (Eukaryotic ribosomes are presumed, in the
absence of definitive evidence, to have molar amounts of most, if not all,
proteins.) Moreover, the formation of the rRNAs and of the ribosomal proteins
has to be precisely balanced. The coordinate synthesis in the nucleolus of
5.8S, 18S, and 28S rRNAs is easily accomplished since their genes are in a
single large transcription unit.[52, 53] However, the synthesis of 5S rRNA occurs
outside of the nucleolus so its transcription must be coordinated with that of
the other rRNAs and, in addition, 5S rRNA must be delivered to the nucleolus.

scription of even these two is functionally coupled, indeed, they are transcribed in opposite directions.[64]

More than 20 yeast ribosomal protein genes have been sequenced and the 5' flanking regions searched for common nucleotide elements that might account for coordinate regulation.[63] A striking finding is of two sequences of 12 to 15 nucleotides each about 300 bases upstream of the start of transcription; they were originally designated HOMOL 1 and RPG box. Not all yeast ribosomal protein genes contain these two elements and they are also associated with rare nonribosomal protein genes, for example, elongation factor (EF)-1α, RNA polymerase I, and mating-type. Deletion experiments provide compelling evidence that HOMOL 1 and RPG box are *cis* acting transcription activation sites (now designated UAS$_{rpg}$). In addition to the conserved boxes there is a pyrimidine stretch (TC-rich) downstream of the UAS$_{rpg}$ which also enhances transcription efficiency; however, this sequence is found in many other highly expressed yeast genes. There is a TATA-like region 40 to 100 nucleotides upstream that presumably positions RNA polymerase II for initiation of transcription. The many elements in the extended promoter region of yeast ribosomal protein genes are presumptive binding sites for *trans* acting factors. One of these has been identified and designated TUF; it is a protein of 150,000 Da that binds to UAS$_{rpg}$. TUF is probably a general factor for the coordination of the transcription of ribosomal protein genes.

The structures of ribosomal protein genes in mammals differs from those in yeast.[27–29, 65–69] One distinctive feature is the lack of a canonical TATA box. In the region where one would be expected there is a six to seven base pair element that contains five or six AT pairs. Presumably, some aspect of this sequence pattern, or the novel organization of the cap region, or both, assumes the function usually served by the TATA box, i.e., to position RNA polymerase II for the initiation of transcription.[70] Although precise initiation of transcription in the absence of a canonical TATA box is rare it does occur; it has been described for a few viral and cellular genes (cf. Reference 27 for a discussion). The latter are, like the ribosomal protein genes, members of the "housekeeping" class. Conceivably, novel promoter structure may confer special regulatory properties on this class of genes.

Another striking characteristic of mammalian ribosomal protein genes is the structure of the cap site. It is embedded in a ≥12 nucleotide stretch of pyrimidines flanked by blocks of greater than 80% GC content. The cap site pyrimidine tract has the motif 5'-CTTCCYTYYTC-3'; initiation of transcription is at the C at position 4 or 5.[27–29]

Mammalian ribosomal protein genes contain a number of *cis* acting control elements. One is unusual if not unique; the 3' end of exon I and the 5' end of intron 1 are necessary for full expression of the gene, indeed, they increase transcription five- to tenfold.[68] This transcriptional regulatory element binds a nuclear factor. The binding site embraces an inverted repeat located near the exon I-intron 1 junction. This region does not function as a typical enhancer (it is not active when transplanted), indeed, ribosomal protein genes do not have typical enhancers. In addition to its special regulatory function intron 1 has a general role in that intron-splicing is necessary for full expression of the gene. The latter function can be assumed by other introns in the gene or even an intron from a foreign gene.

Maximal expression of ribosomal protein genes requires 200 bp in the 5′ flanking region.[67–69] Within this region there are at least five discrete elements that affect expression and which bind nuclear factors. Although the general pattern of the structure of the promoters of the three mouse ribosomal protein genes that have been analyzed are similar, they also have distinct characteristics. For example, the location and distribution of the elements in the ribosomal protein S16 gene are different than those in the L30 and L32 genes.[69] An example of an S16 distinctive element is the one that binds the Sp1 transcription factor and stimulates promoter activity two- to threefold. Finally, the nuclear factors that bind to *cis* acting sequences in the L30 and L32 genes are not the same as those that bind to S16.[69]

As has already been indicated, it seems reasonable to search the sequence of nucleotides in the promoters of mammalian ribosomal protein genes for clues as to how their transcription is regulated and, more importantly, how the transcription of the 70 to 80 ribosomal protein genes is coordinated. The assumption, until proven otherwise, has to be that coordination is mediated by a set of common or overlapping *trans* acting factors affecting in some complex way a pattern of *cis* acting sequences in the promoters. The analysis of the three mouse ribosomal protein genes[27–29, 67–69] has provided an enormous amount of information on the architecture of the promoters but the number is not sufficiently large to establish (1) the exact pattern of the control elements; (2) whether there is likely to be a common architecture for all mammalian ribosomal protein gene promoters; or (3) the identity of the factors that bind to *cis* acting sequences. Clearly, what is required is the structure and the functional organization of additional mammalian ribosomal protein genes.

1.5. RNA-PROTEIN INTERACTION: AN ANALYSIS OF THE RECOGNITION BY α-SARCIN OF A RIBOSOMAL DOMAIN CRITICAL FOR FUNCTION

α-Sarcin is a small, basic, cytotoxic protein produced by the mold *Aspergillus giganteus* (References 71 and 72 and Part III, Chapter 15 of this book) that inhibits protein synthesis by inactivating ribosomes.[73–75] The inhibition is the result of the hydrolysis of a phosphodiester bond[76] on the 3′ side of G-4325[24, 77–78] which is in a single-stranded loop 459 residues from the 3' end of 28S rRNA[24, 79] (cf. Figure 1 also). The cleavage site is embedded in a purine-rich, single-stranded segment of 14 nucleotides that is near universal.[79, 80] This is one of the most strongly conserved regions of rRNA and, indeed, the ribosomes of all the organisms that have been tested, including the producing fungus[81] are sensitive to the toxin. α-Sarcin catalyzes the hydrolysis of only the one phosphodiester bond and this single break accounts entirely for its cytotoxicity.[79] This remarkable specificity is peculiar to α-sarcin; treatment of ribosomes with other ribonucleases causes extensive digestion of rRNA. If, however, the substrate is naked RNA rather than 80S ribosomes or 60S subunits and the amount of the toxin is large it cuts on the 3' side of nearly every purine without regard to whether the nucleotides are in single- or double-stranded regions.[78]

The finding that cleavage of a single phosphodiester bond in the α-sarcin domain inactivates the ribosome implies that this sequence is crucial for function since ribosomes ordinarily survive mild treatment with nucleases despite many nicks in their RNA;[82] indeed, some organisms physiologically divide their 28S rRNA into domains; in *Trypanosomes* the fragmentation is into two large and four small molecules.[83] Thus, intact rRNA *per se* is not essential for protein synthesis. The presumption that the α-sarcin region of 28S rRNA is critical for ribosome function has gained considerable reinforcement from the elucidation of the mechanism of action of ricin. Ricin, which is amongst the most toxic substances known, is a RNA N-glycosidase and the single base in 28S rRNA that is depurinated is A-4324, i.e., the nucleotide adjacent to the α-sarcin cut site (References 84 through 87 and Part III, chapter 15 of this book).

Until recently little was known of the function of individual ribosomal components or even of ribosomal domains. None of the ribosomal proteins or nucleic acids have activity when separated from the particle. To circumvent this impediment to the analysis of function advantage has been taken of the activity of toxins and antibiotics. The value that derives from an analysis of their mechanism of action is in concentrating our attention on regions of the ribosome where our efforts to comprehend functional correlates of structure are likely to be rewarded. Although, the molecular details of the function of the α-sarcin/ricin domain are not known there are good reasons to suspect that it is involved in EF-1 dependent binding of aminoacyl-tRNA to ribosomes and EF-2 catalyzed GTP hydrolysis and translocation. This supposition follows from the findings that these are the partial reactions most adversely affected by α-sarcin[73] and by ricin,[88] respectively, and that cleavage at the α-sarcin site in *E. coli* 23S rRNA interferes only with the binding of elongation factors EF-Tu and EF-G.[89] The most convincing evidence for this interpretation comes from the demonstration[90] that EF-Tu and EF-G footprint in the α-sarcin/ricin domain. EF-Tu protects only four of the nucleotides in prokaryotic 23S rRNA against chemical modification and these correspond in eukaryotic 28S rRNA to A-4324 (ricin), G-4325 (α-sarcin), and G-4319 and A-4329 of which the latter two are also in the universal sequence. EF-G also protects only four nucleotides and three are the same as the ones protected by EF-Tu; the bases that correspond to G-4319, A-4324, and G-4325.[90]

An effort is being made to determine how α-sarcin recognizes a single phosphodiester bond in a particular domain in rRNA.[91] This is part of an effort, unfortunately but necessarily oblique, to understand how ribosomal proteins, which like α-sarcin are small and basic, recognize specific sites in rRNA. To accomplish this we prepared a RNA oligonucleotide[92] using a synthetic DNA template and phage T7 RNA polymerase.[93] The oligoribo-nucleotide, a 35-mer, has the sequence and the secondary structure (a stem and a loop), of the region of 28S rRNA attacked by α-sarcin and, indeed, the toxin hydrolyzes this synthetic oligoribonucleotide at a position that cor-responds precisely to where it cleaves the nucleic acid in intact ribosomes.[92]

Nucleotides in the α-sarcin domain RNA have been systematically altered (Figure 2). The aim is a more precise definition of the sequence of nucleotides and of the higher order structure that prescribes the binding of the protein to the RNA and allows the catalysis of hydrolysis. The determinants of recog-nition that have been examined are (1) nucleotide specificity at the site of covalent modification; (2) the necessity for a bulged nucleotide in the helical stem; (3) the requirement for the stem itself; (4) the influence of context, i.e., the importance of the nucleotides in the universal sequence in the loop in which the α-sarcin guanosine is located; and, finally, (5) an aspect of rec-ognition that is referred to, for want of a better term, as geometry.

The nucleotide specificity at the site of covalent modification of RNA by α-sarcin—The oligoribonucleotide that reproduces the α-sarcin domain in 28S rRNA is referred to as wild type (Figure 2, Structure I). The guanosine at position 21 in this RNA (which corresponds to G-4325 in 28S rRNA) was mutated. A transition of the G to an A (II) produced a less suitable substrate for α-sarcin; cleavage was decreased to 35% of the control (Table 1-3). Transversions of the wild type G to U (III) or to C (IV) reduces hydrolysis to approximately 17 and to 1%, respectively (Table 1-3). Thus, there is strong but by no means absolute dependence on preservation of the G at the site of covalent modification; the preference is G ≫ A > U ≫ C. The results prejudice one to consider structure rather than sequence as the more important deter-minant of specificity.

The requirement for a bulged nucleotide in the stem of the α-sarcin domain RNA—The α-sarcin domain RNA has a canonical protein binding structure: a stem, a loop, and a bulged nucleotide. The last occurs in a number of ribosomal protein binding sites, albeit the bulged nucleotide is usually an A rather than the U that is found here.[94] It was suspected that the bulged U at position 6 in the substrate (position 4310 in 28S rRNA) is not necessary

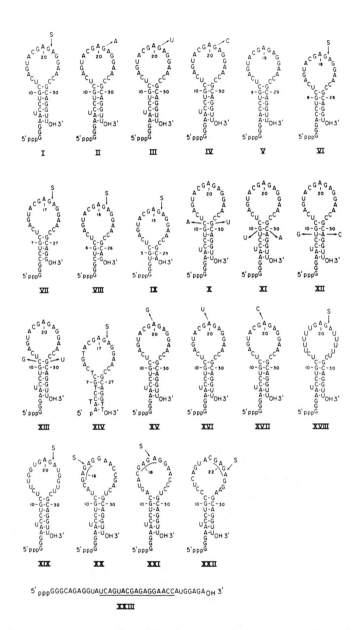

Figure 2. The structure of an oligoribonucleotide that mimics the α-sarcin domain in 28S rRNA and of a series of mutants.

TABLE 1–3.
Recognition by α-Sarcin of
Oligoribonucleotide Mutants of a
Domain in 28S rRNA

Structure[a]	Recognition[b]
I	+ + + +
II	+ +
III	+
IV	±
V	+ + + +
VI	+ + + +
VII	+ + + +
VIII	+ + + +
IX	+ + + +
X	+ + +
XI	+ +
XII	+ + + +
XIII	+ + +
XIV	0
XV	+ + + +
XVI	+ + + +
XVII	+ + + +
XVIII	0
XIX	0
XX	0
XXI	0
XXII	0
XXIII	0

[a]The structures referred to are in Figure 2.
[b]Recognition is graded on the basis of the extent
of the specific cleavage of the oligoribonucleotide by α-sarcin; the wild type response is taken
as + + + +.

for α-sarcin action since it does not occur at the comparable site in *E. coli*
23S rRNA,[1] although the bacteria's ribosomes are sensitive to the toxin.[76, 77]
Still the possibility was tested directly by synthesizing a variant oligoribonucleotide (a 34-mer) that lacked the bulged U in the stem (Figure 2, Structure
V). Just as was suspected the variant is as sensitive to the toxin as the wild
type substrate (Table 1-3). Thus, the bulged nucleotide is not required for
recognition by α-sarcin, i.e., neither for binding to the RNA substrate nor
for covalent modification.

**The requirement for a helical stem for recognition of the α-sarcin
domain RNA**—To test the importance of the stem for α-sarcin action a linear
molecule (35-mer) was constructed that retained the order of the 17 nucleotides
in the loop, including the universal purine-rich sequence of 14 bases, but

with the 5' and 3' ends engineered so that they would not pair (Figure 2, Structure XXIII). This linear molecule was not a substrate for α-sarcin (Table 1-3).

Having established an absolute necessity for the helical stem, we inquired as to the number of base pairs that are needed by removal of successive additional ones (Figures 2, Structures VI through IX). The number can be reduced to three without loss of specificity of recognition (Table 1-3). Thus, it would appear that the helix is in nature longer and hence more stable than is required for recognition by α-sarcin; it is presumed it is neither longer nor more stable than is required for its contribution to the function of this ribosomal domain in protein synthesis.

There is evidence that the nature of the Watson-Crick base pairs in a RNA helix, i.e., GC as opposed to AU, can affect the structure of an included loop.[95] For this reason base pairs in the wild type oligoribonucleotide are being systematically changed (Figure 2, Structures X through XIII). All those analyzed to date are competent substrates for the specific activity of α-sarcin; however, the change in the second base pair (XI) does reduce sensitivity suggesting it may change the structure of the loop (Table 1-3).

Synthetic oligodeoxynucleotides corresponding to *E. coli* tRNA[Phe] or tRNA[Lys] are recognized by their cognate synthetases.[96] However, a DNA oligomer (32-mer) that corresponds to the sequence of the wild type α-sarcin RNA (Figure 2, Structure XIV) was not cleaved specifically by α-sarcin.

The effect of the context of the α-sarcin site guanosine on the recognition of the toxin domain RNA—The purpose here was to evaluate the contribution to recognition of the nucleotides in the single-stranded region surrounding the α-sarcin site guanosine. It was anticipated that alterations in the 5' adjacent adenosine would not have an appreciable effect since depurination of A-4324 by pretreatment with ricin did not affect subsequent cleavage by α-sarcin at G-4325 in the same ribosomes.[97] Nonetheless, a series of variants were constructed (Figure 2, Structures XV through XVII) with alterations of the ricin site A to G, U, or C. All of these variants were recognized by α-sarcin (Table 1-3).

The context was changed in another way: the tetranucleotide GAG(sarcin)A was left intact for reasons that will be apparent shortly and the remainder of the universal portion of the loop sequence was engineered (Figure 1) so it was entirely uridines (XVIII) or in a second mutant uridines and guanosines (XIX). Neither oligonucleotide is a competent substrate for α-sarcin (Table 1-3). Thus, the context is an essential feature of the recognition of the substrate by α-sarcin.

The contribution of the geometry of the sequence GAGA in the loop to the recognition by α-sarcin of the toxin domain RNA—*E. coli* ribosomes are not sensitive to ricin,[98] the ribosomes are not inactivated and A-2660 in the α-sarcin/ricin domain is not depurinated.[84] However, naked *E. coli* rRNA is a substrate for the *N*-glycosidase activity of the toxin.[87] The two sites of covalent modification, one each in 16S and 23S rRNAs, have stems with seven base pairs and loops that have the sequence GAGA. Not all of the sites in rRNA with this structure, however, are modified by ricin. A comparison of the two subsets, modified and unmodified by ricin, indicated that the tetranucleotide GAGA in the loop had to have a particular geometric relationship to, or orientation with, the stem; that the ricin-sensitive adenosine had to be centered over the axis of the helix at least as the structure is depicted in two dimensions.

In agreement with the prediction that comes from this observation, if the tetranucleotide GAGA was moved either four (Figure 2, Structure XX) or two (XXI) bases closer to the 5′ end or two bases closer to the 3′ end (XXII) the specific response to α-sarcin is entirely lost (Table 1-3). Thus recognition is not merely of a guanosine at the correct position in the sequence but of a particular conformation and, in addition, there is a strong requirement for a particular geometry. This latter is an aspect of the recognition of RNA by a protein that to our knowledge has not been defined before.

Recognition by α-sarcin of the toxin domain RNA requires a stem (but not a bulged nucleotide) and a single-stranded loop in which the sequence of at least 14-nucleotides (the universal sequence) affect binding and enzymatic activity; the stem needs only three base pairs and the identity of the Watson-Crick interactions does not have a large influence. Perhaps most important is the observation that the tetranucleotide GAG(sarcin)A in the loop has to have the correct geometry with respect to the stem.

It has been proposed[15, 91] that the structure of the α-sarcin domain in 28S rRNA is more complex than is depicted in the usual two-dimensional cartoons and, furthermore, that it is capable of undergoing reversible alterations.

What it is that suggests that the structure of this domain is complex is the enormous amount of ricin required to cleave the glycosidic bond in naked 28S rRNA or in the synthetic RNA. Depurination of A-4324 in 28S rRNA resident in the ribosome occurs at a ricin:substrate ratio of 0.001, whereas with naked 28S rRNA or with the wild type oligoribonucleotide the ratio is 10, i.e., 10,000 times greater.[84, 92] We do not know the reason for this incredible discrepancy but the most likely explanation is that the ordered structure of the domain is different in ribosomes than it is in naked RNA. We take note that the single-stranded loop is large (17 nucleotides) and, hence, unlikely to exist as such; it might well participate in a tertiary interaction with other regions of 28S rRNA.

What it is that suggests that the domain might undergo transitions in its structure is an observation of the effect of α-sarcin on the association of 5.8S rRNA with 28S rRNA.[99] In the large subunit of eukaryotic ribosomes these two nucleic acids are noncovalently but stably associated.[25] This interaction, which involves approximately 40 hydrogen bonds in two separate contact regions, is destabilized by the α-sarcin catalyzed cleavage which is at a site more than 4000 nucleotides away. Dissociation of 5.8S rRNA from 28S rRNA, which ordinarily requires treatment with 4 to 6 M urea, occurs spontaneously after α-sarcin action on ribosomes.[99]

It is difficult to provide a coherent physical chemical explanation for this phenomenon. The cleavage of G-4325 appears to initiate a propogated change in the secondary, or tertiary structure, or both that in some way is transmitted through the molecule and leads to the collapse of 28S rRNA. This collapse could account for the destabilization of the 5.8S·28S rRNA complex and to the loss of the function of the ribosomes. We suggest further that it interferes with the operation of a rRNA switch, i.e., with the ability to disrupt and reestablish a crucial tertiary interaction that involves the single-stranded region of the α-sarcin domain.

There are other observations that support this conjecture. Oligodeoxynucleotides complimentary to the universal sequence in the loop of the α-sarcin domain will not bind to either *E. coli*[100] or rat ribosomes (Endo Y and Wool IG: unpublished) suspended in buffer; suggesting, but by no means proving, that the structure is not simply single stranded. Occlusion by ribosomal proteins could account for the failure of the cDNA to bind to the site; however, this seems less likely since α-sarcin and ricin have access to the domain. It is most important in this regard that if ribosomes are catalyzing protein synthesis they will bind the complimentary oligodeoxynucleotide (manifest as sensitivity to ribonuclease H) suggesting a reversible change in the structure of the domain (Endo Y: unpublished data). A similar observation has been made with *E. coli* ribosomes (Merryman C and Hill WE: personal communication). Further support for the proposal comes from experiments with inhibitors of protein synthesis (cycloheximide, GMPPNP, and sparsomycin) in which α-sarcin is used as a probe of structure (Endo Y: unpublished data). The results of these experiments also indicate that there is a conformational change in the RNA during translocation since ribosomes are sensitive to α-sarcin only when peptidyl-tRNA is in the A-site prior to translocation. The conformational transition, perhaps the making and breaking of a tertiary interaction, could provide the motive force for changes in the structure of ribosomal 60S subunits and might underlie the movement required for the translocation of peptidyl-tRNA from the A- to the P-site and for the movement of mRNA one codon after each of the reiterative rounds in translation. In this paradigm it is the elongation factors that initiate the reversible transition or switch in rRNA structure that propels translocation either directly or indirectly through the binding or the hydrolysis of GTP. Cleavage at G-4325 in 28S rRNA by α-sarcin or depurination at A-4324 by ricin might abolish the capacity to reversibly switch structures and in this way account for the catastrophic effect of the toxins on ribosome function.

1.6. CODA

All that remains are the difficult tasks: the tertiary folding patterns of the rRNAs, the three-dimensional structure of the proteins, and ultimately the structure of the subunits at atomic resolution. These are formidable problems but progress in studies of macromolecular structure has been so astonishing in the last several years as to encourage optimism. As to the function of the organelle we *only* lack information on how a peptide bond is made and how movement of mRNA and of peptidyl-tRNA is catalyzed. Anyone who is optimistic that we will soon have a solution to these problems is uninformed. But we live in hope.

ACKNOWLEDGMENTS

The experimental work was supported by grants (GM 21769 and GM 33702) from the National Institutes of Health. I am indebted to the colleagues with whom I have had the pleasure of working on the research reported here: To Yuen-Ling Chan, Yaeta Endo, Anton Glück, Joe Olvera, and Veronica Paz. I am grateful to Arlene Timosciek for assistance in the preparation of the manuscript.

REFERENCES

1. Noller HF: Structure of ribosomal RNA. *Annu. Rev. Biochem.* 53:119 (1984).
2. Wittmann-Liebold B: Primary structure of *Escherichia coli* ribosomal proteins. *Adv. Protein chem.* 36:56 (1984).
3. Woese CR, Fox GE, Zablen L, Uchida T, Bonen L, Pechman K, Lewis BJ, and Stahl D: Conversation of primary structure in 16S ribosomal RNA. *Nature* 254:83 (1975).
4. Woese CR, Magrum LJ, Gupta R, Siegal RB, Stahl DA, Kop J, Crawford N, Brosius J, Gutell RR, Hogan JJ, and Noller HF: Secondary structure model for bacterial 16S ribosomal RNA: phylogenetic, enzymatic and chemical evidence. *Nucleic Acids Res.* 8:2275 (1980).
5. Brimacombe R: The emerging three-dimensional structure and function of 16S ribosomal RNA. *Biochemistry* 27:4207 (1988).
6. Stern S, Weiser B, and Noller HF: Model for the three-dimensional folding of 16S ribosomal RNA. *J. Mol. Biol.* 204:447 (1988).
7. Stern S, Wilson RC, and Noller HF: Localization of the binding site for protein S4 on 16S ribosomal RNA by chemical and enzymatic probing and primer extension. *J. Mol. Biol.* 192:101 (1986).
8. Stern S, Changchien LM, Craven GR, and Noller HF: Interactions of proteins S16, S17 and S20 with 16S ribosomal RNA. *J. Mol. Biol.* 200:301 (1988).

9. Svensson P, Changchien LM, Craven GR, and Noller HF: Interaction of ribosomal proteins S6, S8, S15 and S18 with the central domain of 16S ribosomal RNA. *J. Mol. Biol.* 200:301 (1988).

10. Capel MS, Kjeldgaard M, Engleman DM, and Moore P: Positions of S2, S13, S16, S17, S19 and S21 in the 30S ribosomal subunit of *Escherichia coli*. *J. Mol. Biol.* 200:65 (1988).

11. Stöffler G and Stöffler-Meilicke M: Immunoelectron microscopy on *Escherichia coli* ribosomes, in Hardesty B and Kramer G (eds): *Structure, Function, and Genetics of Ribosomes* New York, Springer-Verlag, 1986, 28.

12. Oakes MO, Henderson E, Scheinman A, Clark M, and Lake JA: Ribosome structure, function and evolution: mapping ribosomal RNA, proteins, and functional sites in three dimensions, in Hardesty B and Kramer G (eds): *Structure, Function, and Genetics of Ribosomes* New York, Springer-Verlag, 47 (1986).

13. Yonath A, Frolow F, Shoham M, Müsig J, Makowski I, Glotz C, Jahn W, Weinstein S, and Wittmann HG: Crystallography of ribosomal particles. *J. Cryst. Growth* 90:231 (1988).

14. Warner JR: Synthesis of ribosomes in *Saccharomyces cerevisiae. Microbiol. Rev.* 53:256 (1989).

15. Wool IG, Endo Y, Chan YL, and Glück A: Studies of the structure, function and evolution of mammalian ribosomes, in Hill WE (ed): *The Structure, Function and Evolution of Ribosomes* Washington, D.C., American Society for Microbiology, 203 (1990).

16. Aoyama K, Hidaka S, Tanaka T, and Ishikawa K: The nucleotide sequence of 5S RNA from rat liver ribosomes. *J. Biochem.* 91:363 (1982).

17. Nazar RN, Sitz TO, and Busch H: Structural analyses of mammalian ribosomal ribonucleic acid and its precursors. Nucleotide sequence of 5.8S ribonucleic acid. *J. Biol. Chem.* 250:8591 (1975).

18. Chan YL, Gutell R, Noller HF, and Wool IG: The nucleotide sequence of a rat 18S ribosomal ribonucleic acid gene and a proposal for the secondary structure of 18S ribosomal ribonucleic acid. *J. Biol. Chem.* 259:224 (1984).

19. Torczynski R, Bollon AP, and Fuke M: The complete nucleotide sequence of the rat 18S ribosomal RNA gene and comparison with the respective yeast and frog genes. *Nucleic Acids Res.* 11:4879 (1983).

20. Chan YL, Olvera J, and Wool IG: The structure of rat 28S ribosomal ribonucleic acid inferred from the sequence of nucleotides in a gene. *Nucleic Acids Res.* 11:7819 (1983).

21. Hadjiolov AA, Georgiev OI, Nosikov VV, and Yavachev LP: Primary and secondary structure of rat 28S ribosomal RNA. *Nucleic Acids Res.* 12:3677 (1984).

22. Gutell RR and Fox GE: Compilation of large subunit sequences presented in a structural form. *Nucleic Acids Res.* 16:r175 (1988).

23. Noller HF, Stern S, Moazed D, Powers T, Svensson P, and Changchien LM: Studies on the architecture and function of 16S rRNA. *Cole Spring Harbor Symp. Quant. Biol.* 52:695 (1987).

24. Chan YL, Endo Y, and Wool IG: The sequence of the nucleotides at the α-sarcin cleavage site in rat 28S ribosomal ribonucleic acid. *J. Biol. Chem.* 258:12768 (1983).

25. Wool IG: The structure and function of eukaryotic ribosomes. *Annu. Rev. Biochem.* 48:719 (1979).

26. Proudfoot NJ and Brownlee GG: The 3' non-coding region sequences in eukaryotic messenger RNA. *Nature* 263:211 (1976).

27. Dudov KP and Perry RP: The gene family encoding the mouse ribosomal protein L32 contains a uniquely expressed intron-containing gene and an unmutated processed gene. *Cell* 37:457 (1984).

28. Wiedemann LM and Perry RP: Characterization of the expressed gene and several processed pseudogenes for the mouse ribosomal protein L30 gene family. *Mol. Cell. Biol.* 4:2518 (1984).

29. Wagner M and Perry RP: Characterization of the multigene family encoding the mouse S16 ribosomal protein: strategy for distinguishing an expressed gene from its processed pseudogene counterparts by an analysis of total genomic DNA. *Mol. Cell. Biol.* 5:3560 (1985).

30. Chan YL, Lin A, McNally J, Peleg D, Meyuhas O, and Wool IG: The primary structure of rat ribosomal protein L19: a determination from the sequence of nucleotides in a cDNA and from the sequence of amino acids in the protein. *J. Biol. Chem.* 262:1111 (1987).

31. Chan YL and Wool IG: The primary structure of rat ribosomal protein S6. *J. Biol. Chem.* 263:2891 (1988).

32. Monk RJ, Meyuhas O, and Perry RP: Mammals have multiple genes for individual ribosomal proteins. *Cell* 24:301 (1981).

33. D'Eustachio P, Meyuhas O, Ruddle F, and Perry RP: Chromosomal distribution of ribosomal protein genes in the mouse. *Cell* 24:307 (1981).

34. Davies B, Feo S, Heard E, and Fried M: A strategy to detect and isolate an intron-containing gene in the presence of multiple processed pseudogenes. *Proc. Natl. Acad. Sci. U.S.A.* 86:6691 (1989).

35. Glück A, Chan YL, Lin A, and Wool IG: The primary structure of rat ribosomal protein S10. *Eur. J. Biochem.* 182:105 (1989).

36. Lin A, Chan YL, McNally J, Peleg D, Meyuhas O, and Wool IG: The primary structure of rat ribosomal protein L7: the presence near the amino terminus of L7 of five tandem repeats of a sequence of 12 amino acids. *J. Biol. Chem.* 262:12665 (1987).

37. Chan YL, Lin A, Paz V, and Wool IG: The primary structure of rat ribosomal protein S8. *Nucleic Acids Res.* 15:9451 (1987).

38. Paz V, Olvera J, Chan YL, and Wool IG: The primary structure of rat ribosomal protein L26. *FEBS Lett.* 251:89 (1989).

39. Dingwall C and Laskey RA: Protein import into the cell nucleus. *Annu. Rev. Cell Biol.* 2:367 (1986).

40. Moreland RB, Nam HG, Hereford LM, and Fried HM: Identification of a nuclear localization signal of a yeast ribosomal protein. *Proc. Natl. Acad. Sci. U.S.A.* 82:6561 (1985).

41. Matheson AT: Ribosomes of the archaebacteria, in Woese CR and Wolfe RS (eds): *Bacteria: A Treatise on Structure and Function* Vol. 8, Orlando, Academic Press, 1985, 345.

42. Chan YL, Lin A, McNally J, and Wool IG: The primary structure of rat ribosomal protein L5: a comparison of the sequence of amino acids in the proteins that interact with 5 S rRNA. *J. Biol. Chem.* 262:12879 (1987).

43. Auer J, Lechner K, and Böck A: Gene organization and structure of two transcriptional units from *Methanococcus* coding for ribosomal proteins and elongation factors. *Can. J. Microbiol.* 35:200 (1989).
44. Wool IG and Stöffler G: Structure and function of eukaryotic ribosomes, in Nomura M, Tissieres A, and Lengyel P (eds): *Ribosomes* Cold Spring Harbor, N.Y., Cold Spring Harbor Laboratory, 1974, 417.
45. Lin A, Wittmann-Liebold B, McNally J, and Wool IG: The primary structure of the acidic phosphoprotein P2 from rat liver 60S ribosomal subunits: comparison with ribosomal 'A' proteins from other species. *J. Biol. Chem.* 257:9189 (1982).
46. Wool IG: Studies of the structure of eukaryotic (mammalian) ribosomes, in Hardesty B and Kramer G (eds): *Structure, Function, and Genetics of Ribosomes* New York, Springer-Verlag, 391 (1986).
47. Matheson AT, Möller W, Amons R, and Yaguchi M: Comparative studies on the structure of ribosomal proteins, with emphasis on the alanine-rich, acidic ribosomal 'A' protein, in Chamblis G, Craven GR, Davies J, Davis K, Kahan L, and Nomura M (eds): *Ribosomes: Structure, Function, and Genetics* Baltimore, University Park Press, 297 (1980).
48. Nazar RN: A 5.8S rRNA-like sequence in prokaryotic 23S rRNA. *FEBS Lett.* 119:212 (1980).
49. Barbieri M: *The Semantic Theory of Evolution*. New York, Harwood Academic Publishers, 1985.
50. Cech TR and Bass B: Biological catalysis by RNA. *Annu. Rev. Biochem.* 55:599 (1986).
51. Stern S, Powers T, Changchien LM, and Noller HF: RNA-protein interactions in 30S ribosomal subunits: folding and function of 16S rRNA. *Science* 244:783 (1989).
52. Perry RP: Processing of RNA. *Annu. Rev. Biochem.* 45:605 (1976).
53. Gerbi SA: Evolution of ribosomal DNA, in MacIntyre RJ (ed): *Molecular Evolutionary Genetics* New York, Plenum Press, 1985, 419.
54. Meyuhas O and Perry RP: Construction and identification of cDNA clones for mouse ribosomal proteins: application for the study of r-protein gene expression. *Gene* 10:113 (1980).
55. Meyuhas O, Thompson EA, and Perry RP: Glucocorticoids selectively inhibit translation of ribosomal protein mRNAs in P1798 lymphosarcoma cells. *Mol. Cell. Biol.* 7:2691 (1987).
56. Faliks D and Meyuhas O: Coordinate regulation of ribosomal protein mRNA levels in regenerating rat liver. Study with the corresponding mouse cloned cDNAs. *Nucleic Acids Res.* 10:789 (1982).
57. Pierandrei-Amaldi P, Campioni N, Beccari E, Bozzoni I, and Amaldi F: Expression of ribosomal protein genes in *Xenopus laevis* development. *Cell* 30:163 (1982).
58. Pierandrei-Amaldi P, Bozzoni I, and Cardenali B: Expression of the gene for ribosomal protein L1 in *Xenopus* embryos: alteration of gene dosage by microinjection. *Genes Dev.* 2:23 (1988).
59. Geyer LF, Meyuhas O, Perry RP, and Johnson LF: Regulation of ribosomal protein mRNA content and translation in growth-stimulated mouse fibroblasts. *Mol. Cell. Biol.* 2:685 (1982).

60. Gourse RL, Sharrock RA, and Nomura M: Control of ribosome synthesis in *Escherichia coli* in Hardesty B and Kramer G (eds): *Structure, Function, and Genetics of Ribosomes* New York, Springer-Verlag, 766 (1986).

61. Lindahl L and Zengel JM: Ribosomal genes in *Escherichia coli. Annu. Rev. Genet.* 20:297 (1986).

62. Nomura M, Gourse R, and Banghman G: Regulation of the synthesis of ribosomes and ribosomal components. *Annu. Rev. Biochem.* 53:75 (1984).

63. Mager WH: Control of ribosomal protein gene expression. *Biochim. Biophys. Acta* 949:1 (1988).

64. Leer RJ, van Raamsdonk-Duin MMC, Kraakman P, Mager WH, and Planta RJ: The genes for yeast ribosomal proteins S24 and L46 are adjacent and divergently transcribed. *Nucleic Acids Res.* 13:701 (1984).

65. Nakamichi N, Rhoads D, and Roufa DJ: The Chinese hamster cell emetine resistance gene: analysis of cDNA and genomic sequences encoding ribosomal protein S14. *J. Biol. Chem.* 258:13236 (1983).

66. Chen IT and Roufa DJ: The transcriptionally active human ribosomal protein S17 gene. *Gene* 70:107 (1988).

67. Atchison ML, Meyuhas O, and Perry RP: Localization of transcriptional regulatory elements and nuclear factor binding sites in the mouse ribosomal protein gene rpL32. *Mol. Cell. Biol.* 9:2067 (1989).

68. Chung S and Perry RP: The importance of introns for the expression of the mouse ribosomal protein gene rpL32. *Mol. Cell. Biol.* 9:2075 (1989)

69. Harihan N and Perry RP: A characterization of the elements comprising the promoter of mouse ribosomal protein gene RPS16. *Nucleic Acids Res.* 17:5323 (1989).

70. Breathnach R and Chambon P: Organization and expression of split genes coding for proteins. *Annu. Rev. Biochem.* 50:349 (1981).

71. Olson BH and Goerner GL: Alpha sarcin, a new antitumor agent. I. Isolation, purification, chemical composition, and the identity of a new amino acid. *Appl. Microbiol.* 13:314 (1965).

72. Olson BH, Jennings JC, Roga V, Junek AJ, and Schuurmans DM: Alpha sarcin, a new antitumor agent. II. Fermentation and antitumor spectrum. *Appl. Microbiol.* 13:322 (1965).

73. Fernandez-Puentes C and Vazquez D: Effects of some proteins that inactivate the eukaryotic ribosome. *FEBS Lett.* 78:143 (1977).

74. Conde FP, Fernandez-Puentes C, Montero MTV, and Vazquez D: Protein toxins that catalytically inactivate ribosomes from eukaryotic microorganisms. Studies on the mode of action of alpha sarcin, mitogillin and restrictocin: response to alpha sarcin antibodies. *FEMS Microbiol. Lett.* 4:349 (1978).

75. Hobden AN and Cundliffe E: The mode of action of alpha sarcin and a novel assay of the puromycin reaction. *Biochem. J.* 170:57 (1978).

76. Schindler DG and Davies JE: Specific cleavage of ribosomal RNA caused by alpha sarcin. *Nucleic Acids Res.* 4:1097 (1977).

77. Endo Y and Wool IG: The site of action of α-sarcin on eukaryotic ribosomes. The sequence at the α-sarcin cleavage site in 28S ribosomal ribonucleic acid. *J. Biol. Chem.* 257:9054 (1982).

78. Endo Y, Huber PW, and Wool IG: The ribonuclease activity of the cytotoxin α-sarcin. The characteristics of the enzymatic activity of α-sarcin with ribosomes and ribonucleic acids as substrates. *J. Biol. Chem.* 258:2662 (1983).

79. Wool IG: The mechanism of action of the cytotoxic nuclease α-sarcin and its use to analyze ribosome structure. *Trends Biochem. Sci.* 9:14 (1984).

80. Raué HA, Klootwijk J, and Musters W: Evolutionary conservation of structure and function of high molecular weight ribosomal RNA. *Prog. Biophys. Mol. Biol.* 51:77 (1988).

81. Hobden AN: The Action of α-Sarcin. Ph.D. dissertation. University of Leicester, England, (1978).

82. Cahn F, Schachter EM, and Rich A: Polypeptide synthesis with ribonuclease-digested ribosomes. *Biochim. Biophys. Acta* 209:512 (1970).

83. White TC, Rudenko G, and Borst P: Three small RNAs within the 10 kb trypanosome rRNA transcription unit are analogous to domain VII of other eukaryotic 28S rRNAs. *Nucleic Acids Res.* 14:9471 (1986).

84. Endo Y, Mitsui K, Motizuki M, and Tsurugi K: The mechanism of action of ricin and related toxic lectins on eukaryotic ribosomes. The site and the characteristics of the modification of 28S ribosomal RNA caused by the toxins. *J. Biol. Chem.* 262:5908 (1987).

85. Endo Y and Tsurugi K: RNA N-glycosidase activity of ricin A-chain. Mechanism of action of the toxic lectin ricin on eukaryotic ribosomes. *J. Biol. Chem.* 262:8128 (1987).

86. Endo Y, Tsurugi K, Yutsudo T, Takeda Y, Ogasawara T, and Igarashi K: Site of action of a Vero toxin (VT2) from *Escherichia coli* 0157:H7 and of Shiga toxin on eukaryotic ribosomes. *Eur. J. Biochem.* 171:45 (1988).

87. Endo Y and Tsurugi K: The RNA N-glycosidase activity of ricin A-chain. The characteristics of the enzymatic activity of ricin A-chain with ribosomes and with rRNA. *J. Biol. Chem.* 263:8735 (1988).

88. Montanaro L, Sperti S, Mattioli A, Testoni G, and Stirpe F: Inhibition by ricin of protein synthesis *in vitro*. Inhibition of the binding of elongation factor 2 and of adenosine diphosphate-ribosylated elongation factor 2 to ribosomes. *Biochem. J.* 146:127 (1975).

89. Hausner TP, Atmadja J, and Nierhaus KH: Evidence that the G^{2661} region of 23S rRNA is located at the ribosomal binding sites of both elongation factors. *Biochimie* 69:911 (1987).

90. Moazed D, Robertson JM, and Noller HF: Interaction of elongation factors EF-G and EF-Tu with a conserved loop in 23S RNA. *Nature* 334:362 (1988).

91. Endo Y, Glück A, Chan YL, Tsurugi K, and Wool IG: RNA-protein interaction: An analysis with RNA oligonucleotides of the recognition by α-sarcin of a ribosomal domain critical for function. *J. Biol. Chem.* (in press).

92. Endo Y, Chan YL, Lin A, Tsurugi K, and Wool IG: The cytotoxins α-sarcin and ricin retain their specificity when tested on a synthetic oligoribonucleotide (35-mer) that mimics a region of 28S ribosomal ribonucleic acid. *J. Biol. Chem.* 263:7917 (1988).

93. Milligan JF, Groebe DR, Witherell GW, and Uhlenbeck OC: Oligoribonucleotide synthesis using T7 RNA polymerase and synthetic DNA templates. *Nucleic Acids Res.* 15:8783 (1987).

94. Peattie DA, Douthwaite S, Garrett RA, and Noller HF: A "bulged" double helix in a RNA-protein contact site. *Proc. Natl. Acad. Sci. U.S.A.* 78:7331 (1981).
95. Seong BL and RajBhandary UL: *Escherichia coli* formylmethionine tRNA: Mutations in $\frac{GGG}{CCC}$ sequence conserved in anticodon stem of initiator tRNAs affect initiation of protein synthesis and conformation of anticodon loop. *Proc. Natl. Acad. Sci. U.S.A.* 84:334 (1987).
96. Khan AS and Roe BA: Aminoacylation of synthetic DNAs corresponding to *Escherichia coli* phenylalanine and lysine tRNAs. *Science* 241:74 (1988).
97. Terao K, Uchiumi T, Endo Y, and Ogata K: Ricin and α-sarcin alter the conformation of 60S ribosomal subunits at neighboring but different sites. *Eur. J. Biochem.* 174:459 (1988).
98. Gale EF, Cundliffe E, Reynolds PE, Richmond MH, and Waring MJ: Antibiotic inhibitors of ribosome function, in *The Molecular Basis of Antibiotic Action* 2nd ed., New York, Wiley-Interscience, 1981, 402.
99. Walker TA, Endo Y, Wheat WH, Wool IG, and Pace NR: Location of 5.8S rRNA contact sites in 28S rRNA and the effect of α-sarcin on the association of 5.8S rRNA with 28S rRNA. *J. Biol. Chem.* 258:333 (1983).
100. White GA, Wood T, and Hill WE: Probing the α-sarcin region of *Escherichia coli* 23S rRNA with a cDNA oligomer. *Nucleic Acids Res.* 16:10817 (1988).
101. Devi KRG, Chan YL, and Wool IG: The primary structure of rat ribosomal protein S4. *Biochim. Biophys. Acta* 1008:258 (1989).
102. Tanaka T, Kuwano Y, Ishikawa K, and Ogata K: Nucleotide sequence of cloned cDNA specific for rat ribosomal protein S11. *J. Biol. Chem.* 260:6329 (1985).
103. Lin A, Chan YL, Jones R, and Wool IG: The primary structure of rat ribosomal protein S12: The relationship of rat S12 to other ribosomal proteins and a correlation of the amino acid sequences of rat and yeast ribosomal proteins. *J. Biol. Chem.* 262:14343 (1987).
104. Nakanishi O, Oyanagi M, Kuwano Y, Tanaka T, Nakayama T, Mitsui H, Nabeshima Y, and Ogata K: Molecular cloning and nucleotide sequence of cDNAs specific for rat liver ribosomal proteins S17 and L30. *Gene* 35:289 (1985).
105. Itoh T, Otaka E, and Matsui KA: Primary structures of ribosomal protein YS25 from *Saccharomyces cerevisiae* and its counterparts from *Schizosaccharomyces pombe* and rat liver. *Biochemistry* 24:7418 (1985).
106. Kuwano Y, Nakanishi O, Nabeshima Y, Tanaka T, and Ogata K: Molecular cloning and nucleotide sequence of DNA complementary to rat ribosomal protein S26 messenger RNA. *J. Biochem.* 97:983 (1985).
107. Nakamura H, Tanaka T, and Ishikawa K: Nucleotide sequence of cloned cDNA specific for rat ribosomal protein L7a. *Nucleic Acids Res.* 17:4875 (1989).
108. Devi KRG, Chan YL, Wool IG: The primary structure of rat ribosomal protein L18. *DNA* 7:157 (1988).
109. Aoyama Y, Chan YL, Meyuhas O, and Wool IG: The primary structure of rat ribosomal protein L18a. *FEBS Lett.* 247:242 (1989).
110. Devi KRG, Chan YL, and Wool IG: The primary structure of rat ribosomal protein L21. *Biochem. Biophys. Res. Commun.* 162:364 (1989).

Eukaryotic Ribosomes 33

111. Tanaka T, Kuwano Y, Ishikawa K, and Ogata K: Nucleotide sequence of cloned cDNA specific for rat ribosomal protein L27. *Eur. J. Biochem.* 173:53 (1988).
112. Tanaka T, Kuwano Y, Kuzumaki T, Ishikawa K, and Ogata K: Nucleotide sequence of cloned cDNA specific for rat ribosomal protein L31. *Eur. J. Biochem.* 162:45 (1987).
113. Rajchel A, Chan YL, and Wool IG: The primary structure of rat ribosomal protein L32. *Nucleic Acids Res.* 16:2347 (1988).
114. Aoyama Y, Chan YL, and Wool IG: The primary structure of rat ribosomal protein L34. *FEBS Lett.* 249:119 (1989).
115. Tanaka T, Wakasugi K, Kuwano Y, Ishikawa K, and Ogata K: Nucleotide sequence of cloned cDNA specific for rat ribosomal protein L35a. *Eur. J. Biochem.* 154:523 (1986).
116. Gallagher MJ, Chan YL, Lin A, and Wool IG: Primary structure of rat ribosomal protein L36a. *DNA* 7:269 (1988).
117. Lin A, McNally J, and Wool IG: The primary structure of rat liver ribosomal protein L37: homology with yeast and bacterial ribosomal proteins. *J. Biol. Chem.* 258:10664 (1983).
118. Tanaka T, Aoyama Y, Chan YL, and Wool IG: The primary structure of rat ribosomal protein L37a. *Eur. J. Biochem.* 183:15 (1989).
119. Lin A, McNally J, and Wool IG: The primary structure of rat liver ribosomal protein L39. *J. Biol. Chem.* 259:487 (1984).

Transfer RNAs and Aminoacyl-tRNA Synthetases

Jacques Lapointe
Départment de Biochimie,
Faculté des Sciences et de Génie,
Université Laval, Québec, Canada

Richard Giegé
Laboratoire de Biochimie
Institut de Biologie Moléculaire et Cellulaire du CNRS
Strasbourg Cedex, France

2.1. TRANSFER RIBONUCLEIC ACIDS

2.1.1. The Biological Necessity of tRNA

The existence and function of tRNA molecules were predicted by Crick in 1955[1] before their biochemical discovery.[2] According to the adaptor hypothesis he formulated, specific RNA molecules would serve to accommodate the proper amino acids on the template codons of mRNAs. Originally Crick thought that the adaptor molecules were just trinucleotides corresponding to the anticodons. In fact they are much more sophisticated and contain in addition to the anticodons about 70 to 90 nucleotides folding in a versatile three-dimensional structure. The experimental demonstration that the tRNA moiety of aminoacyl-tRNA is responsible for the selection of the adaptor molecules by the codons of mRNA was performed by Chapeville et al.[3] in their classical experiment where alanine was incorporated into a polypeptide chain from Ala-tRNA^{Cys} obtained after Raney nickel treatment of Cys-tRNA^{Cys}.

2.1.2. Cytoplasmic and Organellar tRNA Structure

Sequences—The first tRNA to be sequenced was that specific for alanine from the yeast Saccharomyces cerevisiae.[4] To date about 1000 sequences of tRNAs or their genes are known (see the sequence compilation in Reference 5). They can all be folded into a cloverleaf secondary structure comprising three stem and loop regions, a variable region and a terminal 3' CCA-end to which amino acids are attached. In that cloverleaf structure some of the residues are conserved or semi-conserved in almost all sequences (U8, A14, G18, G19, U33, G53, T54, ψ55, C56, A58, C61, C74, C75 and A76 for conserved residues and R15 and Y48 for semi-conserved residues); the other residues, mainly those located in stem regions, present a highly variable, although not random, statistical distribution.[6] This variability is likely to be related to the functional identities of the various tRNA species.[7] On the other hand, the sequence similarities, especially those occurring between tRNAs from different phyla belonging to the same anticodon families, account for the early divergence of tRNA species at the very beginning of evolution, much before the emergence of eukaryotes.[8] Thus, most of the structural characteristics of present tRNAs already existed in ancient tRNAs, a conclusion that implies that the functioning of eukaryotic tRNAs must be closely related to that of prokaryotic species. This is the reason why we also discuss in that review some central concepts arising from studies on prokaryotic tRNAs and aminoacyl-tRNA synthetases.

According to the length of the variable region one distinguishes two families of tRNAs, class I with short variable loops of four to five nucleotides and class II with a variable region of ten or more nucleotides. This latter class comprises only leucine, serine, and prokaryotic tyrosine-specific tRNAs. A distinction between elongator and initiator tRNAs which mainly relies on sequence peculiarities in the T-loop (G53AUC56 in initiator and G53TψC56 in elongator tRNAs) is also worth mentioning.

In contrast to other RNAs, tRNAs contain a high number of modified nucleotides (reviewed in Reference 9a and 9b), some of them common to almost all species, such as dihydrouridine located in the so-called D-loop and ribothymine in the T-loop. Some other modified residues are characteristic of specific tRNAs, such as the hypermodified Y residue at position 37 in tRNA[Phe] or the Q residue at the first position in the anticodon of certain tRNA[Asp] species. At present more than 50 modified nucleotides have been discovered in eukaryotic tRNAs, one of the latest being a residue hypermodified on its ribose moiety, *O*-ribosylphosphate-adenosine, which was found at position 65 in the T-stem of initiator tRNA[Met] from yeast.[10]

Bizarre tRNAs in mitochondria—tRNA structures that show considerable variation in both the size and the sequence of D- and T-loops have been found in the mitochondrial genome of metazoa.[11] Because of methodological difficulties in isolation of the gene products, very little is known about the biochemistry of these bizarre tRNAs. However, in the case of bovine heart mitochondrial tRNA^{Ser} lacking the D-arm, it was found that only the mitochondrial synthetase, and not those from *Escherichia coli* or yeast, charge this tRNA.[12]

Three-dimensional structures—Only two crystallographic structures of elongator tRNAs are known at atomic resolution, both from *S. cerevisiae*. This small number most probably reflects the difficulty in obtaining good tRNA crystals. The first structure solved was that specific for phenylalanine (reviewed in Reference 13). It shows the characteristic L-shaped folding of the molecule with the amino acid acceptor CCA-group and the anticodon at both ends of the L-fold and the T- and D-loops at its corner. This structure gives the architectural significance to the conserved and semi-conserved residues of tRNA which mainly cluster near the junction of the arms of the L-conformation.[14, 15] The second structure solved was that of tRNA^{Asp}.[16, 17] As anticipated this tRNA is also L-shaped, but presents some striking conformational peculiarities as compared to tRNA^{Phe}. In fact, its X-ray structure mimics the conformation of a tRNA interacting with messenger RNA. This is the consequence of the crystal packing of tRNA^{Asp} with anticodon/anticodon (GUC/GUC) interactions in the crystal lattice which leads to conformational changes in the D-/T-loop region (opening of the G19-C56 base pair). Solution studies with tRNAs forming dimers confirmed this interpretation.[18] The structure of the initiator tRNAs from *E. coli* and yeast have also been known for a decade[19, 20] and for the yeast species the resolution was markedly improved recently;[21] these initiators present conformations resembling those of elongator tRNAs, which is not unexpected since they possess essentially the same frame of conserved residues.

Noticeable are the conformations of the anticodon and T-loops. Although these loops are built up with seven nucleotides, they fundamentally differ in conformation. The T-loop contains an internal Hoogsteen reverse base pair between residues 54 and 58 which is stacked over the last base pair of the helical T-arm. This gives rise to two quasi-loops of two and five nucleotides which are stabilized by H-bonding between P60 and both C61 and ribose-58. Such a structure facilitates interdigitation of T-loops with D-loops. In contrast, the anticodon loop is designed to make the anticodon residues accessible for interaction with the codons of mRNA. This is achieved by a stacking arrangement of anticodon bases over the 3'-adjacent residue-37, itself stacked over the 32-38 base pair, and a sharp turn of the ribose-phosphate backbone after conserved residue U33. Watson-Crick pairing with codons thus becomes possible as shown in the crystal structure of duplex tRNA^{Asp}.[16, 17] In anticodon loops, residue 37 is always a modified or hypermodified purine which acts as a regulatory signal for codon reading, but might also be involved in other events like tRNA aminoacylation (see below). As a consequence of the particular conformation of the anticodon loop of elongator tRNAs, this domain presents a remarkable electrostatic potential which makes it the most positive region of the molecule, a property which may account for the ability of the anticodon to strongly associate with negatively charged mRNA.[22]

Magnesium ions are an integral part of the tRNA structure and are found in the anticodon loop and P10 region.[13, 17] This divalent ion is an absolute requirement for the crystallization of tRNAs, as are polyamines.[23] However, the localization of the polycationic polyamines in the crystal structures is not clearly defined, a fact which can be explained by multiple and overlapping binding sites as shown by footprinting experiments with affinity photoactivable polyamine derivatives.[25] Finally, tRNAs contain internal structural water (reviewed in Reference 24) and are surrounded by a solvent shell whose structure is modulated by the nature of the monovalent counterions.[26]

Conformations in solution and structural models—The development of chemical structural probes for mapping RNA was particularly useful in the tRNA field for approaching the conformation of a variety of these molecules. Three such probes are particularly worth mentioning: ethylnitrosourea, a phosphate alkylating reagent; dimethylsulfate, specific of N7 of G and of N3 of C residues and diethylpyrocarbonate, specific of N7 of A residues. Studies on both tRNA^{Phe} and tRNA^{Asp} have shown that almost all chemical solution data are explained by the crystal structures, thus demonstrating that chemical mapping reflects the geometry of those tRNAs.[27–29] For instance and interestingly, the lack of reactivity of phosphate 60 in all tRNAs so far mapped with ethylnitrosourea brought the proof of an intrinsic T-loop conformation in tRNAs linked to the presence of the constant C61 residue in the G-C base pair closing the T-loop.[30]

The potentialities of chemical reagents to map RNA conformations gave the impetus to computer-aided modeling on tRNAs reluctant to crystallize. Combining chemical mapping and knowledge of crystallographic informations on tRNA^{Asp} and tRNA^{Phe}, it was possible to model a three-dimensional structure of yeast tRNA^{Ser}, a tRNA of class II with a large variable region.[31] In this model the variable arm is not far from the plane of the common L-shaped structure. Other models of noncanonical tRNAs, also based on chemical modification studies, were built for bovine and human mitochondrial tRNA^{Ser} [32] and valylatable turnip yellow mosaic virus (TYMV) tRNA-like domain.[33, 34]

2.1.3. tRNA Genes and Operons

Because of the high demand for tRNA molecules in eukaryotes, tRNA genes for a given tRNA species can be reiterated ten to several hundred times depending on the organism (reviewed in References 35 and 36). Not all tRNA genes code for functional tRNA isoacceptors. This is because the products of pseudogenes contain mutations in their structural part and thus are not processed correctly, as was shown for the human tRNA^{Val} family.[37] The tRNA genes contain extended 5'- and 3'-flanking sequences,[38] and often introns located 3' of the anticodon sequences.[35] They are often clustered and sometimes transcribed into multimeric tRNA precursors by RNA polymerase III, in a process under the control of intragenic DNA sequences (5'- and 3'- Internal Control Regions at positions 9 to 19 and 52 to 58, respectively), but in which flanking regions also have a role in determining transcription rates (reviewed in Reference 38).

Clustered tRNA genes are also found in organelle genomes. In *Zea mays* chloroplast genome for instance, a sequence coding for duplex tRNA and containing introns of about 1000 nucleotides was found in the 16-23S spacer in a ribosomal RNA operon.[39] Mitochondrial genomes show distinct structural organizations of tRNA genes. Whereas in fungi, echinoderms, and other lower phylae, genes are mainly clustered, they are predominantly dispersed in vertebrates (reviewed in Reference 40).

2.1.4. tRNA Processing

The processing of tRNAs involves several highly specialized steps in both the nucleolar and cytoplasmic compartments of the cells. After transcription pre-tRNAs have to be trimmed at both 5'- and 3'-ends, spliced to remove intervening sequences and finally modified at specific nucleotide positions. Specific enzymes are involved at each step, such as RNAse P, an enzyme containing catalytic active RNA, for the processing of the 5'-end (reviewed in Reference 41). The splicing mechanism of tRNA precursors is rather well understood: in yeast it involves an endonuclease activity removing the intron followed by a multistep ligation process.[42] Since introns are located near the anticodon sequences, it is likely that pre-tRNAs are already folded into the classical three-dimensional tRNA conformation, which suggests that exon structures are major recognition elements for the splicing machinery. Newly synthesized pre-tRNA cannot enter the cytoplasm by diffusion processes and their transport through the nucleolar membrane is coupled with their processing.[43]

An important aspect of tRNA processing concerns the biosynthesis of modified nucleotides.[9, 44–46] For a set of cellular tRNAs, about 100 distinct modification enzymes are involved. In eukaryotic cells these enzymes are found in part in the nucleus and in part in the cytoplasm, and it is now well established that they act in a sequential order.[47] With RNA engineering and microinjection methodologies it was possible to have a deeper insight into the sequence requirements in tRNA of some of these enzymes.[48] For instance, by enzymatic microsurgery it was possible to modify the GUC canonical anticodon of yeast tRNA^{Asp} or to transfer it into other tRNAs. After microinjection of the chimeric tRNAs into *Xenopus laevis* oocytes it could be shown that Q-insertase requires the presence of a U33-G34-U35 sequence for the conversion of G34 into Q34.[49] For Gm34 methylase, it was found that the neighboring nucleotides are not essential, but the structure beyond the anticodon region is.[50] More recently, the *in vitro* preparation of tRNA transcripts deprived of any modified nucleotide gave adequate substrates for studying enzymes specific for positions outside the anticodon loop.[51] With these new tools in hand, it is likely that the biochemistry of nucleotide posttranscriptional modifications will be deciphered in the near future.

Little is known about the incidence of modifications on the overall conformation of tRNA. Some recent studies point to a stabilizing role of modified residues. This is the case of the methylation of G18 (Gm) and A58 (m^{1}A) and thiolation of T54 (s^{2}T) in tRNAs from extreme thermophiles which are responsible for the thermostability of these tRNAs by rigidifying their structure.[52] Consistently, transcripts lacking modified nucleotides possess a relaxed conformation.[53, 54]

2.1.5. tRNA Population and Codon Usage

Transfer RNAs are present in cells in varying amounts. Their relative abundance is correlated with the codon usage in the mRNAs being translated.[55, 56] For isoacceptor tRNA species, structural variations might rely upon differences in posttranscriptional modifications, but in fact, they mainly consist in major sequence differences. Modification in the anticodon loop, however, plays a major regulatory role in the exactness and fidelity of codon reading.[44–46] Moreover, modification of anticodon bases can dictate the choice of synonymous codons by a tRNA.[57] Sometimes minor species are found, which might be involved in regulatory events or in suppression activity. In that case the modification level of the anticodon can modulate suppression, as observed for *Drosophila* tRNA^{Tyr} in which modification of the first position in the anticodon from G to Q abolishes suppression.[58] Suppression might also be regulated by the intracellular amount of suppressors. As illustrated in higher eukaryotes for a glutamine suppressor tRNA, this amount can be specifically increased after viral infection.[59]

2.2. AMINOACYL-tRNA SYNTHETASES

2.2.1. An Overview

Named for their function as catalysts of the first step of protein biosynthesis, the aminoacyl-tRNA synthetases (aaRSs) (or aminoacyl-tRNA ligases, their correct but less often used appellation) form a family of about 20 members whose structures are unexpectedly varied for enzymes catalyzing similar reactions with similar substrates. For reasons not yet clearly understood, several organisms have more than one aaRS specific for a given amino acid, or lack of given aaRS whose function is replaced by the combined action of another aaRS and of another enzyme. The aaRSs have been studied extensively because of their important role in the fidelity of gene expression, because they are a good model system to study protein/RNA interactions, and because it is relatively easy to purify them, taking advantage of affinity chromatography methods[60a] and/or of overproducing strains. So far, most of the information on aaRS structure and biosynthesis has been obtained with prokaryotic cells. Therefore, even in a book on eukaryotic translation, a discussion of the current models for aaRS structure, function(s), biosynthesis, and evolution must refer to results obtained with prokaryotes.

Their quaternary structures are diverse, but are generally conserved throughout evolution among aaRSs specific for the same amino acid. For several aaRSs, deletion of parts of the polypeptide revealed a linear arrangement of functional domains which keep their activity in the absence of other domains; on the other hand, other aaRSs lose their activity after the loss of a short COOH-terminal segment. A significant amino acid sequence similarity is present throughout the lengths of prokaryotic and eukaryotic aaRSs specific for the same amino acid,[60b] except for the presence in the eukaryotic enzymes of additional segments which may have a function in their association into large complexes and in their localization in the cell. In each of two subsets of aaRSs specific for different amino acids, some amino acid sequence motifs involved in substrate binding are also conserved throughout evolution. Significant primary structure similarities over substantial parts of the lengths of smaller subsets of aaRSs specific for different amino acids suggested that at least certain regions of these aaRSs derive from a common ancestor. Domains or regions dispensable for the catalysis of the aminoacylation reaction have been shown in some cases to mediate other functions, or to influence the aminoacylation activity. The domains or regions responsible for other functions, such as the mRNA maturase activity of some mitochondrial aaRSs, have not yet been identified. Considering the large size of several aaRSs compared with the expected size for the smallest polypeptide able to catalyze the aminoacylation reaction (the core aaRS), it is tempting to speculate that several activities or regulatory sites coupling protein biosynthesis to other metabolic pathways will be discovered in these enzymes. The search for core aaRSs is under way in several laboratories, and will be facilitated by the knowledge of the three-dimensional structures of tRNA/aaRS complexes, presently established at high resolution only for the *E. coli* tRNA^{Gln}/GlnRS complex[61] and the yeast tRNA^{Asp}/AspRS complex.[62a] Recently, from the existence of two mutually exclusive sets of amino acid sequence motifs in the known aaRSs, it was proposed that these enzymes partition into two classes;[62b] keeping with this conclusion the first known 3-dimensional structures of class II aaRSs, the *E. coli* SerRS[62c] and yeast AspRS (in tRNA complexed form),[62a] are completely different from the three known structures of class I aaRSs (those specific for methionine, tyrosine, and glutamine). One of the most striking differences is the presence of a Rossman-fold-like domain in these three class I aaRSs but not in the class II SerRS[62c] and AspRS.[62a] On the other hand, SerRS has an N-terminal domain that forms an antiparallel α-helical coiled-coil, stretching 60Å out in the solvent, and stabilized by interhelical hydrophobic interactions; this structure, absent in the three class I aaRSs of known structure, may be involved in tRNA binding.[62c] Further, there appears to be a relation between this structural classification of the aaRSs and their catalytic properties: indeed most of the class I aaRSs are known to aminoacylate their cognate tRNAs on the 2'-OH of the terminal adenosine, whereas those of class II (with one exception) do it on the 34-OH.[62b, d]

Although the intracellular concentrations of the aaRSs increase similarly with increasing exponential growth rates in bacteria (the so called metabolic regulation), different regulatory mechanisms, including transcriptional and translational autoregulation and transcription attenuation, have been found for several prokaryotic genes encoding aaRSs specific for different amino acids (see Reference 63 for a review). The results obtained with a few eukaryotic aaRSs suggest that a similar diversity will be found.

2.2.2. Structural Diversity of the Aminoacyl-tRNA Synthetases and some Unifying Principles

The primary and quaternary structures are known for all aaRSs from *E. coli*, for several aaRSs from other bacteria (particularly from thermophilic ones because of the advantages of thermostable enzymes for physicochemical studies), for several cytoplasmic and mitochondrial aaRSs from lower eukaryotes and for a few mammalian ones. A general impression of diversity emerges from such a large family picture which includes monomers, α_2 dimers, α_4 and $\alpha_2\beta_2$tetramers, and where the molecular weights of the native enzymes range from about 50 Da to 380 kDa[64, 65] (reviewed in References 66 and 67), but some unifying principles are also apparent:

1. The quaternary structure of aaRSs specific for a given amino acid is conserved[68] with few exceptions (e.g. Reference 60), from bacteria to mammalian cells, suggesting that it is important for their role as catalyst of tRNA aminoacylation, or for their involvement in another important function acquired at a very early stage of evolution. The quaternary structure of the *E. coli* AlaRS appears to be an example of the latter model since its aminoacylation activity is not apparently altered by the removal of a COOH-terminal domain responsible for the association of four identical subunits,[68] whereas the role of this

enzyme in the transcriptional regulation of its own structural gene might be lost.[69]An example of the quaternary structure requirement for the catalysis of the aminoacylation reaction is the *E. coli* GlyRS where the association of the subunits may generate the active site,[70] as found in many enzymes with more than one substrate.[71]

2. aaRSs specific for different amino acids generally share very little amino acid sequence similarity, in the form of a few short conserved motifs. The first motif to be noticed was the signature sequence,[66] (consensus HIGH, specific for enzymes of class I), found near the amino-terminal end of 10 out of 20 aaRSs of known primary structure (those specific for Arg, Cys, Gln, Glu, Ile, Leu, Met, Tyr, Trp, and Val.[72a-c] These 10 aaRSs also contain the KMSKS motif (see below), and are the members of class I. The signature sequence is present in these aaRSs in all the organisms where they have been studied. The three-dimensional structure of one of them, the TyrRS from *B. Stear-othermophilus,* has been determined at high resolution. Differences between the electron densities of the TyrRS and of its complexes with ATP and with tyrosinyl-adenylate (a stable analog of tyrosyl-adenylate) showed the involvement of several conserved residues in and around this motif in the binding of several atoms of ATP and of tyrosinyl-adenylate (reviewed in Reference 73). These results were confirmed and the model was extended by the characterization of kinetic properties of mutants of the TyrRS where these residues were substituted, using site-specific mutagenesis (reviewed in Reference 74). Model building suggested that two neighboring residues, Thr40 and His45, also present in the *E. coli* TyrRS, were involved in binding the γ-phosphate of ATP, making its α-phosphate more accessible for an attack by the carboxyl group of tyrosine. Site-specific mutagenesis of these residues gave "facticious" enzymes with much lower turn-over numbers (k_{cat}), but virtually unchanged for tyrosine and ATP binding. Model building indicated that these residues are essential for catalysis because they interact with the γ-phosphate group of ATP only in the transition state,[75, 76a] and stabilize that state. In class II aaRSs, "motif 3",[62b] the only motif reported so far to be shared by these 10 aaRSs (those specific for Ala, Asn, Asp, Gly, His, Lys, Phe, Pro, Ser, and Thr), appears to be involed in ATP binding since a single point mutation generated in this region of *E. coli* AsnRS, has a 15-fold increase in its Km for ATP, while all the other kinetic parameters remain unchanged.[76b]

Another motif, whose consensus is KMSKS, is present in segments of the *E. coli* TyrRS and MetRS cross-linked to the 3'-end of their cognate periodate-oxidized tRNAs.[77] It is present in each class I aaRS, in the COOH-terminal half of several aaRSs and was shown to be in a structural domain distinct from the one (NH_2-terminal in prokaryotic aaRSs) containing the HIGH motif. The same approach used with the *E. coli* AlaRS, which does not carry the signature sequence, led to the cross-linking of the 3′-end of tRNA^{Ala} to the Lys73 residue in the sequence RAGGKHNDL located in the adenylate synthesis domain.[78] No sequence similar to this site has yet been reported in synthetases that do not have the KMSKS motif.

A less evident motif is in the process of being discovered in the amino acid binding site of some aaRSs, starting from the structure of the tyrosine binding site of the *B. Stearothermophilus* TyrRS, determined by X-ray diffraction studies of a deletion mutant of this enzyme complexed with tyrosine.[79] All the residues involved in this site are in the α/β NH_2-terminal domain, and one of them, Asp78 is separated from the signature sequence by 26 residues consisting of an α-helix and a β-strand bringing the tyrosine binding site into proximity with the signature sequence. At the same distance from the signature sequence of the IleRS, the amino acid sequence similar to that around Asp78 of the TyrRS was mutated at random, which led to the isolation of a Gly94 → Arg mutant whose K_m for isoleucine is elevated by 6000-fold, whereas the affinities for ATP and tRNA, and the k_{cat} are not altered appreciably;[80] (for a review see Reference 72).

Finally, several small groups of synthetases share low but significant levels of similarity over a large proportion of their lengths. We will discuss later the possibility of closer evolutionary linkages within these groups.

3. aaRSs specific for the same amino acid, and isolated from evolutionarily distant organisms, have a substantial amino acid sequence homology over the length of the bacterial enzymes. This homology is not regularly spread over this length, but defines a limited number of conserved motif (e.g. References 80 and 81). The characterization of mutant enzymes altered in these motifs is likely to identify residues involved in substrate binding sites or in regulatory properties of these aaRSs. Such a motif in the carboxy-terminal domain of the trypsin-modified *E. coli* MetRS, the motif $L^F/_Y\text{-}^R/_K\,^I/_L D$ conserved in the MetRS of *B. Stearothermophilus* and of yeast at the same distances from the HIGH and KMSK motifs, is located in a short C-terminal extension shown by X-ray diffraction studies to fold back towards the N-domain and to link it to the C-domain.[82] The kinetic properties of mutants altered in this extension show that it is involved in the interaction with tRNA^{Met}.

This structural information gathered mostly with prokaryotic aaRSs will be useful to construct eukaryotic aaRS mutants with expected properties, by site-specific mutagenesis of evolutionarily conserved residues in motifs of known function. Conversely, the determination of intron positions in eukaryotic aaRS genes may be useful to identify the regions containing active site residues, since both are frequently found in the same loop regions in α/β domains, at the carboxyl edge of the β-sheet,[83] as found for residues of *B. Stearothermophilus* TyrRS involved in binding tyrosyl-adenylate.[84] The determination of the three-dimensional structure of most eukaryotic aaRSs by X-ray diffraction studies may be difficult because of their large sizes. The use of deletion mutants may facilitate the crystallization and yield more ordered crystals. Even with prokaryotic aaRSs, much structural information was gathered with this approach. Fully active fragments of the *E. coli* MetRS[85] and *B. Stearothermophilus* TyrRS were more amenable to crystallization and structure determination than the native enzymes.[84] With a genetically engineered TyrRS able to activate tyrosine with kinetic properties similar to those of the wild-type enzyme but no longer able to bind tRNA^{Tyr}, the residues (all present in the NH_2-terminal α/β domain) forming hydrogen bonds with tyrosine were identified.[79]

2.2.3. Search for the Core Aminoacyl-tRNA Synthetases and Evolutionary Relationships Between Them

Among aaRSs specific for the same amino acid, the major difference between eukaryotic and prokaryotic enzymes is the presence of amino-terminal and/or carboxy-terminal extensions in the former. Therefore, prokaryotic aaRSs are a more favorable material to search for the core aaRSs, i.e., the shortest polypeptides able to catalyze the aminoacylation reaction.

The *E. coli* AlaRS can be shortened by about half its length, from its carboxy-terminus, without losing its aminoacylation activity; a polypeptide containing the NH_2-terminal 300 residues will activate alanine, but cannot transfer it to tRNA (for a review see Reference 86). On the other hand, the *E. coli* GlnRS and the *S. cerevisiae* AspRS and IleRS are inactivated by the removal of a few C-terminal amino acids;[64, 87, 88] interestingly, the amino acid activation step is not always suppressed (e.g. Reference 64). Similarly, limited proteolytic treatment of yeast AspRS generates a truncated species, with full enzymatic activity, that has lost the first 64 NH_2-terminal residues.[89]

The primary structure of aaRSs from extreme thermophiles may be useful to identify elements of the core enzymes, since the smallest known polypeptide endowed with aminoacylation activity is the 29 kDa amino-terminal structural domain of the MetRS from *Thermus thermophilus,* obtained by limited proteolysis of the native enzyme.[90]

Another promising approach, which resulted in the finding that the core enzyme is not present as a contiguous sequence, is to generate internal deletions in an aaRS in a segment corresponding to a gap in its amino acid sequence when aligned with that of a partially homologous aaRS. Homology between the primary structures of *E. coli* IleRS, MetRS and ValRS, and yeast mitochondrial LeuRS, strongest in the region corresponding to the first 150 amino acids of the amino-terminal domain of the MetRS, suggest that the aaRSs specific for branched-chain amino acids, and the MetRS may have evolved from a common ancestor.[91, 92] Deletions of segments spanning collectively 145 contiguous amino acids of the IleRS corresponding to gaps in its alignment with the MetRS,[93] and known to be parts of connective peptides between three regions of the nucleotide binding fold of the MetRS,[84] yielded active IleRS. Comparison of the primary structure of these related aaRSs also suggests that the metal binding domains containing the motif $Cys-X_2-Cys-X_{9-16}$-$Cys-X_2-Cys$, present near the carboxy-terminus of the IleRS and in the amino-terminal third of the MetRS,[95, 96] but absent in the ValRS,[91] was probably not present in the ancestral progenitor of these four synthetases.

Other subsets of aaRSs specific for different amino acids but having a substantial amino acid sequence similarity, include those specific for glutamate and glutamine, where the greatest similarity covers the first 100 amino acids of the *E. coli* GluRS corresponding to its amino-terminal structural domain, identified by limited proteolysis,[97, 98] and those specific for aspartate and lysine.[99, 100] Analysis of internal deletion mutants of these evolutionary linked aaRSs will certainly contribute to the identification of their cores. Inversely, the evolutionary linkages between the 20 aaRSs will be easier to determine once the core of each member of this family will have been defined. It is important to note that the sorting of the 20 aaRSs into two classes on the basis of the presence of a few structural motifs involved (all or most) in the catalytic reaction, does not exclude the possibility that aaRSs of two different classes may have homologous regions,[62b] such as those involved in the interactions with their cognate tRNAs (for a discussion see 10 and references therein).

2.2.4. Distinctive Features of Eukaryotic Aminoacyl-tRNA Synthetases

The association of mammalian aaRSs into high molecular weight complexes has been first reported about 20 years ago;[102] recent results and models are reviewed in References 103 through 105. Complexes containing all or most of the 20 aaRSs have been obtained in extracts from cells lysed under gentle conditions. Purification of such complexes from various mammalian tissues led to the isolation of a relatively stable ubiquitous multienzyme complex containing 9 aaRSs, specific for Arg, Asp, Glu, Gln, Ile, Leu, Lys, Met, and Pro. This complex is retained by hydrophobic resins. Controlled proteolysis of the complex releases smaller aaRSs that are still active for tRNA aminoacylation but have lost their hydrophobic property. To account for these results, Deutscher[103] proposed a model in which hydrophobic regions of individual enzymes would keep them together and possibly also bound to membranes or other lipids, while the hydrophilic catalytic regions would be exposed to the environment. Eukaryotic aaRSs are larger than the corresponding bacterial enzymes, and include extensions on the amino-terminal and/or on the carboxy-terminal sides of the part homologous to the bacterial aaRSs. The presence of these additional domains is not necessary for the catalysis of the aminoacylation reaction. The N-terminal region of the human AspRS contains a segment likely to be a neutral amphiphilic helix.[106] Such helices are known to have high hydrophobic moments, and are found also in apolipoproteins, peptide hormones, toxins and the signal peptides for import of mitochondrial proteins; moreover, the other mammalian aaRSs whose amino acid sequence is known, the HisRS, does not contain a neutral amphiphilic helix and is not present in the stable complex. Therefore, this helix is a good candidate for the hydrophobic region responsible for the formation of the high molecular weight complex(es). It is not possible, however, to rule out the possibility that all aaRSs are present in a high molecular weight complex *in vivo*, because the purification of the above mentioned complex of the nine synthetases can remove lipids that are responsible for a variety of functional and structural properties of the individual synthetases.[107]

The aaRSs from yeast or bacteria do not form such complexes, but those from all eukaryotes differ from bacterial enzymes by their affinity *in vitro* for polyanions such as immobilized ribosomal RNA or heparin. In fact, chromatography on heparin-Ultrogel is an efficient method to purify them. Deletion of an N-terminal polycationic domain from the yeast AspRS or LysRS does not destroy their aminoacylation activity[89, 108] but eliminates their affinity for polyanions. Interestingly, the truncated domain in AspRS as well as the N-terminal domain of other yeast aaRSs are lysine rich and can be represented in form of amphiphilic α-helices.[89] Moreover, this deletion makes LysRS less stable *in vivo*. In the context of the observation[109, 110] by immunofluorescence microscopy of an association between the high molecular weight complexes of mammalian aaRSs with the endoplasmic reticulum, Cirakoglu et al.[104] suggested that a structural domain carrying a cationic net charge contiguous to the functional domain may be a common feature of the structural organization of the lower eukaryotic aaRSs and may serve to anchor them to polyanionic structures at the site of protein synthesis.

2.2.5. Other Functions of Aminoacyl-tRNA Synthetases

Autoregulation—Regulatory functions have been clearly identified for a few prokaryotic aaRSs. The *E. coli* AlaRS was shown by footprinting studies to bind specifically to a palindromic sequence flanking its structural gene's transcription start site. This binding is enhanced by high concentrations of alanine.[69] It is likely that the part of the AlaRS involved in this interaction with DNA is not essential for the aminoacylation activity; indeed, its tetra- meric structure, probably required for its interaction with the palindromic sequence is due to a carboxy-terminal domain whose presence is not necessary for the AlaRS activity *in vivo*.[69]

The *E. coli* ThrRS controls the translation of its mRNA by binding to a tRNA^{Thr}-like structure at the 5'-end of this mRNA, competing with ribosome binding. The enzyme apparently recognizes similar features on both its tRNA substrate and the tRNA-like translational operator since synthetase mutants that repress to an extreme level the translation of their mRNA also have an increased affinity for the tRNA substrate, whereas mutant ThrRS unable to repress translation binds tRNA^{Thr} with a decreased affinity.[111] This functional intepretation was confirmed by structural studies on the mRNA demonstrating the existence of a conformation mimicking that of the anticodon arm of tRNA^{Thr}.[112a, b]

RNA splicing—The involvement of mitochondrial aaRSs in the splicing of group 1 introns has been found from the homology between an open- reading frame in the locus of nuclear cyt-18 mutant of *Neurospora crassa* defective in splicing, and the amino acid sequences of bacterial TyrRS.[113] Characterization of the TyrRS activity of these mutants and of partial re- vertants provided evidence that the same protein is endowed with TyrRS and splicing activities, but that these two functions are perfomed by separate domains. The role of yeast (*S. cerevisiae* and *S. douglasii*) mitochondrial LeuRS in the splicing of group I and group II introns has been demonstrated by a similar approach;[114, 115] their amino acid sequences are very similar to that of the *E. coli* LeuRS, except for significant features (such as a sequence which could form a "zinc-finger") which may be important for their splicing function and are absent in the *E. coli* LeuRS.

Role in chlorophyll biosynthesis—The GluRS is involved in chlorophyll biosynthesis in higher plants, algae, cyanobacteria, and several photosynthetic and nonphotosynthetic bacteria since the universal precursor for tetrapyrrole compounds, 8-aminolevulinate, is synthesized from a glutamate aminoacy- lated at the 3-terminal adenosine of tRNA^{Glu}.[116] This pathway may allow the coordinate regulation of chlorophyll and protein biosyntheses during plastid development.[117]

Regulation of amino acid biosynthesis—Via their aminoacyl-tRNA product, some aaRSs are involved in the regulation of amino acid biosynthesis in bacteria (for a review, see Reference 118), in yeast (for a review, see Reference 119) and in hamsters.[120] In *S. cerevisiae,* derepression of several genes encoding enzymes in multiple amino acid biosynthetic pathways occurs when mutants, whose IleRS or MetRS are thermosensitive, are grown at semipermissive temperatures, suggesting that either the presence of uncharged tRNA^{Ile} or tRNA^{Met} or a reduced rate of protein synthesis is a more proximal signal for derepression than the depletion of an amino acid pool. Using a thermosensitive mutant of the LeuRS, the same approach was used to show that the system L of amino acid transport in Chinese hamster ovary cells is influenced by the aminoacylation level of tRNA^{Leu} (for a review, see Reference 121).

Multiple aaRSs specific for the same amino acid—Certain cells carry two distinct aminoacyl-tRNA synthetases specific for the same amino acid and encoded by distinct genes. For instance, *E. coli* contains two LysRS which differ by their amino acid sequences and by the regulation of their biosynthesis;[122] one is constitutively expressed, whereas the other is under the control of a heat-shock promoter. Two ThrRSs were recently found in *Bacillus subtilis;* each is sufficient for cell growth, but one of them, is present mostly during vegetative growth, whereas the other is present in higher concentration during sporulation.[123] It remains to be shown what regulatory functions, if any, these cryptic synthetases have in the heat-shock response and in sporulation.

Alarmone biosynthesis—Several aaRSs from prokaryotes and eukaryotes catalyze the formation of Ap$_4$A, a product of the reaction of ATP with an aaRS·aminoacyl-adenylate, the intermediate complex in tRNA aminoacylation. The level of Ap$_4$A increases 50- to 100-fold when mammalian cells in culture pass from the growth arrested phase into the early S phase, prior to cell division.[124] Ap$_4$A and the TrpRS are found tightly associated with DNA polymerase α, suggesting that it is involved in the control of DNA replication (for reviews, see References 125 and 126). Overproduction of LysRS, MetRS, PheRS, or ValRS in *E. coli* increases significantly the intracellular Ap$_4$A concentration,[127] indicating that aaRSs are involved in its *in vivo* synthesis. The Ap$_4$A synthesis by AlaRS, LysRS, MetRS, and PheRS from *E. coli,* yeast, and sheep liver are strongly stimulated by zinc.[128] In this context, it is interesting to note that removal of zinc ions from bovine TrpRS by chelation with a phosphonate analog of Ap$_4$A, destroyed the ability of the enzyme to form tryptophanyl-adenylate, but revealed a hydrolytic activity (ATP $\rightarrow$ ADP + Pi $\rightarrow$ AMP + Pi) which is fully suppressed after incubation of this modified TrpRS with zinc.[129] Rat liver LysRS dissociated from the multi-aaRS complex is sixfold more active than free LysRS in Ap$_4$A synthesis.[130] This and related adenylylated bis (5'-nucleosidyl) tetraphosphates (Ap$_4$A, where N is G, C, or U), called alarmones, occur naturally in prokaryotes and eukaryotes in response to various stresses, and appear to be involved in cellular homeostasis after heat-shock or oxidizing treatment. AppppGpp, formed by the reaction of ppGpp with aaRS·aminoacyl-adenylate complexes, as a response to cross-linking of 4-thiouridine in tRNA by near UV, appear to induce the synthesis of proteins necessary for resistance to near UV irradiation.[131] In this context, it is worth mentioning the correlation between Ap$_4$A accumulation and DNA repair processes.[132]

Phosphorylation of aaRS—Autoantibodies against human HisRS were used to show that this enzyme is phosphorylated in HeLa cells. Subsequently, analysis of the core complex of aaRSs from reticulocytes previously incubated with [^{32}P] revealed that the AspRS, GlnRS, GluRS, LysRS and MetRS are phosphorylated, in all instances on serine. Phosphorylation appeared to have no influence on the integrity of the complex or on the aminacylation activity (for a review, see Reference 133). Phosphorylation of reticulocyte ThrRS and SerRS (not present in the core complex) has no effect on the aminoacylation activity, but increases Ap$_4$A synthesis.[134a] Exposure of reticulocytes to tumor promoting phorbol esters stimulates the phosphorylation of one synthetase in the complex, the GluRS.[134b] The phosphorylation of aaRSs is also linked to physiological changes in *E. coli* since heat shock proteins DnaK and DnaJ participate in the phosphorylation of ThrRS and GlnRS.[135]

Autoantibodies against human aaRS—Autoantibodies against AlaRS, GlyRS, HisRS, IleRS, and ThrRS have been found in the serum of patients afflicted with myositis, a human idiopathic inflammatory myopathy (Reference 136 and references therein). Each patient produces antibodies to a single aaRS.[137] The relation between the presence of these autoantibodies and this disease is not known.

2.2.6. Biosynthesis of Aminoacyl-tRNA Synthetases

The control of aaRS biosynthesis is important not only to optimize the growth rate of cells, but also to ensure the accuracy of *in vivo* aminoacylation. For example, the overproduction of the GlnRS in *E. coli* led to the mischarging of supF amber suppressor tRNA^{Tyr} with glutamine. When the intracellular concentration of tRNA$_2$^{Gln} was increased in these cells, this mischarging was abolished.[138]

The levels of bacterial aaRSs studied so far increase about 2.5-fold for a 7-fold increase in exponential growth rate.[139] It was shown, by quantitating the levels of ten aaRSs purified on the O'Farrell two-dimensional gel system from cells uniformly labeled with [^{14}C] amino acids, that these changes reflect new synthesis of these enzymes. Starvation for the cognate amino acids causes transient derepression of the levels of the aaRS specific for His, Leu, Met, Pro, Ser, Thr, and Val, and a long-term derepression for those specific for Arg, Ile, and Phe;[140] (for a review, see Reference 141). The mechanism regulating the expression of aaRS genes was studied in detail for only a few of them. The tetrameric AlaRS of *E. coli* binds to a palindromic sequence overlapping the -10 region of the alaS promoter, autoregulating its transcription; this repression is enhanced by the binding of alanine to AlaRS.[69] As mentioned above, the *E. coli* ThrRS regulates the translation of its own mRNA.[111] Finally, the expression of *E. coli* pheST operon encoding the PheRS is controlled by the phenylalanyl-tRNA[Phe]-dependent attenuation of its transcription.[142]

In yeast, the genes encoding the IleRS and the LysRS are under the general control of amino acid biosynthesis, which is a transcriptional control for several genes, and leads to a coordinate derepression of several biosynthetic pathways during a limitation of Arg, His, Ile, Leu, Lys, Thr, and Val.[143, 144] On the other hand, the biosynthesis of the yeast MetRS is not under this type of control.[143] The yeast aaRS genes studied so far exist in a single copy in the genome, and are not clustered.[145]

Yeast mitochondria contain a complete set of aaRSs,[146] all encoded by nuclear genes which are distinct from the genes encoding the corresponding cytoplasmic enzymes, for all but two of the mitochondrial aaRSs studied to date. The cytoplasmic and the mitochondrial HisRS and ValRS are produced from common genes, through differential transcription leading to the presence of an NH_2-terminal targeting signal in the latter (e.g. Reference 147 and reviewed in Reference 148). In *N. crassa,* the genes for cytoplasmic and mitochondrial LeuRS are regulated independently: the cytoplasmic enzyme is controlled by the "cross-pathway control system", analogous to the general amino acid control in *S. cerevisiae,* whereas the mitochondrial enzyme is not induced by amino acid limitation.[149]

Little is known yet at the molecular level about the mammalian aaRSs. cDNA clones encoding rat liver AspRS,[150] human ThrRS,[151] the core region of human GlnRS,[152] and hamster and human HisRS[153] were obtained and sequenced. In Chinese hamster ovary cells, methionine restriction leads to a twofold increase in MetRS level, as assessed by specific activity measurement and immunotitration, but the levels of the eight other aaRSs with which it forms a multienzyme complex are not altered. The excess of MetRS is not present in the free form, but the complex isolated under these conditions displayed twice the amount of MetRS relative to the other components, revealing flexibility in the structure of this multienzyme complex.[154] Finally, it is interesting to note that the extremely high content (0.2 to 0.5% of total proteins) of TrpRS measured by radioimmunoblotting in the pancreas of ruminants but not of other mammals, may be connected with unknown functions of this enzyme.[155, 156]

2.3. tRNA AMINOACYLATION

2.3.1. Background—Specific tRNA Aminoacylation is a Kinetic Process

For functional necessity aminoacylation of tRNA has to be very specific, otherwise it would lead to erroneous incorporations of amino acids into proteins. It was therefore a great surprise to find that both the recognition between tRNAs and aaRS[157] and the activation of amino acids (reviewed in Reference 158) are not highly specific, and that mischarging of tRNA can be catalyzed by synthetases.[159] This apparent paradox is resolved if one accepts that the specificity of tRNA aminocylation relies more on kinetic[160] and proofreading or editing[158, 161, 162] events than on strictly specific interactions. Here we will not discuss all aspects of tRNA/aaRS interactions (see References 66 and 67 for reviews on earlier work) but will restrict the discussion to selected topics mainly promoted by recent methodological breakthroughs (see also critical reviews and comments in References 163 through 166).

2.3.2. Specific tRNA/aaRS Interactions

From a low to a high resolution view of tRNA/aaRS complexes— Complexes between tRNAs and aaRSs are usually characterized by dissociation constants in the 10^{-6} to 10^{-7} M^1 range, much higher than for DNA/ protein complexes. This is indicative of rather weak interactions which facilitate, during protein synthesis, a rapid turnover of the tRNA in the aminoacylation systems. The tRNA/aaRS recognition is entropy driven[167] and involves both electrostatic and hydrophobic interactions as shown by the existence of catalytically active complexes in the presence of high concentrations of ammonium sulfate.[168, 169] The unexpected property of ammonium sulfate to permit tRNA/aaRS interactions led to the first crystallization of a complex between a tRNA and its cognate aaRS, that specific for aspartic acid in yeast[3, 170, 171] and later to that of the glutaminyl complex.[172] Complex formation is accompanied by conformational changes of the two types of macromolecules, as seen for instance by small-angle neutron scattering with the valine system from yeast.[173] Beside an important conformational change of the synthetase, this study showed in addition the location of the tRNA which is buried within the enzyme. A more precise view came from a low-resolution X-ray analysis of the yeast tRNA^{Asp}/AspRS complex, which shows a rather elongated shape of the particle with the two tRNA molecules lying on the surface of the aaRS with their faces comprising the variable region.[174a] Shortly after, the structure of the prokaryotic tRNA^{Gln}/GlnRS complex from *E. coli* was solved from a 2.8 Å resolution electron density map; it has an elongated shape with the tRNA spanning on the enzyme from the acceptor stem to the anticodon by the inside of the L. The tRNA retains its overall conformation as seen in free tRNA^{Asp} or tRNA^{Phe}, but with a disruption of the terminal U1-A72 base pair of the acceptor stem, a distortion of the 3'-GCCA end, and an alteration in the anticodon loop conformation.[61] At the same time, new crystals of the eukaryotic tRNA^{Asp}/AspRS complex diffracting to high resolution were obtained[174b] and the structure solved soon later[62a] shows striking differences with that of the glutaminyl system, a fact not surprising since AspRS and GlnRS belong to different synthetase classes.[62b] Among them, the 3'-GCCA end exhibits a regular helical conformation and the terminal U1-A72 base pair of the acceptor stem is not disrupted. However, like in the glutaminyl system, the anticodon loop undergoes a large conformation change. Of interest to recall are the different topologies of GlnRS and AspRS,[61, 62a] and particularly in the catalytic site domain the presence of a five-stranded parallel β-pleated dinucleotide binding fold in GlnRS replaced by an antiparallel β-sheet flanked by 3 α helices in AspRS.

Contacts between tRNA and aaRS—Ultraviolet light-induced cross-linking of tRNA to synthetases was the first successful approach which permitted direct definition of contact points between the two macromolecules. The method, pioneered by Schimmel and co-workers, led to a schematic model of interaction between yeast tRNA[Phe] and its cognate synthetase in which the major part of the binding site for the synthetase is along and around the diagonal side of the tRNA structure which contains the acceptor-, the D-, and anticodon stems.[175] More precise information was furnished by enzymatic and chemical mapping experiments. Digestions of complexed tRNAs by the higher-order specific RNase from cobra venom, revealed in all systems thus far studied, contacts of the anticodon stem with the aaRS (e.g., References 176 and 177a). The most straightforward results were obtained with ethylnitrosourea, the phosphate alkylating reagent. Protected phosphates were found all over the tRNA molecules with stretches located in the anticodon stem, in the variable or the P10 region. When mapping data are displayed in the three-dimensional structures of the tRNAs, quite different pictures emerge for different tRNAs.[28] While the contacts of tRNA[Phe] fit well with the UV cross-linking derived model (see Reference 175), those obtained for tRNA[Asp] are completely different. In the latter case the tRNA is lying on the synthetase by the face comprising the variable region, a model consistent with the crystal structure of the complex.[62a] Worth mentioning is a new footprint method using phosphorothioate-containing tRNA transcripts and iodine cleavage of the RNA. This technique that probes phosphate groups was recently applied to identify contacts between *E. coli* tRNA[Ser] and its cognate synthetase[177b] and will certainly become in future a rapid and convenient tool for mapping aaRS bound tRNAs.[28]

Detection by biochemical means of regions of synthetases in contact with tRNA is more difficult from an experimental point of view, and therefore only limited information is available. However, the cross-linking reagent, trans-diaminodichloro-platinum, shows an interesting potential to map both oligonucleotides and peptides in close proximity in tRNA/aaRS complexes.[178] Affinity labeling of aaRSs by periodate-oxidized substrates is another way to search for peptides in the vicinity of the active sites on the enzymes, provided they contain a lysine residue. As mentioned before, labeling by oxidized tRNA occurs in the consensus sequence KMSKS of several class I aaRSs,[77] but this is not a general property of all synthetases as was shown for yeast AspRS, a class II enzyme which does not possess such a residue near to the binding site of the CCA-terminus of tRNA^{Asp}.[179] Other chemical methods permitted definition of the region in *E. coli* MetRS, around Lys465, interacting with the anticodon of tRNA^{Met}.[180] Also worth mentioning here is an attempt by protein engineering combined with docking experiments to define the basic amino acids in prokaryotic TyrRS in contact with the tRNA.[181a] It would be interesting to search whether tRNA molecules from eukaryotic systems would interact on both subunits of an aaRS as was found in the prokaryotic tyrosine system.[181a, b]

More direct information on tRNA/aaRS contacts became available for the *E. coli* glutaminyl and the yeast aspartic acid systems on the basis of crystallographic studies. In the glutaminyl system the aaRS interacts with the acceptor stem via the tRNA minor groove[61] in contrast to the aspartic acid system where the tRNA is approached from the major groove side.[62a] In both systems the discriminator base (G73), the first U1-A72 base pair in the acceptor stem, and the anticodon residues are in close contact with the enzymes.[61, 62a] Based on sequence and crystallographic homologies between *E. coli* GlnRS and MetRS it is suggested a similar overall orientation of tRNA^{Gln} and tRNA^{Met} to their respective class I synthetases.[181c] This seems not to be the case of tRNA^{Tyr} that spans over the two TyrRS subunits,[181b] suggesting that TyrRS belongs to a class I subclass.

Identity elements—Two novel methodologies recently provided the experimental tools to rapidly determine identity elements in tRNAs. The first one, restricted to date for the study of prokaryotic systems, permits study of the specificity of tRNA variants *in vivo*, under conditions where the variant molecules are screened in the presence of the 20 aaRSs (reviewed in Reference 163). It consists in *de novo* synthesis of suppressor tRNA genes which are inserted into a proper plasmid, expressed *in vivo* and screened. The aminoacylation specificity of the suppressors is checked after analysis of the amino acids inserted at position 10 of *E. coli* dihydrofolate reductase synthesized from a gene carrying an amber mutation at codon 10.[182] Worthy of note is the theoretical limitation of this method for screening anticodon residues. The second method, easily applicable to eukaryotic tRNAs, permits classical biochemical analysis of tRNA variants obtained by *in vitro* transcription by bacteriophage T7 RNA polymerase of synthetic genes starting with the phage promoter.[183] These tRNAs do not contain the modified nucleotides but represent nevertheless good substrates for aaRSs. To mention here are chemical methods for the total synthesis of tRNA molecules that could represent original ways to prepare tRNA variants; they were first pioneered for the sequence of yeast tRNA^{Ala}[184] and much improved recently, allowing synthesis of functional molecules on automated DNA synthesizers.[185]

With these tools, it could be shown that the sites that determine the specificity of tRNAs correspond to small sets of nucleotides. For *E. coli* tRNA[Ala] this set is limited to only the G3–U70 base pair;[186a, b] this simple identity set seems to be conserved in evolution from bacteria to man.[187a] However, the effect of the lateration of this major identity determinant can be compensated *in vivo* by a mutation in AlaRS.[187b] The picture which emerges in other systems appears more complex, and in the case of yeast tRNA[Phe] five nucleotides located in the anticodon, the D-loop and the so-called discriminator position (nucleotide 73) are crucial for phenylalanylation.[188] Interestingly, the 3–70 base pair (G3–C70) has also been identified as one of the identity elements of tRNA[Gln] (in addition to other elements from the amino acid acceptor stem and the anticodon) as well as its counter part on the interacting protein (D235) which has been shown to contribute to the specificity of GlnRS. Indeed, changing the contact with the G3–U70 base pair by a mutation at D235 relaxes the specificity of GlnRS.[189]

It is noteworthy that anticodon residues are found to be important in many aminoacylation systems (reviewed in Reference 163). As was already proposed in the past on the basis of indirect experiments (reviewed in 190), anticodon residues have been identified as specificity determinants in many aminoacylation systems (reviewed in 163 and 166b). In the valine, methionine and threonine systems from *E. coli*, anticodon alone is sufficient to confer upon tRNA[Val], tRNA[Met] or tRNA[Thr] their identity.[191a, b] In the system specific for phenylalanine[188, 192a] or aspartic acid,[192b] additional nucleotides, including the discriminator base, are required. A striking difference between the two systems concerns the presence of a determinant position in the D-loop (at G20) in tRNA[Phe] and in the D-stem (at base pair G10-U21) in tRNA[Asp]. In contrast, anticodon residues are not required at all in a series of other systems. As shown above, that is the case for alanylation,[187a] but also for the charging of tRNAs recognizing degenerate codons (e.g., specific for serine).[163] Footprinting experiments on yeast tRNA[Ser] complexed to SerRS agree with that view.[193] At the other extremity of the tRNA molecule, residue 73 is found important for aminoacylation in an increasing number of tRNAs (e.g. References 61, 188, 192b, 194, and 195), in agreement with earlier proposals based on indirect evidences.[160, 196]

The theoretical consequence of the concept of identity determinants is the possiblity of changing the specificity of a tRNA by transplantation of its identity set into another tRNA species. This prediction was verified in several systems, either *in vivo* or *in vitro.*. Transplanting *in vivo* the G3–U70 base pair from tRNA[Ala] to tRNA[Phe] makes the latter an alanine acceptor.[186b] Similarly, switching the anticodons in valine and methionine specific tRNAs, or from threonine in methionine tRNA, switches their specificities[191a, b] and substituting two base pairs in the acceptor stem changes a serine suppressor tRNA to an efficient glutamine acceptor.[198a] Also *E. coli* tRNA[Tyr] can be converted to tRNA[Ser] by the insertion of only two nucleotides into the variable stem of tRNA.[Tyr 198b] Other examples concern the transplantation of yeast tRNA[Phe] determinants into yeast tRNA[Tyr 188] or tRNA[Asp 198c] and *vice versa* of yeast aspartic acid determinants into yeast tRNA.[Phe 192b] While the transplanted tRNAs have lost their old canonical specificity, they generally do not acquire an optimal new specificity (e.g. 198b, c) because of non-optimal presentation of determinant nucleotides to the new synthetase. Altogether these facts indicate that tRNA identity for aminoacylation is the result of a subtle interplay between effects brought by identity nucleotides and conformational features expressed at the three-dimensional level. An unexpected observation relies on a specificity switch of an *E. coli* tRNA[Tyr] suppressor that in the yeast *S. cerevisiae* becomes a leucine specific tRNA.[198d]

A remarkable result concerns mini- and even microhelices mimicking the amino acid acceptor branch of the L-shaped tRNA, and containing the alanine determinant G3–U70 base pair, that retain their capability to be charged by AlaRS.[199a] This result is a direct confirmation of old results having indicated alanylation of an amino acid accepting arm formed by the association of the 3′- and 4′-quarters of yeast tRNA[Ala].[199b] A recent extension of such kind of studies lead to the alanylation of 3′-fragments in the presence of complementary and limiting 5′-oligomers.[199c] This phenomenon is due to transient duplex formation followed by charging and opens new routes to the evolution of early aminoacylation systems.

An interesting aspect of the *in vitro* approach relies on the possibility of checking for the influence of modified nucleotides on the aminoacylation capacity of tRNAs. With yeast tRNA[Asp] transcribed *in vitro* from a synthetic gene it was shown that the transcripts possess an increased propensity to be mischarged by noncognate ArgRS, while keeping their aspartylation properties unchanged.[199d] This can be rationalized if one admits that posttranscriptional modifications in tRNA contribute to negative selections by preventing a tRNA from being recognized by noncognate aaRS. In perfect agreement with this view is the finding that a single posttranscriptional modification in the gene product of *E. coli* tRNA[Ile]2 changing C34 to Lysidine 34 prevents this tRNA from being mischarged by MetRS and thus ensures specificity of isoleucylation.[200]

In conclusion, it should be noticed that residues in tRNA in direct contact with aaRS are not necessarily identity determinants. This is best illustrated with *E. coli* tRNA[Ala], whose identity is given by the G3–U70 base pair, but which interacts wtih AlaRS at several additional sites.[201] The inverse might be true as well, although the functional consequences of mutations at positions outside the contact area with an aaRS and which would introduce too strong structural alterations in the tRNA, should be interpreted with caution. However, as seen in the crystal structures of the glutamine and aspartic acid tRNA/aaRS complexes, major identity determinants are in contact with the synthetases.[61, 192b]

The special case of tRNA-like molecules—The genomic RNAs of a number of RNA plant viruses can be aminoacylated at their 3'-CCA termini by host or heterologous aaRSs (reviewed in Reference 202a). So, the RNA of TYMV can be valylated and that of brome mosaic virus (BMV) tyrosylated. The 3'-ends of these viral RNAs, however, are lacking several primary structural features of tRNAs, such as strategic D- or T-loop sequences and modified bases. As a consequence they cannot be folded into a canonical cloverleaf. The question which then arises is: how do structures as different as tRNAs and 3'-regions of viral RNAs behave in such a similar fashion in the presence of aaRSs? Most likely, similar structural domains are recognized by the aaRSs. For TYMV RNA, the simplest tRNA-like structure, a three-dimensional L-shaped conformation mimicking tRNA with a new RNA folding principle in the amino acid accepting limb, the pseudo-knot, was proposed.[33] Recently, chemical mapping combined with melting and NMR studies gave further support to this model, but showed in addition the great sensitivity of the pseudo-knotted structure to solvent conditions.[202b, 203, 204] Pseudo-knotted foldings were found in a variety of other plant viral tRNA-like molecules (e.g., Reference 205) and seem to represent an original alternative conformation for the interaction with components of the translation appartus.[206] They involve base pairing between residues in a hairpin loop with a single-stranded region of the RNA which gives rise to a quasi-helix mimicking a RNA double helix.[207, 208]

A deeper understanding of the molecular recognition between tRNA-like molecules and aaRSs came from dissection, mutational analysis, and footprinting experiments on the RNAs. For TYMV RNA, the sequence corresponding to the L-shaped folding represents the minimal structure acylatable by ValRS,[209] a fact corroborated by comparative mapping data that indicate protections by ValRS of the tRNA-like structure at positions homologous to those protected in tRNA^{Val}, in particular in the anticodon region.[210] However, regions outside the L-fold are also protected by ValRS, in agreement with the better valylation properties of RNA fragments of lengths greater than the minimal active structure.[211a] Some of these structural conclusions are supported by functional studies that have identified the middle position of the anticodon as a major determinant for valylation. Interestingly, while the variants mutated at the determinant position have quasi-completely lost their valylation activity, their affinity for ValRS is not decreased, indicating that the interaction between the anticodon determinant and ValRS controls the kinetics of valylation.[211b] With the BMV tRNA-like structure an extensive mutational analysis[212, 213] showed that only alterations at the 3'-terminus, the pseudo-knot, and in two stem regions (that might be compared to the anticodon branch of tRNA^{Tyr}),[177a] give strong decreases in the tyrosylation rates. The positions of these alterations overlap with those found protected by TyrRS at topologically similar positons in the strategic residues in yeast tRNA^{Tyr}.[177a]

Obviously, the tRNA-like structure mimics that of tRNA and the aaRS can accomodate such a mimicry.

2.3.3. Mechanism of tRNA Charging, Proofreading, and Editing

The present view on tRNA aminoacylation clearly shows (1) that the specificity of tRNA recognition by aaRSs is ensured by a limited number of positive determinants, (2) that negative discriminations might play a major role in the mechanisms leading to specificity, and (3) that aaRSs can accomodate major conformational alterations in the tRNA structure provided the determinants are present. This last point is particularly well illustrated by the potential of tRNA fragments,[214] tRNA minihelices[199a] and tRNA-like structures,[202a] as well as of synthetic DNA corresponding to a tRNA sequence[215] to be aminoacylated. However, recognition alone does not provide the level of specificity required for the correct biological functioning of protein synthesis. This level is caused by kinetic effects as already discussed in the early 1970s.[160] As to the enzymatic mechanism of tRNA charging, it is at present definitely established that it is a two-step event, with first the activation of the amino acid, and second the transfer of the activated amino acid to the CCA-end of the tRNA (reviewed in Reference 216). This general mechanism is valid even for those aaRSs (ArgRS, GlnRS, GluRS) requiring tRNA for the PPi-ATP exchange reaction (e.g., Reference 217).

$$\text{aaRS + amino acid + ATP} \underset{\longleftarrow}{\overset{Mg^{2+}}{\longrightarrow}} \text{(aa-AMP)aaRS + PPi} \qquad (1)$$

$$\text{(aa-AMP)aaRS + tRNA} \rightleftarrows \text{aa-tRNA + AMP + aaRS} \qquad (2)$$

Subtle mechanistic differences, however, appear in the various aminoacylation systems. This is for instance the case for the stability of the (aa-AMP)aaRS complex, which in most systems is stable and can be isolated, but which sometimes dissociates into free aminoacyl-adenylate and enzyme when tRNA is absent as was shown with *S. cerevisiae* AspRS.[218] Other differences, related with the ranking of the aaRSs into structural families,[71b] concern the site of primary esterification of the amino acids either on the 3'-, or on the 2'-, or indiscriminately on the 2'- or 3'-hydroxyl residue of the terminal 3'-ribose of tRNA, before (if needed) the transesterification leading to the significant 3'-aminoacyl-tRNAs.[62d, 219–221] Finally, the rate-limiting step of the aminoacylation reaction is not necessarily the charging of the tRNA but can be the release of the aminoacylated tRNA from the enzyme as observed with *S. cerevisiae* ValRS.[222]

Under certain experimental conditions (for instance at low enzyme concentration, at high ionic strength or in incorrect aminoacylation systems) the aminoacylation of a tRNA may be incomplete (e.g., Reference 159). This can be explained by the aminoacylation plateau theory[223, 224] according to which the level of aminoacylation reflects the balance between the aminoacylation rate and the rates of the various deacylation reactions (spontaneous and enzymic deacylations of aminoacyl-tRNAs and the reverse of the aminoacylation reaction). The existence of such low aminoacylation plateaus was often misinterpreted, and in present studies on tRNA variants with reduced charging k_{cat}, the kinetic interpretation of incomplete aminoacylation should not be overlooked.

The mechanisms of tRNA aminoacylation involves conformational changes on both partners of the reaction (e.g., References 61, 167, 173, and 225). Some of them are induced by the body of the tRNA interacting with an aaRS. This was convincingly demonstrated by the aminoacylation of 3′-terminal CCA fragments or even of adenosine that occur only upon binding to aaRSs of tRNAs lacking their 3′-end, that activate their catalytic centers.[226, 227] In several systems these conformational changes concern the anticodon region of the tRNA (e.g., References 61 and 225).

As mentioned above, the specificity of both activation of amino acids by aaRSs and recognition of tRNA during the amino acid transfer step is not absolute (reviewed in References 157 through 159). To prevent the deleterious effects of misactivations of amino acids, nature has selected discrimination and correction mechanisms. For instance, with yeast PheRS discrimination between phenylalanine and tyrosine occurs through kinetic effects favoring the activation of phenylalanine[228] according to a proofreading-type mechanism.[161, 162] Specificity can also be brought by stereochemical effects as was rationalized by Fersht in the double-sieve editing mechanism.[158] According to this mechanism, experimentally proven for several aaRSs,[158, 229] natural amino acids that are larger than the specific substrates are rejected by steric exclusion in the first sieve while smaller or isosteric amino acids are accepted; in the second sieve a hydrolytic site accepts all the smaller amino acids but rejects the cognate amino acid. The evolutionary aspects of accuracy as studied with PheRS and LeuRS from *E. coli* and from several eukaryotes should be mentioned. These enzymes appear to use different strategies with a better initial discrimination in the aminoacylation and less elaborated proofreading for the prokaryotic enzymes and the inverse strategy for the eukaryotic synthetases.[230, 231] Finally, TrpRSs from higher organisms may use an additional strategy for increasing their specificity by directing the amino acid activation along a nonproductive pathway for tryptophan analogs in a process coupled with pyrophosphorylation of the TrpRSs.[232]

Are correction mechanisms needed for tRNA? Corrections could occur via the aaRS catalyzed deacylation of aminoacyl-tRNAs in the absence of AMP and PPi since the deacylation rates of mischarged tRNAs are sometimes highly increased (reviewed in Reference 233), but this mechanism does not apply in all cases (e.g., the deacylation by yeast ValRS of Val-tRNA^{Val} is ten times faster than that of Val-tRNA^{Phe}.[234] As to a tRNA-proofreading, this possibility remains open, although the kinetic discrimination mechanism for tRNA charging may be sufficient to keep the level of mischarged tRNAs low.[160]

2.3.4. Associations of Aminoacylated tRNAs with Metabolic Processes

Phosphoseryl- and selenocysteinyl-tRNA—The selenium is used to modify a minor suppressor ser-tRNA^{Ser} into selenocysteinyl-tRNA. This selenocysteine can then be incorporated into proteins in response to UGA codons recognized by the minor tRNA^{Ser} isoacceptor.[235, 236] In eukaryotes, selenocysteine formation appears to proceed through a phosphoseryl-tRNA intermediate.[237] In *E. coli,* the selenocysteine-inserting tRNA is highly unusual in structure and modification,[238] and selenocysteyl-tRNA recognized by a novel translation elongation factor.[239]

Glutaminyl-tRNA^{Gln}—In all organelles studied so far, no GlnRS was detected.[116] Gln-tRNA^{Gln} is synthesized by two enzymes: a GluRS which acylates (incorrectly) tRNA^{Gln} with glutamate, and a specific amido-transferase which catalyzes the glutamine-dependent modification of Glu-tRNA^{Gln} into Gln-tRNA^{Gln}. The same situation had been previously observed in all the Gram-positive microorganisms studied.[240, 241]

Other examples and evolutionary implications—Beside the above involvements of tRNAs in metabolic processes one can also mention the participation of particular aminoacylated tRNA species in the ribosome independent synthesis of cell wall peptides,[242] in the synthesis of lipids,[243] or in other metabolic processes not related to translation (see Reference 244 for a review). These facts, together with Wächtershäuser's theory of surface metabolism in the origin of life,[245] lead recently to the concept of homeotopic metabolic transformation. According to this view, tRNA molecules would have been used in early evolution as carriers for such transformations of metabolic precursors, well before translation.[243] Thus, some of the odd functions of certain tRNAs would be cryptic remains of early evolutionary processes. Because such functions are not well known and often were fortuitously discovered, it may well be that other cryptic relations betrween tRNAs and intermediary metabolic pathways have been conserved through evolution. Similarly, it has been proposed that 3'-terminal tRNA-like structures in viruses are molecular fossils of the original RNA world that functioned as primitive telomers and could have been aminoacylated by aberrant activities of the early replicases.[246] These views open new routes in tRNA research.

REFERENCES

1. Crick FHC: On Degenerate Template and the Adaptor Hypothesis, a note for the RNA Tie Club, 1955, unpublished; discussed by Crick in *What Mad Pursuit, A Personal View of Scientific Discovery* New York, Basic Books, 95 (1988).
2. Hoagland MB, Stephenson ML, Scott JF, Hecht LL, and Zamecnik PC: *J. Biol. Chem.* 231:241 (1958).
3. Chapeville F, Lipman F, Von Ehrenstein G, Weisblum B, Kay WJ, Ray WJ, and Benzer Z: *Proc. Natl. Acad. Sci. U.S.A.* 48:1086 (1962).
4. Holley RW, Apgar J, Everett GA, Madison JT, Marquisse M, Merrill SH, Penswick JR, and Zamir A: *Science* 147:1462 (1965).
5. Sprinzl M, Hartmann T, Meissner F, Moll J, and Vorderwülbecke T: *Nucleic Acids Res.* 17:rl (1989).
6. Grosjean H, Cedergren RG, and McKay W: *Biochimie* 64:387 (1982).
7. Nicholas HB Jr. and McClain WM: *Comput. Appl. Biol. Sci.* 3:177 (1987).
8. Eigen M, Lindemann BF, Tietze M, Winkler-Oswatitsch R, Dress A, and von Haeseler A: *Science* 244:673 (1989).
9a. Björk GR, Ericson JU, Gustafsson CED, Hagervall TG, Jönsson YH, and Wikström PM: *Annu. Rev. Biochem.* 56:263 (1987).
9b. Gehrke CW and Kuo KCT (eds.) *J. Chromat.* Library, Vol. 45 (1990).
10. Desgrès J, Keith G, Kuo KC, and Gehrke C: *Nucleic Acids Res.* 17:868 (1989).
11. Garey JR and Wolstenholme DR: *J. Mol. Evol.* 28:374 (1989).
12. Ueda T, Ohta T, and Watanabe K: *J. Biochem.* 98:1275 (1985).
13. Rich A and RajBhandary UL: *Annu. Rev. Biochem.* 45:805 (1976).
14. Kim SH, Suddath FL, Quigley GJ, McPherson A, Sussman JL, Wang AHJ, Seeman N, and Rich A: *Science* 185:435 (1974).
15. Robertus JD, Ladner JE, Finch JT, Rhodes D, Brown RS, Clark BFC, and Klug A: *Nature* 250:546 (1974).
16. Moras D, Thierry JC, Comarmond MB, Fischer J, Weiss R, Ebel JP, and Giegé R: *Nature* 288:669 (1980).
17. Westhof E, Dumas P, and Moras D: *J. Mol. Biol.* 184:119 (1985).
18. Moras D, Dock AC, Dumas P, Westhof E, Romby P, Ebel JP, and Giegé R: *Proc. Natl. Acad. Sci. U.S.A.* 83:932 (1986).
19. Woo NA, Roe B, and Rich A: *Nature* 286:346 (1980).
20. Schevitz RW, Podjarny AD, Krischnamachar N, Hugues TT, Sigler P, and Susman JL: *Nature* 278:188 (1979).
21. Basavappa R and Sigler PB: 13th Int. tRNA Workshop, Abstr. TU-37 (1989).
22. Sharp KA, Honig B, and Harvey SC: *Biochemistry* 29:340 (1990).
23. Dock AC, Lorber B, Moras D, Pixa G, Thierry JC, and Giegé R: *Biochimie* 66:179 (1984).
24. Garcia A, Giegé R, and Behr JP: *Nucleic Acids Res.* 18:89 (1990).
25. Westhof E: *Annu. Rev. Biophys. Chem.* 17:125 (1988).
26. Li ZQ, Giegé R, Jacrot B, Oberthur R, Thierry JC, and Zaccaï G: *Biochemistry* 22:4380 (1983).
27. Peattie DA and Gilbert W: *Proc. Natl. Acad. Sci. U.S.A.* 77:4679 (1980).
28. Romby P, Moras D, Bergdoll M, Dumas P, Vlassov VV, Westhof E, Ebel JP, and Giegé R: *J. Mol. Biol.* 184:455 (1985).

29. Romby P, Moras D, Dumas P, Ebel JP, and Giegé R: *J. Mol. Biol.* 195:193 (1987).
30. Romby P, Carbon P, Westhof E, Ehresmann C, Ebel JP, Ehresmann B, and Giegé R: *J. Biomolec. Struct. Dyn.* 5:669 (1987).
31. Dock-Bregeon AC, Westhof E, Giegé R, and Moras D: *J. Mol. Biol.* 206:707 (1989).
32. de Bruijn MHL and Klug A: *EMBO J.* 2:1309 (1983).
33. Rietveld K, Van Poelgeest R, Pleij CWA, Van Boom JH, and Bosch L: *Nucleic Acids Res.* 10:1929 (1982).
34. Dumas P, Moras D, Florentz C, Giegé R, Verlaan P, Van Belkum A, and Pleij CWA: *J. Biomol. Struct. Dyn.* 4:707 (1987).
35. Clarkson SG: In McLean N, Gregory SP, and Flavell RA (Eds): *Eukaryotic genes: Their Structure, Activity and Regulation,* Butterworth Press, London,239 (1983).
36. Sharp SJ, Schaak J, Cooley L, Johnson Burke D, and Söll D: *Crit. Rev. Biochem.* 19:107 (1985).
37. Thomann H-U, Schmutzler C, Hüdepohl U, Blow M, and Gross HJ: *J. Mol. Biol.* 209:505 (1989).
38. Geiduschek EP and Tocchini-Valentini GP: *Annu. Rev. Biochem.* 57:873 (1988).
39. Koch W, Edwards K, and Kössel H: *Cell* 25:203 (1981).
40. Jacobs HT, Asakawa S, Araki T, Miura K, Smith MJ, and Watanabe K: *Curr. Genet.* 15:193 (1989).
41. Altman S, Baer MF, Guerrier-Takada C, and Vioque A: *Trends Biochem. Sci.* 11:515 (1986).
42. Greer CL, Peebles CL, Gegenheimer P, and Abelson P: *Cell* 32:537 (1983).
43. Zasloff M: *Proc. Natl. Acad. Sci. U.S.A.* 80:6436 (1983).
44. Nishimura S: *Prog. Nucleic Acid Res. Mol. Biol.* 28:49 (1983).
45. Dirheimer G: *Rec. Results Cancer Res.* 84:15 (1983).
46. Kersten H: *Prog. Nucleic Acid Res. Mol. Biol.* 31:59 (1984).
47. Nishikura K and deRobertis EM: *J. Mol. Biol.* 145:405 (1981).
48. Grosjean H and Kubli E: In Celis JE et al. (eds): *Microinjection and Organelle Transplantation Techniques,* New York, Academic Press, 304 (1986).
49. Carbon P, Haumont E, Fournier M, deHenau S, and Grosjean H: *EMBO J.* 2:1093 (1983).
50. Droogmans L, Haumont E, deHenau S, and Grosjean H: *EMBO J.* 5:1105 (1986).
51. Grosjean H, Droogmans L, Giegé R, and Uhlenbeck H: *Biochim. Biophys. Acta* 1050:267 (1990).
52. Yokoyama S, Watanabe K, and Miyazawa T: *Adv. Biophys.* 23:115 (1987).
53. Hall KB, Sampson JR, Uhlenbeck OC, and Redfield AG: *Biochemistry* 28:5794 (1989).
54. Perret V, Garcia A, Puglisi J, Grosjean H, Ebel JP, Florentz C, and Giegé R: *Biochimie* 72:735 (1990).
55. Ikemura T and Ozeki H: *Cold Spring Harbor Symp. Quant. Biol.* 47:1087 (1982).
56. Grosjean H and Fiers W: *Gene* 18:199 (1982).
57. Meir F, Suter H, Grosjean H, Keith G, and Kubli E: *EMBO J.* 4:823 (1985).

58. Bienz M and Kubli E: *Nature* 294:188 (1981).

59. Kuchino Y, Nishimura S, Schröder HC, Rottmann M, and Müller WEG: *Virology* 165:518 (1988).

60a. Schwob E, Sanni A, Fasiolo F, and Martin RP: *Eur. J. Biochem.* 178:235 (1988).

60b. Mirande M: *Aminoacyl-tRNA Synthetase Family from Prokaryotes and Eukaryotes: Structural Domains and Their Implications. Prog. Nucleic Acid Res. Mol. Biol. 40:95 (1991)*

61. Rould MA, Perona JJ, Söll D, and Steitz TA: *Science* 246:1135 (1989).

62a. Ruff M, Krishnaswamy S, Boeglin M, Poterzman A, Mitschler A, Podjarny A, Rees B, Thierry JC, and Moras D: *Science* (1991) in press.

62b. Eriani G, Delarue M, Poch O, Gangloff J, and Moras D: *Nature* 347:249 (1990).

62c. Cusak S, Berthet-Colominas C, Härtlein M, Nassar N, and Leberman R: *Nature* 347:249 (1990).

62d. Hecht S: In Schimmel PR, Söll D, and Abelson JN (Eds.) *Transfer-RNA structure, properties and recognition* Cold Spring Harbor Laboratory, Cold Spring Harbor, NY pp. 345–360 (1979).

63. Grunberg-Manago M: In Neidhardt FC (Ed. in chief): *Escherichia coli and Salmonella typhimurium Cellular and Molecular Biology* Washington, D.C., American Society for Microbiology, 1386 (1987).

64. Eriani G, Dirheimer G, and Gangloff J: *Nucleic Acids Res.* 17:5725 (1989).

65. Prevost G, Eriani G, Kern D, Dirheimer G, and Gangloff G: *Eur. J. Biochem.* 180:351 (1989).

66. Schimmel P: *Annu. Rev. Biochem.* 56:125 (1987).

67. Schimmel P and Söll D: *Annu. Rev. Biochem.* 48:601 (1979).

68. Jasin M, Regan L, and Schimmel P: *Nature* 306:441 (1983).

69. Putney SD and Schimmel P: *Nature* 291:632 (1981).

70. McDonald T, Breite L, Pangbrun KLW, Hom S, Manser J, and Nagel GM: *Biochemistry* 19:1402 (1980).

71. Chothia C: *Annu. Rev. Biochem.* 53:537 (1984).

72a. Burbaum JJ, Starzyk RM, and Schimmel P: *Proteins* 7:99 (1989).

72b. Eriani G, Dirheimer G, and Gangloff J: *Nucleic Acids Res.* 19:265 (1991).

72c. Hou Y-M, Shiba K, Mottes C, and Schimmel P: *Proc. Natl. Acad. Sci. U.S.A.* 88:976.

73. Blow DM and Brick P: In Jurnak F and McPherson A (Eds): *Biological Macromolecuels and Assemblies* vol. 2, New York, John Wiley & Sons, 442 (1985).

74. Ward WHJ and Fersht AH: In Brenner SA (Ed): *Redesigning the Molecules of Life,* Berlin, Springer-Verlag, 5a (1988).

75. Leatherbarrow RJ, Fersht AR, and Winter G: *Proc. Natl. Acad. Sci. U.S.A.* 82:7840 (1985).

76a. Fersht AR, Leatherbarrow RJ, and Wells TNC: *Trends Biochem. Sci.* 11:321 (1986).

76b. Anselme J and Härtlein M: *Febs Letters* 280:163 (1991).

77. Hountondji C, Lederer F, Dessen P, and Blanquet S: *Biochemistry* 25:16 (1986).

78. Hill K and Schimmel P: *Biochemistry* 28:2577 (1989).

79. Brick P and Blow DM: *J. Mol. Biol.* 194:287 (1987).
80. Clarke ND, Lien DC, and Schimmel P: *Science* 240:521 (1988).
81. Breton R, Watson D, Yaguchi M, and Lapointe J: *J. Biol. Chem.* 265:18248 (1990).
82. Mellot P, Mechulam Y, Le Corre D, Blanquet S, and Fayat G: *J. Mol. Biol.* 208:429 (1989).
83. Bränden CI: In Moras D, Drenth J, and Strandberg B (Eds): *Crystallography in Molecular Biology* Plenum Press, New York 359 (1987).
84. Blow DM, Brick P, Brown KA, Fersht AR, and Winter G: In Moras D, Drenth J, and Strandberg B (eds): *Crystallography in Molecular Biology* New York, Plenum Press, 241 (1987).
85. Walker J-P, Risler J-L, Monteilhet C, and Zelwer C: *FEBS Lett.* 16:186 (1971).
86. Schimmel P: *Adv. Enzymol.* 63:233 (1990).
87. Conley JG: Ph.D. Thesis, Yale University, New Haven, CT (1988).
88. English-Peters S and Cramer F: 13th Int. tRNA Workshop, Abstr. Mo-am-9 (1989).
89. Lorber B, Mejdoub H, Reinbolt J, Boulanger Y, and Giegé R: *Eur. J. Biochem.* 174:155 (1988).
90. Kohda D, Yokoyama S, and Miyazawa T: *J. Biol. Chem.* 262:558 (1987).
91. Heck JD and Hatfield GW: *J. Biol. Chem.* 263:868 (1988).
92. Tzagoloff A, Akai A, Kurkulos M, and Repetto B: *J. Biol. Chem.* 263:850 (1988).
93. Starzyk RM, Webster TA, and Schimmel P: *Science* 237:1614 (1987).
94. Brunie S, Mellot P, Zelwer C, Risler J-L, Blanquet S, and Fayat G: *J. Mol. Graphics* 5:18 (1987).
95. Webster T, Tsai H, Kula M, Mackie G, and Schimmel P: *Science* 226:1315 (1984).
96. Freedman R, Gibson B, Donovan D, Biemann K, Eisenbeis S, Parker J, and Schimmel P: *J. Biol. Chem.* 260:10063 (1985).
97. Breton R, Sanfaçon H, Papayannopoulos I, Biemann K, and Lapointe J: *J. Biol. Chem.* 261:10610 (1986).
98. Brisson A, Brun YV, Bell AW, Roy PH, and Lapointe J: *Biochem. Cell Biol.* 67, 404–410 (1989).
99. Gampe A and Tzagoloff A: *Proc. Natl. Acad. Sci. U.S.A.* 86, 6023–6027 (1989).
100. Gatti DL and Tzagoloff A: Structure and evolution of a group of related aminoacyl-tRNA synthetases. *J. Mol. Biol.* 218:557 (1991).
101. Schimmel P: *Trends Biochem. Sci.,* 16:1 (1991).
102. Bandyopadhyay AK and Deutscher MP: *J. Mol. Biol.* 60, 113–122 (1971).
103. Deutscher MP: *J. Cell. Biol.* 99, 373–377 (1984).
104. Cirakoglu B, Mirande M, and Waller J-P: *FEBS Lett.* 183, 185–190 (1985).
105. Yang DCH and Jacobo-Molina A: In Srere PA, Jones MA, and Mathews CK (Eds): *UCLA Symposium on Structural Organizational Aspects of Metabolic Regulation* New York, Alan R. Liss Inc. 199 (1990).
106. Jacobo-Molina A, Peterson R, and Yang DCH: *J. Biol. Chem.* 264:16608 (1989).
107. Sivaram P, Vellekamp G, and Deutscher MP: *J. Biol. Chem.* 263:18891 (1988).

108. Martinez P, Waller JP, and Mirande M: 13th Int. tRNA Workshop, abstr. TU-23 (1989).

109. Dang CV, Yang DCH, and Pollard TD: *J. Cell. Biol.* 96:1138 (1983).

110. Mirande M, Le Corre D, Louvard D, Reggio H, Pailliez JP, and Waller JP: *Exp. Cell Res.* 156:91 (1985).

111. Springer M, Graffe M, Dondon J, and Grunberg-Manago M: *EMBO J.* 8:2417 (1989).

112a. Moine H, Romby P, Springer M, Grunberg-Manago M, Ebel JP, Ehresmann C, and Ehresmann B: *Proc. Natl. Acad. Sci. U.S.A.* 85:7892 (1988).

112b. Moine H, Romby P, Springer M, Grunberg-Manago M, Ebel JP, Ehresmann B, and Ehresmann C: *J. Mol. Biol.* 216:299 (1990).

113. Akins RA and Lambowitz AM: *Cell* 50;331 (1987).

114. Herbert CJ, Labouesse M, Dujardin G, and Slonimski PP: *EMBO J.* 7:473 (1988).

115. Labouesse M: personal communication.

116. Schön A, Kannangara CG, Gough S, and Söll D: *Nature* 331:187 (1988).

117. Kannangara CG, Gough SP, Bruyant P, Hoober JK, Kahn A, and von Wettstein D: *Trends Biochem. Sci.* 13:140 (1988).

118. Lanrick R and Yanofsky C: In Neidhardt FC (Ed. in chief): *Escherichia coli and Salmonella typhimurium Cellular and Molecular Biology* Washington, D.C., American Society for Microbiology, 1276 (1987).

119. Hinnebusch AG: *Microbiol. Rev.* 52:248 (1988).

120. Andrulis IL, Hatfield GW, and Arfin SM: *J. Biol. Chem.* 254:10629 (1979).

121. Shotwell MA and Oxender DL: *Trends Biochem. Sci.* 9:314 (1983).

122. Neidhardt FC and Van Bogelen RA: In Neidhardt FC (Ed. in chief): *Escherichia coli and Salmonella typhimurium Cellular and Molecular Biology* Washington, D.C., American Society for Microbiology, 1334 (1987).

123. Putzer H, Brakhage A, and Grunberg-Manago M: *J. Bacteriol.* 172:4593 (1990).

124. Rapaport E and Zamecnik PC: *Proc. Natl. Acad. Sci. U.S.A.* 73:3984 (1976).

125. Zamecnik P: *Anal. Biochem.* 134:1 (1983).

126. Varshavsky A: *Cell* 34:711 (1983).

127. Brevet A, Chen J, Lévêque F, Plateau P, and Blanquet S: *Proc. Natl. Acad. Sci. U.S.A.* 86:8275 (1989).

128. Blanquet S, Plateau P, and Brevet A: *Mol. Cell. Biochem.* 52:3 (1983).

129. Kovaleva GK, Tarussova NB, and Kisselev LL: *Mol. Biol. (Moscow)* 22:1307 (1988).

130. Wahab SZ and Yang DCH: *J. Biol. Chem.* 260:12735 (1985).

131. Kramer GF, Baker JC, and Ames BN: *J. Bacteriol.* 170:2344 (1988).

132. Gilson G, Ebel JP, and Remy P: *Exp. Cell Res.* 177:143 (1988).

133. Traugh JA and Pendergast AM: *Prog. Nucleic Acid Res. Mol. Biol.* 33:195 (1986).

134a. Dang CV and Traugh JA: *J. Biol. Chem.* 264:5861 (1989).

134b. Venema RC and Traugh JA: *J. Biol. Chem.* 266:5298 (1991).

135. Itikawa H, Wada M, Sekine K, and Fujita H: *Biochimie* 71:1079 (1989).

136. Ramsden DA, Chen J, Miller FW, Misener V, Bernstein RM, Siminovitch KA, and Tsui FWL: *J. Immunol.* 143:2267 (1989).

137. Targoff IN, Arnett FC, and Reichlin M: *Arthrit. Rheumat.* 31:515 (1988).

138. Swanson R, Hoben P, Sumner-Smith M, Uemura H, Watson L, and Söll D: *Science* 242:15487 (1988).
139. Neidhardt FC, Parker J, and McKeever WG: *Annu. Rev. Microbiol.* 29:215 (1975).
140. Neidhardt F, Bloch P, Pedersen S, and Reeh J: *J. Bacteriol.* 129:378 (1977).
141. Morgan SD and Söll D: *Progr. Nucleic Acid Res.* 21:181 (1979).
142. Springer M, Mayaux JF, Fayat G, Plumbridge JA, Graffe M, Blanquet S, and Grunberg-Manago M: *J. Mol. Biol.* 181:467 (1985).
143. Meussdoerffer F and Fink GR: *J. Biol. Chem.* 258:6293 (1983).
144. Mirande M and Waller JP: *J. Biol. Chem.* 263:18443 (1988).
145. Kolman CJ, Snyder M, and Söll D: *Genomics* 3:201 (1988).
146. Martin NC and Rabinowitz M: *Biochemistry* 17:1628 (1978).
147. Chatton B, Walter P, Ebel JP, Lacroute F, and Fasiolo F: *J. Biol. Chem.* 263:52 (1988).
148. Tzagoloff A, Vambutas A, and Akai A: *Eur. J. Biochem.* 179:365 (1989).
149. Chow CM and RajBhandary UL: *Mol. Cell. Biol.* 9:4645 (1989).
150. Mirande M and Waller JP: *J. Biol. Chem.* 264:842 (1989).
151. Cruzen ME and Arfin SM: 13th Int. tRNA Workshop, Abstr. TH-am-17 (1989).
152. Thömmes P, Fett R, Schray B, Kunze N, and Knippers R: *Nucleic Acids Res.* 16:5391 (1988).
153. Tsui FWL and Siminovitch L: *Nucleic Acids Res.* 15:3349 (1987).
154. Lazard M, Mirande M, and Waller JP: *J. Biol. Chem.* 262:3982 (1987).
155. Sallafranque ML, Garett M, Benedetto JP, Fournier M, Labouesse B, and Bonnet J: *Biochem. Biophys. Acta* 882:192 (1986).
156. Favorova OO, Zargarova TA, Rukosuyev VS, Beresten SF, and Kisselev LL: *Eur. J. Biochem.* 184:583 (1989).
157. Bonnet J and Ebel JP: *Eur. J. Biochem.* 58:193 (1975).
158. Fersht AR: In Schimmel P, Söll D, and Abelson JN (Eds): *Transfer RNA: Structure, Properties, and Recognition* Cold Spring Harbor, New York, Cold Spring Harbor Laboratory, 247 (1979).
159. Giegé R, Kern D, Ebel JP, Grosjean H, De Henau S, and Chantrenne H: *Eur. J. Biochem.* 45:351 (1974) and references included.
160. Ebel JP, Giegé R, Bonnet J, Kern D, Befort N, Bollack C, Fasiolo F, Gangloff J, and Dirheimer G: *Biochimie* 55:547 (1973).
161. Hopfield JJ: *Proc. Natl. Acad. Sci. U.S.A.* 71:4135 (1974).
162. Ninio J: *Biochimie* 57:587 (1975).
163. Normanly J and Abelson J: *Annu. Rev. Biochem.* 58:1029 (1989).
164. Schimmel P: *Biochemistry* 28:2747 (1989).
165. DeDuve C: *Nature* 333:117 (1988).
166. Yarus M: *Cell* 55:739 (1988).
167. Krauss G, Riesner D, and Maass G: *Eur. J. Biochem.* 68:81 (1976).
168. Giegé R, Lorber B, Ebel JP, Moras D, Thierry JC, Jacrot B, and Zaccaï G: *Biochimie* 64:357 (1982).
169. Florentz C, Kern D, and Giegé R: *FEBS Lett.* 261:335 (1990).
170. Giegé R, Lorber B, Ebel JP, Moras D, and Thierry JC: *C.R. Acad. Sci. Paris D-2* 291:393 (1980).

171. Lorber B, Giegé R, Ebel JP, Berthet C, Thierry JC, and Moras D: *J. Biol. Chem.* 258:8429 (1983).
172. Perona JJ, Swanson R, Steitz TA, and Söll D: *J. Mol. Biol.* 202:121 (1988).
173. Zaccaï G, Morin P, Jacrot B, Moras D, Thierry JC, and Giegé R: *J. Mol. Biol.* 129:483 (1979).
174a. Podjarny A, Rees B, Thierry JC, Cavarelli J, Jésior JCX, Roth M, Lewit-Bentley A, Kahn R, Lorber B, Ebel JP, Giegé R, and Moras D: *J. Biomol. Struct. Dyn.* 5:187 (1987).
174b. Ruff M, Cavarelli J, Mikol V, Lorber B, Mitschler A, Giegé R, Thierry JC, and Moras D: *J. Mol. Biol.* 210:235 (1988).
175. Rich A and Schimmel P: *Nucleic Acids Res.* 4:1649 (1977).
176. Favorova OO, Fasiolo F, Keith G, Vassilenko SH, and Ebel JP: *Biochemistry* 20:1006 (1981).
177a. Perret V, Florentz C, Dreher T, and Giegé R: *Eur. J. Biochem.* 185:331 (1989).
177b. Schatz D, Leberman R, and Eckstein F: *Proc. Nat. Acad. Sci. U.S.A.* (1991) in press.
178. Tukalo MA, Kubler MD, Kern D, Mougel M, Ehresmann C, Ebel JP, Ehresmann B, and Giegé R: *Biochemistry* 26:5200 (1987).
179. Théobald A, Kern D, and Giegé R: *Biochimie* 70;205 (1988).
180. Leon O and Schulman LH: *Biochemistry* 26:5416 (1987).
181a. Bedouelle H and Winter G: *Nature* 320:371 (1986).
181b. Ward HJW and Fersht AR: *Biochemistry* 27:5525 (1988).
181c. Perona JJ, Rould MA, Steitz TA, Risler JL, Zelwer C, and Brunie S: *Proc. Nat. Acad. Sci. U.S.A.* 88:2903 (1991).
182. Normanly J, Ogden RC, Horvath SJ, and Abelson J: *Nature* 321:213 (1986).
183. Sampson JR and Uhlenbeck OC: *Proc. Natl. Acad. Sci. U.S.A.* 85:1033 (1988).
184. Wang Y: *Acc. Chem. Res.* 17:393 (1984).
185. Ogilvie KK, Usman N, Nicoghosian K, and Cedergren RJ: *Proc. Natl. Acad. Sci. U.S.A.* 85:5764 (1988).
186a. Hou YM and Schimmel P: *Nature* 333:140 (1988).
186b. McClain WH and Foss K: *Science* 240:793 (1988).
187a. Hou YM and Schimmel P: *Biochemistry* 28:6800 (1989).
187b. Miller WT, Hou YM, and Schimmel P: *Biochemistry* 30:2635 (1991).
188. Sampson JR, DiRenzo AB, Behlen LS, and Uhlenbeck OC: *Science* 243;1363 (1989).
189. Perona JJ, Swanson RN, Rould MA, Steitz TA, and Söll D: *Science* 246:1152 (1989).
190. Kisselev LI: *Prog. Nucleic Acids Res. Mol. Biol.* 32:237 (1985).
191. Schulman LH and Pelka H: *Nucleic Acids Res.* 18:285 (1990).
192a. McClain WH and Foss K: *J. Mol. Biol.* 202:697 (1988).
192b. Pütz J, Puglisi JD, Florentz C, and Giegé R: *Science* (1991) in press.
193. Dock-Bregeon AC, Garcia A, Giegé R, and Moras D: *Eur. J. Biochem.* 188:283 (1990).
194. Hasegawa T, Himeno H, Ishikura H, and Shimizu M: *Biochem. Biophys. Res. Commun.* 163:1534 (1989).
195. Himeno H, Hasegawa T, Ueda T, Watanabe K, Miura K, and Shimizu M: *Nucleic Acids Res.* 17:7855 (1989).

196. Crothers DM, Seno T, and Söll D: *Proc. Natl. Acad. Sci. U.S.A.* 69:3063 (1972).
197. McClain WH and Foss K: *Science* 240:793 (1988).
198a. Rogers MJ and Söll D: *Proc. Natl. Acad. Sci. U.S.A.* 85:6627 (1988).
198b. Himeno H, Hasegawa T, Ueda T, Watanabe K, and Shimuzu M: *Nucleic Acids Res.* 18:6815 (1990).
198c. Giegé R, Florentz C, Garcia A, Grosjean H, Perret V, Puglisi J, Thèobald-Dietrich A, and Ebel JP: *Biochimie* 72:453 (1990).
198d. Edwards H, Trézéguet V, and Schimmel P: *Proc. Nat. Acad. Sci. U.S.A.* 88:1153 (1991).
199a. Francklyn C and Schimmel P: *Nature* 337:478 (1989).
199b. Chambers RW: *Progr. Nucleic Acid Res. Mol. Biol.* 11:489 (1971).
199c. Musier-Forsyth K, Scaringe S, Usman N, and Schimmel P: *Proc. Nat. Acad. Sci. U.S.A.* 88:209 (1991).
199d. Perret V, Garcia A, Grosjean H, Ebel JP, Florentz C, and Giegé R: *Nature* 344:787 (1990).
200. Muramatsu T, Nishikawa K, Nemoto F, Kuchino Y, Nishimura S, Miyazawa T, and Yokoyama S: *Nature* 336:179 (1988).
201. Park SJ and Schimmel P: *J. Biol. Chem.* 263:16527 (1988).
202a. Haenni AL, Joshi S, and Chapeville F: *Prog. Nucleic Acids Res. Mol. Biol.* 27:85 (1982).
202b. Mans RMW, Pleij CWA, and Bosch L: *Eur. J. Biochem.* (1991) in press..
203. Puglisi JD, Wyatt JR, and Tinoco I Jr.: *Nature* 331:283 (1988).
204. van Belkum A, Wiersema PJ, Joordens J, Pleij C, Hilbers CW, and Bosch L: *Eur. J. Biochem.* 183:591 (1989).
205. Rietveld K, Pleij CWA, and Bosch L: *EMBO J.* 2:1079 (1983).
206. Schimmel P: *Cell* 58:9 (1989).
207. Pleij CWA, Rietveld K, and Bosch L: *Nucleic Acids Res.* 13:1717 (1985).
208. Pleij CWA and Bosch L: *Methods Enzymol.* 180:289 (1989).
209. Joshi S, Chapeville F, and Haenni AL: *Nucleic Acids Res.* 10:1947 (1982).
210. Florentz C and Giegé R: *J. Mol. Biol.* 191:117 (1986).
211a. Dreher T, Florentz C, and Giegé R: *Biochimie* 70:1719 (1988).
211b. Florentz C, Dreher TW, Rudinger J, and Giegé R: *Eur. J. Biochem.* 195:229 (1991).
212. Dreher TW, Bujarski JJ, and Hall TC: *Nature* 31:171 (1984).
213. Dreher TW and Hall TC: *J. Mol. Biol.* 201:41 (1988).
214. Renaud M, Ehrlich R, Bonnet J, and Remy P: *Eur. J. Biochem.* 100:157 (1979).
215. Khan AS and Roe BA: *Science* 241:74 (1988).
216. Freist W: *Biochemistry* 28:6787 (1989).
217. Kern D and Lapointe J: *Eur. J. Biochem.* 106:137 (1980).
218. Kern D, Lorber B, Boulanger Y, and Giegé R: *Biochemistry* 24:1321 (1985).
219. Fraser TH and Rich A: *Proc. Natl. Acad. Sci. U.S.A.* 72:3044 (1975).
220. Sprinzl M and Cramer F: *Proc. Natl. Acad. Sci. U.S.A.* 72:3049 (1975).
221. Chladeck S and Sprinzl M: *Angew. Chem. Int. Ed. Engl.* 24:371 (1985).
222. Kern D and Gangloff J: *Biochemistry* 20:2065 (1981).
223. Bonnet J and Ebel JP: *Eur. J. Biochem.* 31:335 (1972).

224. Dietrich A, Kern D, Bonnet J, Giegé R, and Ebel JP: *Eur. J. Biochem.* 70:147 (1976).

225. Lefevre JF, Bacha H, Renaud M, Ehrlich R, Gangloff J, von der Haar F, and Remy P: *Eur. J. Biochem.* 117:439 (1981).

226. Renaud M, Bacha H, Remy P, and Ebel JP: *Proc. Natl. Acad. Sci. U.S.A.* 78:1606 (1981).

227. Bacha H, Renaud M, Lefevre JF, and Remy P: *Eur. J. Biochem.* 127:87 (1982).

228. Lin SX, Baltzinger M, and Remy P: *Biochemistry* 22:681 (1983).

229. Igloi GL, von der Haar F, and Cramer F: *Biochemistry* 16:1696 (1977).

230. Gabius HJ, von der Haar F, and Cramer F: *Biochemistry* 22:2331 (1983).

231. English S, English U, von der Haar F, and Cramer F: *Nucleic Acids Res.* 14, 7529 (1986).

232. Kisselev LL, and Kovaleva GK: *Sov. Sci. Rev. D. Physicochem. Biol.* 7:137 (1987).

233. Schimmel P: *Acc. Chem. Res.* 6:299 (1973).

234. Bonnet J, Giegé R, and Ebel JP: *FEBS Lett.* 27:139 (1972).

235. Zinoni F, Birkman A, Stadtman TC, and Bock A: *Proc. Natl. Acad. Sci. U.S.A.* 83:3156 (1986).

236. Leinfelder W, Zehelein E, Mandrand-Berthelot MA, and Bock A: *Nature* 331:727 (1988).

237. Lee BJ, Wordand PJ, Davis JN, Stadtman TC, and Hatfield DL: *J. Biol. Chem.* 264:9724 (1989).

238. Schön A, Böck A, Ott G, Sprinzl M, and Söll D: *Nucleic Acids Res.* 17:7159 (1989).

239. Forchhammer K, Leinfelder W, and Böck A: *Nature* 342:453 (1989).

240. Wilcox M and Nirenberg M: *Proc. Natl. Acad. Sci. U.S.A.* 61:229 (1968).

241. Lapointe J, Duplain L, and Proulx M: *J. Bacteriol.* 165:88 (1986).

242. Roberts RS: *J. Biol. Chem.* 249:4787 (1974).

243. Nesbitt JA and Lennarz WJ: *J. Biol. Chem.* 243:3088 (1968).

244. Danchin A: *Prog. Biophys. Mol. Biol.* 54:81 (1990).

245. Wächtershäuser G: *Proc. Natl. Acad. Sci. U.S.A.* 87:200 (1990).

246. Weiner AM and Maizels N: *Proc. Natl. Acad. Sci. U.S.A.* 84:7383 (1987).

Chapter

3

mRNA and mRNP

J. Bag
University of Guelph
Department of Molecular Biology and Genetics
Guelph, Ontario, Canada

3.1. INTRODUCTION

Regulation at the level of transcription of a gene will permit the cells to control the production of the mRNA precursor. It is, however, not sufficient for expression of the polypeptide product in the cytoplasm. Beside posttranscriptional processing of the mRNA, the level of gene expression may also depend on how efficiently a specific mRNA is translated and how long this mRNA is stable in the cytoplasm. To help us understand how these processes may be regulated I will first discuss briefly certain structural features of eukaryotic mRNA that are involved in its translation and modulating its stability. The major focus of this chapter will be the unique ability of eukaryotic mRNA to complex with specific RNA-binding proteins forming mRNA-protein complexes (mRNPs). Furthermore, the present state of our knowledge regarding the possible role of interaction of mRNA with the cytoskeletal framework (CSK) and small cytoplasmic RNA-protein complexes (scRNP) in its translation will be discussed.

3.2. TRANSLATION AND STRUCTURE OF mRNA

A number of features of eukaryotic mRNAs separate it from the prokaryotic mRNAs. Prokaryotic mRNAs are usually polycistronic, whereas eukaryotic mRNAs are almost invariably monocistronic. Therefore, only one polypeptide is produced from most eukaryotic mRNAs. There are, however, occasional deviations of the monocistronic nature of eukaryotic mRNAs. The 5' end of the normal cellular mRNA is characterized by a terminal cap structure consisting of 7-methyl guanine linked via a 5'-5' triphosphate with the first transcribed nucleotide of the mRNA. A few animal and plant viral mRNAs are naturally uncapped. In addition to the terminal m^7G moieties the penultimate and subpenultimate nucleotides ($m^7GpppNpN'$) are often methylated in higher eukaryotes. This (cap) feature is absent in prokaryotic mRNAs and is considered to be characteristic of eukaryotes. Like prokaryotic mRNAs the eukaryotic mRNAs also have a 5' untranslated region (UTR). The length of the ''leader'' sequence between the initiator AUG and cap may vary in eukaryotic mRNAs from three to several hundred. However, the extremely short and extremely long 5' UTRs are exceptions rather than the rule. Generally, in most cellular mRNAs the length of 5' UTR varies form 40 to 100 nucleotides. Recent evidences support a role of 5' UTR in regulating translation of some mRNAs by sequence specific interaction with mRNA-binding proteins.[1] Earlier studies using prokaryotic mRNAs have shown that prokaryotic ribosomes are capable of using in addition to AUG a number of other triplet codons for initiation. A small number of genes have been shown to utilize either GUG or UUG as initiation codons (reviewed in Reference 2). Until recently all eukaryotic mRNAs have been shown to utilize AUG as the only codon for initiation. Analysis of *in vitro* translation products of *in vitro* transcribed RNA from a cDNA clone of human basic fibroblast growth factor (bFGF) showed synthesis of 24.2, 23.1, 22.5, and 17.8 kDa polypeptides. Three longer versions of these polypeptides were shown by nucleotide sequencing and studies using *in vitro* mutagenesis to be initiated with CUG start codons, whereas the 17.8 kDa polypeptide was initiated with an AUG codon. Furthermore, transfection of COS cells with bFGF cDNA clone from which the AUG initiator was eliminated produced the three longer versions of the polypeptides.[3] Presence of non-AUG start codon has also been reported for myc protooncogene.[4] Once the ribosome stops and selects the correct initiation codon, it starts the process of elongation of polypeptide chain. Recent evidences suggest that ribosomes may not elongate polypeptides at a uniform rate. Analysis of how ribosomes bind and stack along the length of the mRNA showed that there are regions of mRNA where ribosomes tend to stall and therefore stacking of ribosomes occurs.[5] The secondary structure of the coding region may be responsible for this stacking phenomenon. Other factors like abundance of a specific tRNA, codon usage, methylation of ribosomal RNA (rRNA) could also influence how ribosomes would proceed along the length of a specific mRNA.

3.3. TRANSLATIONAL FRAMESHIFTING

Ribosomes in both eukaryotes and prokaryotes can jump from one reading frame to the other at a discrete position on a few unique mRNAs to synthesize a single protein using two separate reading frames. Frameshifting by eukaryotic ribosomes have been described for synthesis of gag-pol polyprotein by a number of retroviral templates, and Ty transposon transcripts of yeast *Saccharomyces cerevisiae*. In many retroviruses the pol gene product is synthesized as a carboxyterminal fusion protein with the gag gene product. Posttranslational cleavage of the fusion protein produces several active enzymes. The pol gene is, however, in -1 reading frame with respect to the gag gene. Since the pol gene product does not arise from independent initiation of translation of the mRNA, synthesis of pol end of the gag-pol polyprotein may depend on ribosomal frameshift. Using translation of *in vitro* synthesized RNA from Roussarcoma virus (RSV) genomic template in a reticulocyte lysate cell free system, it was shown that gag-pol polyprotein was synthesized by a single mRNA. Furthermore, it was shown that synthesis of polyprotein depends on the template alone and not on the presence of protein factors from RSV infected cells. The mechanism of frameshifting is still unclear. It may depend on the primary or secondary structure of mRNA. A stretch of uracils in RNA may be involved in influencing frameshifting. Certain codons may have shifty natures. Unusual behavior of certain tRNAs may also be responsible for frameshifting (reviewed in Reference 6).

3.4. THE 3'-UNTRANSLATED REGION AND TRANSLATION

The 3' untranslated region appears to be less exciting than the 5'-end of mRNA in the context of protein synthesis. The sequence between the translation termination and the poly(A) addition sites may vary from 50 to 150 nucleotides. In some mRNAs 3' UTR of more than 1000 nucleotides have been found. The mRNA for a single protein sometimes differ markedly in the length of 3' UTR. There are, for example, mRNAs of four different sizes for dihydrofolate reductase. Each of these mRNAs differ only in the length of their 3' UTR. It is not certain whether this difference has any influence on the translation of mRNAs. The 3' UTR, however, has been shown to have a role in the stability of mRNAs and therefore may indirectly influence the level of polypeptide produced from a single mRNA (reviewed in Reference 2). In addition to 3' UTR, the poly(A) region of eukaryotic mRNA also influences the stability of mRNA. furthermore, a polypeptide that specifically binds to the poly(A) region of mRNA may also influence the translation of mRNA.[7] The precise mechanism of this effect is, however, unknown.

3.5. ORGANELLAR mRNAs

This discussion of mRNA structure will remain incomplete without some comments about the structure of organellar mRNAs. The mitochondrial mRNA is uncapped at the 5' end but has a 3' terminal poly(A) segment. The initiation codon of mammalian mitochondrial mRNA is frequently AUA or AUU and is generally found very close to the 5' end of the mRNA. The yeast mitochondrial mRNA, however, utilizes AUG as initiator codon and has long 5' untranslated region which is normally AU-rich. On the other hand, the 3' terminal poly(A) segment of yeast mitochondrial mRNA is only eight nucleotides long. Chloroplast mRNAs lack 3' terminal poly(A) and 5' untranslated regions. Chloroplast mRNAs still use AUG as initiation codon which is preceeded by purine-rich sequences similar to the Shine and Dalgarno sequence found in bacteria. In addition to these differences between nuclear genome coded cellular mRNAs and mitochondrial and chloroplast mRNAs, with regard to initiation of translation, there are also exceptions to the universal code for mitochondrial mRNAs (reviewed in Reference 2).

3.6 STABILITY AND mRNA STRUCTURE

In contrast to prokaryotes eukaryotic mRNAs generally possess much longer half-lives. Since different mRNA decay at different rates it is evident that certain structural features are involved in regulating mRNA decay. It is conceivable that the 5' and 3' untranslated regions of mRNA may be involved in the selective decay process. Special structural features of 5' and 3' ends may provide protection of these ends from exonuclease(s). On the other hand, recognition of special structural features at 3' and 5' UTRs or within the coding region by endonucleases or other factors could also be involved in the selective decay process.

The stability of an mRNA may be selectively modified in response to various stimuli. Stabilization of general population of mRNA was shown to occur during differentiation of muscle cells.[8] More recently, it has been shown that mRNA for creatine phosphokinase is selectively stabilized following differentiation of rat myoblast cells in culture. Approximately tenfold increase in the stability of heat shock protein hsp70 occurred in HeLa cells after heat shock treatment. Similarly, human transferrin receptor mRNA showed a selective 20-fold increase in its stability when cells were treated with iron chelator desferrioxamine. After the addition of iron salts, rapid degradation of this ferritin mRNA occurs with a half-life of only 90 min. Synthesis of tubulin is similarly regulated by selectively degrading the tubulin mRNA. Increase in the tubulin subunit level caused by either drug-induced dissociation of the microtubule or by microinjection of exogenous tubulin subunits results in reduced levels of tubulin synthesis. This was shown to be achieved by a five- to tenfold reduction in the stability of tubulin mRNAs (reviewed in Reference 9). One of the earliest examples of selective degradation of mRNA is the histone mRNA, its degradation was shown to be coupled with the cell cycle. In proliferating cells the half life of histone mRNA is approximately 45 min. Withdrawal of cells from the cell cycle during differentiation or serum starvation or inhibition of DNA synthesis by a variety of drugs have shown to shorten the half-lives of various histone mRNAs to only 5 to 10 min.[10]

Earlier studies used histone mRNA as a model to examine how decay rates of mRNAs are controlled. Since histone mRNAs lack 3' poly(A) region, it was postulated that poly(A) structure plays an important role in stabilization of mRNAs. Shortening of 3' poly(A) structure was believed to be the signal for decay of poly(A)-containing mRNAs. To test the role of poly(A) in stability of mRNAs, poly(A) was added *in vitro* to histone mRNA. Injection of polyadenylated histone mRNA into *Xenopus* oocyte showed that poly(A) containing histone mRNA was more stable than the nonpolyadenylated species.[11] Furthermore, degradation of deadenylated globin mRNA in *Xenopus* oocyte was shown to be linked to translation.[12] In keeping with the results of these epxeriments, naturally occurring poly(A)-containing H_4 histone mRNAs have been found to be stable and are not coordinately regulated with other histone mRNAs during the cell cycle.[13] Rapid decay of nonpolyadenylated histone H_3 mRNA (and probably others) in nondividing cells requires the presence of a unique stem and loop structure at the 3' end of mRNA. This 3' end structure is believed to be the primary target for exonuclease attack. Since histone mRNAs decay at slower rates in cells in S-phase it is likely that interaction of a protein factor with the 3' UTR could regulate the accessibility of this stem and loop structure from exonuclease attack (reviewed in Reference 9). Studies introducing a number of derivatives of histone H_3 and H_{2a} genes in mouse L cells have shown that in fact two features of these mRNAs are necessary for regulation of degradation. In addition to the 3' stem and loop structure, the histone mRNAs must be capable of being translated to within 300 nucleotides of the 3' end of the mRNA.[14]

The iron-dependent stability of human transferrin receptor (HTR) mRNA also requires the presence of the 3' UTR. Analysis of a series of deletion mutants of HTR mRNA have delineated the regulatory region in the 3' UTR of this mRNA. Analysis of the nucleotide sequence in this region suggest the ability of forming a stem and loop structure of 60 nucleotides. It also contains five repeats of a palindromic sequence element. Deletion of more than one repeat of the palindromic element or point mutations in the stem and loop region of the HTR mRNA completely abolished iron regulation of its stability. A protein factor that binds to the palindrome of HTR mRNA has been identified and partially characterized.[15] Furthermore, this binding activity is inversely related to the intracellular level of iron. Therefore, this iron responsive factor (IRF) may be working as a cytoplasmic repressor of degradation of a specific mRNA. The IRF does not, however, bind to the stem and loop structure of the HTR mRNA. It has been proposed that the stem and loop region is the inital binding site for an exonuclease which could be sterically hindered by the binding of IRF to the adjacent palindrome.[15] The IRF also appears to be involved in regulating translation of ferritin mRNA by iron. Ferritin mRNA contains an iron responsive signal at 5′ UTR which is similar to the one described above for the 3′ UTR of HTR mRNA.[1] These observations also suggest that depending on whether IRF binds to the 3′ end or 5′ end of mRNA, it may produce different results. Binding of IRF to the 5′ UTR of ferritin mRNA is probably involved in blocking binding of ribosomes to this mRNA. On the other hand, binding of IRF to the 3′ UTR of HTR mRNA prevents degradation by blocking movement of a ribonuclease. Therefore, sequence specific binding of an mRNA binding protein could play a pivotal role in coordinating the physiological response to an increase of intracellular iron level in higher eukaryotes.

Degradation of tubulin mRNAs like histone and HTR mRNAs are also regulated. The decay of β-tubulin mRNA is specified by the nucleotide sequences of the mRNA that lie within the first 13 translated nucleotides. This is a unique example where mRNA decay is not controlled by sequences present in the 3′ UTR. In the regulation of tubulin mRNA decay the nascent tubulin polypeptide probably plays an important role.[16] It is not, however, certain if a cytoplasmic mRNA binding protein is also involved in the destabilization of tubulin mRNA.

Translating ribosomes probably have a role in mRNA decay as well. Decay of both histone and tubulin mRNAs require translation. The rapid degradation of histone mRNA after cessation of DNA synthesis was not seen when premature termination sites were inserted in the coding region.[14] Autoregulation of tubulin mRNA stability has been shown to occur only with translationally active polysomal population. Determination of β-tubulin mRNA stability after slowing the elongation of polypeptide chains with various inhibitors showed that destabilization of β-tubulin mRNA requires ongoing ribosome translocation beyond at least ninety codons of the coding region.[17]

One possible explanation of requirement of translation for histone mRNA decay is that degradation of mRNA is initiated by a ribosome linked nuclease which recognizes the stem and loop structure of the 3′ UTR of histone mRNA. Since there is no requirement for the 3′ UTR in the decay of β-tubulin mRNA this may not be a general model. To explain the decay of tubulin mRNA it has been proposed that a contranslational binding event between tubulin subunit (or cellular mRNA binding protein factor), the nascent polypeptide and ribosome-associated nuclease is required.[16]

A variety of mechanisms for regulation of mRNA decay have been described here. A number of putative target sequences and in one case its interaction with a cytoplasmic protein have also been described. Future studies will delineate the molecular mechanism of these interactions. It is not yet known whether the enzyme that attacks the 3' loop structure is endonuclease or exonuclease. Characterization of these enzymes will be necessary before we can fully understand how their action is regulated by physiological stimuli in the cell. Furthermore, like bacterial mRNAs a role of 5' the UTR in mRNA decay may also be found for some eukaryotic mRNAs. Some basic features of mRNA structure that can control mRNA decay appear to be similar between eukaryotes and prokaryotes.

3.7. mRNA-PROTEIN COMPLEXES

More than two decades have passed since presence of rapidly labeled mRNA-like molecules as RNA-protein complexes in fish and sea urchin embryos have been demonstrated.[18, 19] Analysis of the sedimentation behavior of these complexes in sucrose gradients and buoyant densities in CsCl gradients suggested that the mRNA-like molecules were not present as part of the polyribosome where the majority of mRNAs are normally found. The result of these analyses showed that the mRNA-like molecules are complexed with protein complexes and are comprised for approximately 75% proteins and 25% RNAs. Subsequently, the presence of translatable mRNA in these complexes were demonstrated by the ability of the deproteinized RNAs from these complexes to direct protein synthesis *in vitro*.[20–23] Since these mRNA-protein complexes were found in the post-polysomal cytoplasmic fraction which is not actively engaged in protein synthesis *in vivo*, they were thought to represent a population of repressed mRNAs. It was, therefore, proposed by Spirin that the post-polysomal mRNA-protein complexes may represent stored information for protein synthesis which could later be activated for translation. Therefore, gene expression could potentially be regulated at the level of mRNA translation by storing mRNA in the form of a nontranslatable mRNA-protein complex or ''informosome'' and selectively activating the ''informosome'' of a specific mRNA under appropriate conditions.[24, 25]

The buoyant density of mRNA-protein complexes after fixing the RNA and protein moieties by formaldehyde was found to be 1.42 to 1.45 g/ml, whereas similarly treated ribosomal subunits showed buoyant densities between 1.52 to 1.58 g/ml in CsCl gradients.[19, 22] This feature of informosomes was subsequently used as an important criterion for defining and characterizing ribonucleoprotein complexes. Using this technique it was later found that translationally active mRNA from polyribosomes could also be released as complexes between mRNA and proteins, following dissociation of polyribosomes by puromycin or EDTA treatment.[26, 27] It was soon established that in the cytoplasm both polysomal and nontranslated free mRNAs (post-polysomal) exist as mRNA-protein complex (mRNP). Within a short period studies using a wide variety of eukaryotic cells showed that cytoplasmic mRNA is distributed in a wide variety of eukaryotes between polysomal and nonpolysomal populations as mRNP complexes.[25] Even in muscle cells from developing chick embryos which are normally very active in protein synthesis, as much as 30% of cytoplasmic mRNAs were found as nonpolysomal or free mRNP complexes.[22] Analysis of the *in vitro* translation products of deproteinized actin and myosin heavy chain mRNAs isolated from both polysomal and free fractions (informosomes) showed that mRNAs present in both populations are equally effective in synthesizing authentic polypeptides in a cell-free system.[22, 23] Similar conclusions were reached from analysis of translation products of polysomal and free globin mRNAs from duck erythroblasts.[28] It is now generally accepted that mRNAs present in the informosome particles are not per se defective in protein synthesis.[25] Therefore, it became necessary to investigate the reason for failure of informosomes to take part in protein synthesis *in vivo*.

To examine the cause of the repressed state of free mRNP complexes *in vivo*, a number of laboratories focused their attention to the polypeptides associated wtih cytoplasmic mRNPs. Initial studies on the polypeptide complements of polysomal and free mRNP complexes showed that in addition to a number of common polypeptides between the free mRNPs and the polysome-released mRNPs each share a set of different polypeptides.[29] It was thought that translation of a group of mRNAs was inhibited by association with repressors of protein synthesis with these mRNAs.[28] However, attempts to clearly demonstrate the presence of repressor in free mRNPs which can inhbit translation of a specifc mRNA *in vitro* were unsuccessful. Furthermore, results of translation of informosomes in cell-free systems showed that these free mRNP complexes can be translated *in vitro*.[30, 31] Results of other studies indicated that *in vitro* translation of free mRNPs depends on the method of their isolation.[32] The ability of informosomes to participate in protein synthesis *in vivo* was further established by using pulse and chase experiments.[33, 34] These studies showed that mRNAs can shuttle back and forth between the polysomal and the free fractions. In addition, kinetics of hybridization between mRNA and cDNA prepared from either polysomal and free mRNAs showed that the same mRNAs are present in both cytoplasmic compartments.[35] A majority of mRNAs which can be found in the polysomes are also abundant in the free or apparently repressed population.[36–38] These observations raised the question whether free mRNP complexes represent mRNAs in transit to polyribosomes for translation.

3.8 PROTEIN COMPOSITION OF CYTOPLASMIC mRNP COMPLEXES

To understand the physiological significance of mRNP complexes one needs to know what is the true protein composition of mRNP complexes. As long as no functional assay is available for these complexes, the true composition of mRNP proteins could always be questioned. The cytoplasm contains large quantities of proteins that can bind to negatively charged RNAs.[39] How can one eliminate all fortuitous binding of proteins to mRNAs and at the same time preserve its native structure? If functional assays for polysomal or free mRNP complexes are available one could determine the presence of polypeptide complements necessary to preserve this function. In absence of such assays, various approaches were taken to obtain mRNP complexes free from adventitiously bound proteins. It was shown that both polysomal and free mRNPs were stable in the presence of 0.5 M NaCl.[22] By using high-ionic strength buffers during isolation of these complexes, it was possible to decrease the likelihood of nonspecific association between mRNA and proteins.[22, 23] This treatment, however, carried the risk of dissociating some genuine mRNP proteins from these complexes.

Studies from a number of laboratories using various eukaryotic cells and a number of different methods of isolating mRNPs showed that polypeptide composition of these complexes are very similar. Two polypeptides of 52,000 and 72,000 kDa appeared to be consistent mRNP components in many different cell types (for a review see Reference 40). The 72,000 kDa polypeptide was shown to complex with poly(A) regions of mRNAs.[41, 42] In addition to these two polypeptides, a large number of polypeptides of 25 to 150 kDa were reported in mRNP complexes by various laboratories. At least 15 different polypeptides were detected by one-dimensional gel electrophoresis, whereas two-dimensional gel electrophoretic analysis showed the presence of approximately 30 different polypeptides in mRNP complexes.[31, 43, 44]

Attempts were made to improve the method for preparing mRNPs. It was shown by a number of laboratories that the presence of the polypeptides did not prevent mRNPs to bind to oligo(dT)-cellulose column.[31, 45, 46] Therefore, a simple isolation procedure using affinity chromatography was developed for obtaining mRNPs. The most important improvement in the method for isolating mRNPs was achieved by using *in vitro* photo cross-linking of RNA and protein moieties. Irradiation of polysomal and free fractions *in vitro* to UV was shown to covalently link the RNA-protein moieties. The cross-linked mRNPs were then purified by affinity chromatography using oligo(dT)-cellulose.[47, 48] These various isolation procedures yielded mRNPs which showed a similar and consistent pattern of prominent mRNP polypeptides.

The covalent bond formation between mRNA and proteins following UV treatment of cell extract does not, however, prove that similar interaction takes place in intact cells. Therefore, Van Venrooij and co-workers applied this technique to intact cells.[49] The result of their studies confirmed that both free and polysomal mRNAs exist as mRNP complexes in the cytoplasm of intact cells. Analysis of the polypeptide composition of polysomal and free mRNP complexes in chicken muscle cells following UV-induced cross-linking of RNA and proteins *in vivo* showed that both free (nontranslated) and po-lysomal (translated) mRNPs contain very similar polypeptides.[50] The poly-somal mRNPs showed a slightly higher protein content than free mRNPs. The major polypeptides of M_r 150, 130, 95, 85, 73, 65, 60, 54, 46, 43, 32, and 28 kDa were present in both free and polysomal mRNP complexes of *in vivo* cross-linked preparations. The free mRNP complexes were, however, found to be deficient in polypeptides of Mr 28, 38, 46, and 65 kDa. While a greater proportion of M_r 43 and 130 kDa polypeptides were found in the free mRNPs than in the polysomal mRNPs.[50] These results are different from the results of an earlier study on the polypeptide complements of free and polysomal mRNPs using non-cross-linked preparations from chick embryonic muscle[29] which showed the presence of six unique polypeptides of M_r between 40,000 to 100,000 kDa in free mRNP complexes. This difference may, how-ever, result from failure to completely eliminate fortuitous binding of proteins to mRNA in the non-cross-linked preparations used in the studies of Jain and Sarkar.[29] This fortuitous binding may be more intense in the free mRNP preparation due to the presence of a large pool of RNA-binding proteins in this fraction.[39] It is also possible that *in vivo* cross-linking approach[50] only showed the proteins that were in direct contact with mRNA (core proteins) and, therefore, did not show some true mRNP proteins which were present in these complexes through their interaction which the core complexes formed by direct mRNA-protein interaction.

In addition to these observations, *in vivo* cross-linking studies using cells in culture showed that ^{35}S-methionine-labeled proteins produced after inhi-bition of RNA synthesis by actinomycin D could be assembled into the existing polysomal and free mRNP complexes.[48, 50] This is possible only if the majority of polypeptides of existing complexes could undergo exchange with a pool of newly synthesized mRNP polypeptides. To explain the result of these studies it was suggested that free mRNPs are intermediates of mRNA trans-lation,[48, 50, 51] thus, free mRNPs represent the mRNAs in transit to polysomes. Several lines of evidence support this postulate:

1. Recent and more accurate measurements of the distribution of mRNA into the two cytoplasmic compartments have shown that only 10 to 15% of the majority of mRNAs are in the free pool.[50, 52] This value

is lower than the previous estimate of the presence of 30 to 40% mRNA in the free pool.[22]

2. Pulse and chase studies have shown that mRNA from the free pool can enter the polysomal population.[33, 34]

3. Translation of polysomal and free mRNAs in cell-free systems have shown that mRNAs which are predominant in the polysomal fraction are also abundant in the free fraction.[38, 52]

4. Furthermore, analyses of hybridization kinetics of mRNA and cDNA revealed the presence of mRNAs with similar complexities in both populations.[35] However, a number of studies have also revealed that a few species of mRNAs can be more abundant in the free population[38, 53, 54] (the significance of this observation will be discussed later).

5. It was found that the mRNA binding proteins are necessary for translation of mRNA in a cell-free system. Removal of mRNA-binding proteins from a cell-free extract of HeLa cells by affinity chromatography using mRNA-Sepharose column rendered this extract unsuitable for supporting mRNA translation. It was also found that the protein synthesizing ability of this extract could be restored by the addition of a bound protein fraction from the mRNA affinity column which resembled the polypeptides of mRNP complexes.[55]

Is it then possible that *in vivo* mRNA binds to a number of protein factors in a sequence-specific manner forming the mRNP complex? The mRNP then interact with the 43S preinitiation complex for protein synthesis. Furthermore, the kinetics of association with these proteins may be different for different mRNAs, which could explain why some mRNAs are more abundant in the free pool while others are more abundant in the polysomal population.

3.9. SEQUENCE SPECIFIC BINDING OF mRNP PROTEINS

Studies of Blobel[42] showed that the 3′/poly(A)-rich region of eukaryotic mRNA is complexed with a 78 kDa polypeptide. Recent studies using UV light to cross-link RNA and proteins showed that a 78 kDa polypeptide can be found cross-linked to the 3′-poly(A) region of mRNA in diverse cells (see Reference 40 for a current review). Setyono and Greenberg[56] investigated the proteins associated with the poly(A) region of free and polysomal mRNAs in mouse L cells. In these studies, cross-linking of RNA and proteins by UV light was carried out and poly(A)-protein complexes were isolated by digesting the mRNP complexes with RNase A and T_1 followed by chromatography on oligo(dT)-cellulose. The proteins recovered in the ribonuclease-protected poly(A) region of both polysomal and free mRNAs was analyzed by peptide mapping. A single 78 kDa polypeptide found in both preparations gave identical peptide maps. This observation suggested that the same protein was associated with the poly(A) of both polysomal and free mRNAs. It is not certain, however, whether the 78 kDa poly(A) binding polypeptide could be different in free and polysomal mRNPs. Adam et al.[57] examined the structure and function of poly(A)-binding protein in yeast. The poly(A) segment of yeast mRNA was found to cross-link with a 72 kDa polypeptide. Antibodies to the 72 kDa polypeptide was used to screen a yeast genomic library in λ gt11 expression vector. A recombinant clone producing β-galactosidase-72 kDa fusion protein was isolated. Analysis of the DNA sequence of this clone showed that this polypeptide has distinct domains and consists of repeating structural elements. A sequence of 11 to 13 amino acids is repeated three times in this poly(A)-binding polypeptide. Furthermore, this repeating structure is highly homologous to a sequence found in a major mammalian heterogeneous nuclear RNP protein and other nucleic acid binding proteins.[58, 59]

The 5' end of mRNA and particularly the 5'-cap structure can bind to several proteins. These have molecular weights of 24, 50, 80, and 220 kDa.[40] The 24 kDa polypeptide is known as cap-binding protein.[60] This polypeptide may be transiently associated with mRNA *in vivo* and therefore UV cross-linking *in vivo* have failed to detect this polypeptide.[40] Studies from several laboratories, however, have also shown that eukaryotic initiation factors like eIF-2, eIF4A, and eIF4B could directly bind to mRNA (reviewed in Reference 61). Furthermore, a 62 kDa polypeptide that shares a common epitope with eukaryotic elongation factor Tu has also been reported in mRNP complexes.[62] It is not clear whether in all mRNAs sequences other than poly(A) tail and 5' cap can bind to proteins. The major mRNP proteins of approximate molecular mass of 72 to 75 kDa, 65 to 68 kDa, 53 to 54 kDa, and 46 to 50 kDa have been found in a variety of eukaryotic cells. A large number of other polypeptides have also been described.[40] Attempts have to be made for enumeration of these polypeptides. Furthermore, understanding of the mRNA sequences involed in binding to these polypeptides will be helpful in elucidating the function of the polypeptides of mRNP complexes.

3.10. mRNP COMPLEXES OF SPECIFIC mRNAs

The majority of results on the polypeptide complements of both free and polysomal mRNP complexes came from studies using the global population of mRNA. Analysis of the proteins associated with a large number of different mRNAs have been useful during the initial phase of these studies. It is, however, important to analyze the proteins that interact with a specific mRNA. Furthermore, important insights can be obtained from analysis of the proteins associated with a specific mRNA when it is translated (polysomal mRNP) and also under a different circumstance when it is not translated (free mRNP). Until recently, only partial purification of mRNP complexes for some mRNAs was achieved. This includes globin[63] actin,[22] myosin heavy chain,[23] and protamin.[64] Several recent studies used UV irradiation of cells to cross-link RNA and proteins so that specific mRNPs can be isolated by hybrid selection without losing polypeptide moieties of this complex during hybridization of mRNP with the immobilized DNA probe. van Venrooij et al.[65] reported that in infected cells adenovirus 2 mRNAs are bound to host mRNP proteins. Ruzdijic et al.[66] reported that the polyribosomal histone H_4 mRNA was cross-linked to 49 and 52 kDa polypeptides, whereas the free histone H_4 mRNA was cross-linked to 43 and 57 kDa polypeptides. Additional studies by Ruzdijic et al.[67] showed that a subspecies of H_4 histone mRNA which contains poly(A) binds to same polypeptidees as myosin heavy chain (MHC) mRNA. The MHC mRNA is, however, significantly larger in size than that of histone mRNA and binds to a number of addiitional polypeptides. Comparison of the polypeptide complements of free and polysomal histone and myosin heavy chain mRNPs suggested that a specific mRNA binds to a different set of proteins depending upon the subcellular location of the mRNA. It is not certain whether this difference in mRNP proteins is involved in translational control. The major limitation of these studies is that it has not been shown whether translation of histone or myosin heavy chain mRNAs is regulated under the conditions used in these studies. Recent studies from our laboratory have shown that an mRNA coding for a 40 kDa polypeptide (P40) is translated in proliferating myoblasts, but is not translated in differentiated myotubes. The polypeptide complements of both repressed and translated P40 mRNP complexes were analyzed following hybrid selection of UV cross-linked mRNPs. The result of these studies showed that two polypeptides of 55 and 72 kDa were less abundant in the repressed P40 mRNP than in the translated P40 mRNP. It was therefore suggested that inefficient binding of these two polypeptides to P40 mRNA may be responsible for the repressed state of P40 mRNA in myotubes.[68]

Adam et al.[69] studied the polypeptides associated with vesicular stomatitis virus (VSV) mRNAs in infected cells and found that the major polypeptides that bind to host mRNAs also bind to the five VSV mRNAs. Furthermore, reconstitution experiments using globin mRNA and extracts from rabbit reticulocytes showed that a number of polypeptides can be cross-linked to globin mRNA.[70] The polypeptides associated with globin mRNA were similar to those normally found when the global mRNA population was examined. While hybrid selection of mRNP complexes for serum albumin using non-cross-linked native mRNP complex showed that proteins associated with albumin mRNA is only a subset of global poly(A)$^+$ mRNA-associated proteins.[71]

It therefore seems that even from the result of these studies the only general conclusion that can be reached is that a group of mRNP proteins are common to different mRNAs and, therefore, must represent binding to common features of mRNA. These may include the 5'-cap structure, the 3'-poly(A) region, and the polyadenylation signal (AAUAAA) of mRNAs.

3.11. mRNP COMPLEXES AND TRANSLATION

What is, therefore, the function of cytoplasmic mRNP complexes? It is known that mRNAs which are translationally active are complexed with proteins. However, we do not know if these proteins are actually involved in the translation of mRNA. Among the several omnipresent mRNP polypeptides a role for the poly(A)-binding proteins is, however, emerging. Biochemical studies using antibody to this polypeptide to deplete a cell-free system from the poly(A)-binding polypeptide has shown that absence of this polypeptide facilitates degradation of poly(A)-containing mRNA. Results of studies in yeast using a number of conditional mutants of the gene for poly(A)-binding polypeptide suggested that this polypeptide is involved in poly(A) shortening and binding of 60S ribosomal subunit to mRNA.[7] The role of other mRNP polypeptides may also be in facilitating the translation process. This is supported by the inability of cell-free extracts depleted of mRNA-binding proteins to support protein synthesis.[55] More direct proof, however, is necessary to unequivocally elucidate the role of other mRNP polypeptides. Furthermore, it is also not clear if poly(A)-binding polypeptide of higher eukaryotes function in a manner similar to that of yeast. If mRNP complexes are involved in translation, then why are free mRNPs which generally contain the same set of polypeptides as polysomal mRNPs not engaged in protein synthesis? It is possible that in contrast to polysomal mRNPs, free mRNP complexes actually consist of two populations of mRNAs. The majority of free mRNPs may represent mRNAs in transit to polysomes, whereas a small population of mRNAs may be translationally repressed due to interaction with specific repressor proteins.

Studies have shown that mRNA for a 40 kDa polypeptide of mouse ascites cells, mouse erythroleukemia cells and the mRNA for a polypeptide of similar molecular mass of rat L6 muscle cells can be translated in dividing cells but not in the terminally differentiated mouse or rat cells.[38, 54] In our laboratory we have been investigating the molecular mechanism of translational control of P40 mRNA in rat L6 muscle cells. At first it appeared that the translation of this mRNA is regulated with the cell cycle.[38] Results of more recent studies using inhibitors of DNA synthesis like dimethyl sulfoxide, cytosine arabinoside, and aphidicolin to block DNA synthesis without expressing genes for muscle-specific proteins have shown that inhibition of DNA synthesis alone is insufficient to selectively prevent translation of P40 mRNA. It seems that expression and possibly translation of genes for differentiated function is necessary to repress translation of P40 mRNA.[68]

Studies in our laboratory to examine the polypeptide complement of translated and repressed P40 mRNPs suggest that presence of large quantities of muscle specific mRNAs may influence binding of two polypeptides of 55 and 72 kDa to the P40 mRNA. Therefore, reduced level of binding of these two polypeptides is probably responsible for translational repression of P40 mRNA in myotubes. Brawerman's laboratory has investigated the mRNA protein interaction of another similar mRNA which is translated in dividing mouse cells but not in the differentiated cells. It was found that the 5' UTR adjacent to the AUG codon was less accessible to digestion by RNase T_1 in polysomal mRNPs than in the repressed free mRNPs.[72] Therefore, it appears that failure of the 5' UTR of mRNA to bind some critical proteins in non-proliferating cells may result in translational repression of this mRNA. It seems possible that the intrinsic ability of mRNA to complex with certain mRNA-binding proteins to form a pre-initiation complex may differ. Thus, the conformation of a specific mRNA and the relative abundance of this critical mRNA-binding protein in the cytoplasmic pool could determine the distribution of this mRNA between polysomal and free fractions. An mRNA could therefore autogenously regulate its translation. When the conditions are favorable in growing cells all mRNAs may find enough critical mRNA-binding proteins to take part in translation. However, during terminal differentiation it may not be necessary to synthesize housekeeping proteins like mRNA-binding proteins at the same level as in dividing cells. Therefore, the translation of some mRNAs may be regulated by a cascade mechanism. The decrease in the level of mRNA-binding proteins coupled with an increase in level of differentiation-specific mRNAs in the differentiated cells may force some mRNAs at a competitive disadvantage for translation and, therefore, a higher proportion of these mRNAs may be found in the free mRNPs. The intrinsic ability to bind the mRNA binding proteins may differ among mRNAs. It is, therefore, not surprising that different mRNAs may be affected differently by the reduction in the level of a critical mRNA binding protein(s). A great deal of attention was focused in the past on the translational regulation of gene expression for luxury proteins. Results of several studies have, however, shown that gene expression for many luxury proteins is regulated at the level of transcription. On the other hand, transcription may not play a major role in regulating expression of genes for many housekeeping proteins. Control at the level of posttranscriptional processing, stability, and translation may play a greater role than transcriptional control in regulating the synthesis of housekeeping proteins. Although the identity of P40 mRNA studied in our laboratory is still unknown, its presence in all the rat tissues examined suggest that it is a housekeeping protein. The transcription of this gene has been shown to be continued in differentiated myotubes. The stability of P40 mRNA was also the same in both myoblast and myotubes.[73] Therefore, translational control appeared to be the only method of regulating the expression of P40 gene.

In addition to these studies recent attempts to find the molecular basis of translational repression of mRNAs in *Xenopus* oocyte have met with some success. It has been shown that translation of mRNAs in *Xenopus* oocyte is inversely correlated with the presence of a 56 kDa polypeptide on mRNP complexes (reviewed in Reference 74). Similarly, repressors of translation of globin and ferritin mRNAs have been described.[1, 63] Translation of ferritin mRNA is regulated by the intracellular iron level. When cells are treated with iron, ferritin mRNA is translated. On the other hand, in absence of iron, ferritin mRNA enters the free mRNP pool. A 35 nucleotide long region in the 5' UTR of ferritin H mRNA was found to be sufficient and necessary for translational regulation by iron. A 90 kDa repressor polypeptide that binds to this region has been identified and characterized in several laboratories.[1, 75, 76] It is believed that binding of this polypeptide to the 5' UTR prevents translation of ferritin mRNA and this interaction is modulated by iron (see Part II, Chapter 11 of this book).

3.12. CYTOSKELETON AND mRNA TRANSLATION

The cytoplasm of eukaryotic cells is comprised of an elaborate filamentous structure known as the cytoskeletal framework.[77] It is believed to be involved in cellular processes such as maintaining the cellular morphology, cell movement, organization of cellular organelles, and cytoplasmic streaming (reviewed in Reference 78). Penman and co-workers first showed that the majority of polyribosomes with approximately 75% of the poly(A)$^+$ mRNAs are attached to the cytoskeletal (CSK) framework.[77] It was shown that treatment of cells with ribonuclease released all the ribosomes from the CSK framework, whereas dissociation of polyribosomes had no effect on the association between the poly(A)$^+$ mRNAs and the CSK framework.[79] Results of these studies suggested a direct interaction between mRNA and the CSK framework. Further studies using a milder ribonuclease treatment than earlier studies to degrade the mRNA, it was shown that the ribosomes and initiation factors could also remain attached to the CSK framework independent of the presence of mRNA.[80] The physiological significance of the attachment of mRNA to the CSK framework was examined. It was first proposed by Penman that the attachment of poly(A)$^+$ mRNAs to the CSK is obligatory for translation. Infection of HeLa cells with VSV which results in the introduction of a completely new and different set of messages into the cytoplasm was used to examine the role of CSK in mRNA translation. When the fate of the newly synthesized mRNAs was monitored by pulse and chase experiments, it was found that the VSV mRNAs became transiently attached to the CSK framework and that these mRNAs were translated only when they were associated with the CSK framework.[81] From the results of these studies it was concluded that attachment of mRNA to the CSK was required for translation. A number of other studies from various laboratories, however, suggest that attachment of poly(A)$^+$ mRNAs to the CSK framework do not mean that they will be immediately translated. In sea urchin eggs, for example, approximately 42% of the mRNAs which are repressed from translation were associated with the CSK.[82] Upon fertilization of sea urchin eggs, the amount of poly(A)$^+$ mRNA associated with the CSK was increased to 90% of the total poly(A)$^+$ population with a concomitant increase of polysomes in these cells. Similarly, studies from Sonenberg's laboratory using several eukaryotic cellular and viral mRNAs it was shown that following infection of cells with VSV although translation of cellular mRNAs like actin was blocked, but that mRNA was not released from the CSK framework. In contrast, cellular actin and VSV mRNAs were released from the CSK framework and their translation was inhibited following infection of cells with poliovirus.[83] These results, therefore, support the view that association between mRNA and the CSK framework may be helpful but not sufficient for translation. These studies were, however, carried out using either viral infected cells or unfertilized eggs where translation process is generally impaired. The relationship between mRNA translation and its attachment to the CSK framework under conditions permissible for the translation of most cellular mRNAs was examined in differentiating muscle cells. We have used translation of mRNA for a 40 kDa (P40) housekeeping polypeptide in rat L6 cells to re-examine the role of attachment of mRNA to the CSK framework and its translation. This mRNA was previously shown to be translated in proliferating L6 myoblasts but not in the terminally differentiated myotubes.[38] The distribution of this mRNA between the CSK and soluble compartments in both myoblasts and myotubes showed an identical pattern. Both translationally active and repressed P40 mRNA was found to be associated with the CSK. These studies also showed that a small fraction of translationally active polyribosomal actin mRNA, on the other hand, was not associated with the CSK and present in the soluble fraction.

The mRNA appears to directly interact with the CSK framework. It was shown by using a variety of inhibitors of protein synthesis that integrity of polyribosome structure is not necessary for mRNAs attachment to the CSK. Dissociation of polyribosomes with heat shock treatment or inhibition of initiation or elongation of protein synthesis by NaF or cycloheximide, respectively, showed no effect on the distribution of global population of mRNA between the CSK and soluble compartments.[77] Recent studies from our laboratory used the transitionally repressed P40 mRNA and translationally active actin mRNA of myotubes to examine if specific mRNAs with differential ability to participate in protein synthesis would interact differently with the CSK framework.[84] The distribution of both P40 and actin mRNAs between the CSK framework and soluble (Sol) compartments was examined following inhibition of both RNA and protein synthesis. The results of inhibition of RNA synthesis showed that very little exchange of both mRNAs occur between the CSK framework and Sol fractions. The CSK framework-bound mRNAs remained bound in absence of new RNA synthesis and were degraded while they remained attached to the CSK structure. The result of inhibition of protein synthesis by NaF, however, showed that compared to actin mRNA a significant amount of P40 mRNA was released shortly after NaF treatment from the CSK framework. This differential release of P40 mRNA was only evident in proliferating myoblasts but not in the differentiated myotubes. Since both P40 and actin mRNAs were translationally active in myoblasts, it is unlikely that this differential response was related to translation of mRNA. It is possible that this difference is due to the manner in which each mRNA is associated with the CSK framework.

How is mRNA attached to the CSK framework? It is possible that mRNA interacts with a specific protein of the CSK framework. To examine this, mRNA and proteins were cross-linked *in vivo* by irradiating muscle cells with UV. Following this treatment CSK framework-bound and soluble mRNA-protein complexes were isolated. The nature of cross-linked polypeptides to the mRNAs of these two cellular compartments were analyzed after removing the mRNA of the mRNP complexes by ribonuclease treatment. The cross-linked polypeptide complements of both CSK framework-bound and soluble mRNAs appeared to be identical with one exception. A polypeptide of 165 kDa was noticed to be preferentially distributed in the CSK framework-bound mRNA protein complexes. It is not yet known if this polypeptide is involved in binding mRNAs to the CSK[52] framework.

In summary, results of a variety of studies suggest that mRNAs,[77] ribosomes,[80] and cap binding proteins[85] are present in association with the CSK framework. Although most of the translated mRNA is present in the CSK framework evidences so far have failed to show an obligatory relationship between mRNA translation and its attachment to the CSK framework. Since ribosome and a key factor in initiation of protein synthesis, namely, the cap-binding protein is present in the CSK framework, it is possible that the CSK framework acts as a matrix to bring the large number of components necessary for protein synthesis together. Therefore, the CSK framework may enhance the overall rate and efficiency of protein synthesis.

3.13. SMALL CYTOPLASMIC RNA-PROTEIN COMPLEXES

Many years have passed since the presence of a large number of small RNA species have been described in eukaryotic cells. These small RNAs were grouped into two classes based on their cellular locations. One group is predominantly present in the nuclear fraction and referred to as small nuclear or snRNA. The other group is predominantly localized in the cytoplasm and is called small cytoplasmic RNA or scRNAs. Both sn and sc RNAs were found to be complexed with proteins. These protein bound small RNAs are now known as Snurps and Scyrps, meaning small nuclear/or cytoplasmic ribonucleoproteins (reviewed in Reference 86).

These RNAs are abundant and ubiquitous in eukaryotic cells. It was demonstrated that in autoimmune diseases like Sjogren's syndrome (or lupus) the antibody is directed against Snurps or Scyrps. These antibodies are known as Sm, Ro, or La RNPs. The two letters of this designation refer to the names of patients. The SmRNP was found to react with snRNP, whereas Ro is directed against scyrps. The role of snurps in mRNA metabolism has been extensively studied. These RNPs take part in the formation of splice of some complex which is responsible in processing pre-mRNAs in the nucleus. The role of scyrps is, however, uncertain. The ribonucleoproteins precipitated by Ro antibodies are distinct from those associated with Sn and are localized in the cytoplasm. The two small RNAs of these complexes are referred to as Y_1 and Y_2 to distinguish them from the nuclear RNAs (i.e., URNAs). The cytoplasmic small RNAs are less abundant than the URNAs and approximately 10^5 copies are present in a cell. They are not as highly conserved as URNAs. Their role in regulating protein synthesis and mRNA degradation has been speculated. Studies from several laboratories showed that various scRNAs may be involved in modulating mRNA translation. Small RNAs isolated from Artemia Salina embryos or from the dialyzate of chicken muscle initiation factor 3 (eIF-3) showed inhibitory effect on mRNA translation *in vivo*. A small 4.4S RNA was isolated as 10S RNP complex from chick embryonic muscle and showed strong inhibitory effect on binding of mRNA to 43S initiation complex. The 4.4S RNA from this complex was shown to have complementary base sequence to several mRNAs.[87, 88] This small RNA was inhibitory as free RNA as well as RNP complex. The purified proteins of the RNP complex alone was not inhibitory for protein synthesis. As many as 45 distinct polypeptides ranging in sizes from 12 to 150 kDa were detected in this 10S RNP complex (reviewed in Reference 89). Recently, several small RNAs have been identified to contain complementary sequences to several known genes. These include dihydrofolate reductase, chicken myosin heavy chain (MHC), and barley α amylase (reviewed in Reference 90). It is not clear if these RNAs act as naturally occurring antisense genes to regulate gene expression. The small RNA from chicken muscle was shown to have an inhibitory effect on mRNA translation and was referred to as translational control RNA (tcRNA$_{102}$). This RNA has been colocalized with free myosin heavy chain (MHC) mRNP complex. Sequence analysis of this 102 nucleotide long RNA revealed the presence of sequences complementary to both 5' and 3' ends of MHC mRNA. It is currently believed that during chicken embryogenesis expression of various isoforms of MHC is regulated by a family of antisense tcRNAs. Similar to the MHC mRNA family, a family of tcRNA exist which can selectively block translation of a specific MHC mRNA. It is possible that such an interaction between a specific antisense tcRNA and MHC mRNA could lead to selective degradation of the MHC mRNA.[91]

Another group of highly compact small ribonucleoprotein particles has been isolated from several repressed mRNP complexes from duck, mouse, and human cells. A 19S RNP complex of extraordinary stability has been isolated by dissociating it from the repressed globin mRNP complex by high salt or EDTA treatment. The mRNP, however, remained translationally repressed in assays using *in vitro* cell-free systems. It is therefore unlikely that presence of this 19S RNP complex was the cause of repression of globin mRNA translation *in vivo*. The term 'prosome' has been coined to describe a specific group of scRNPs associated with translationally inactive cytoplasmic mRNPs.[92] This structure is ubiquitous to animal cells and has been shown to undergo changes in their protein structure and cellular localization following heat shock treatment in *Drosophila* cells in culture. Analyses of partial nucleotide sequence of several small RNAs associated with prosomes in *Drosophila* revealed a strong sequence homology with the mammalian U6 snRNA. The presence of U6 snRNA in the prosomes, however, argues for a role in mRNA processing.[93]

REFERENCES

1. Leibold EA and Munro HN: Cytoplasmic proteins binds *in vitro* to a highly conserved sequence in the 5' untranslated region of ferritin heavy and light-subunit mRNAs. *Proc. Natl. Acad. Sci. U.S.A.* 85:2171 (1988).
2. Kozak M: Comparison of initiation of protein synthesis in prokaryotes, eukaryotes and organelles. *Microbiol. Rev.* 47:1 (1983).
3. Florkiewcz RZ and Sommer A: Human basic fibroblast growth factor gene encodes four polypeptides: three initiate translation from non AUG codons. *Proc. Natl. Acad. Sci. U.S.A.* 86:3978 (1989).
4. Hann SR, King MW, Bentley DL, Anderson WC and Eisenman RN: A non-AUG translational initiation in C-myc exon 1 generates an N-terminally distinct protein whose synthesis is disrupted in Burkitts Lymphomas. *Cell* 52:185 (1988).
5. Woolin SL and Walter P: Ribosome pausing during translation of a eukaryotic mRNA. *EMBO J.* 7:3559 (1988).

6. Craigen WJ and Caskey CT: Translational frameshifting: where will it stop? *Cell* 50:1 (1987).

7. Sachs AB and Davis RW: The poly(A) binding protein is required for poly(A) shortening and 60S ribosomal subunit-dependent translation initiation. *Cell* 58:857 (1989).

8. Buckingham ME, Caput D, Cohen A, Whalen RG, Gros F: The synthesis and stability of cytoplasmic messenger RNA during myoblast differentiation in culture. *Proc. Natl. Acad. Sci. U.S.A.* 71:1466 (1974).

9. Brawerman G: mRNA decay: finding the right targets. *Cell* 57:9 (1989).

10. Baumbach LL, Marashi F, Plumb M, Stein G, and Stein J: Inhibition of DNA replication coordinately reduces cellular levels of H1 histone mRNAs: requirement for protein synthesis. *Biochemistry* 23:1618 (1984).

11. Huez G, Marbaix G, Gallwitz D, Weinberg E, Devos R, Hubert E, and Cleuter Y: Functional stabilization of HeLa cell histone messenger RNAs injected into xenopus oocytes by 3'-OH polyadenylation. *Nature* 271:572 (1978).

12. Huez G, Marbaix G, Burney A, Hubert E, Leclercq M, Cleuter Y, Chantrene H, Soreq H, and Littauer UZ: Degradation of deadenylated rabbit α-globin mRNA in Xenopus oocytes is associated with its translation. *Nature* 266:473 (1977).

13. Bird RC, Jacobs FA, Sells BH: Coordinate and non-coordinate regulation of histone mRNAs during myoblast growth and differentiation, in Skehan P and Friedman SJ (eds): *Growth, Cancer and the Cell Cycle* Clifton, NJ, The Humana Press, 79 (1984).

14. Graves RA, Pandey NB, Chodchoy J, and Marzluff WF: Translation is required for regulation of histone mRNA degradation. *Cell* 48:615 (1987).

15. Müllner EW, Neupert B, and Kühn LC: A specific mRNA binding factor regulates the iron-dependent stability of cytoplasmic transferrin receptor mRNA. *Cell* 58:373 (1989).

16. Gay DA, Sisodia SS, and Cleveland DW: Autoregulatory control of β-tubulin mRNA stability is linked to translation elongation. *Proc. Natl. Acad. Sci. U.S.A.* 86:5763 (1989).

17. Pachter JS, Yen TJ, and Cleveland DW: Autoregulation of tubulin expression is achieved through specific degradation of polysomal tubulin mRNA. *Cell* 51:283 (1987).

18. Spirin AS and Nemer M: Messenger RNA in early sea urchin embryos: Cytoplasmic particles. *Science* 150:214 (1965).

19. Spirin AS: The second Sir Hans Krebs lecture informosomes. *Eur. J. Biochem.* 10:20 (1969).

20. Jacobs-Lorena M and Baglioni C: Messenger RNA for globin in the postribosomal supernatant of rabbit reticulocytes. *Proc. Natl. Acad. Sci. U.S.A.* 69:1425 (1972).

21. Gross KW, Jacobs-Lorena M, Baglioni C, and Gross PR: Cell-free translation of maternal messenger RNA from sea urchin eggs. *Proc. Natl. Acad. Sci. U.S.A.* 70:2614 (1973).

22. Bag J and Sarkar S: Cytoplasmic non-polysomal messenger ribonucleoprotein particles containing actin messenger RNA in chicken embryonic muscle. *Biochemistry* 14:3800 (1975).

23. Bag J and Sarkar S: Studies on a non-polysomal ribonucleoprotein coding for myosin heavy chains from chick embryonic muscles. *J. Biol. Chem.* 251:7600 (1976).

24. Spirin AS: In Bosh L (Ed): *The Mechanism of Protein Synthesis and Its Regulation* Amsterdam, North Holland Publishing, 515 (1972).

25. Preobrazhensky AA and Spirin AS: Informosomes and their protein components: the present state of knowledge. *Prog. Nucleic Acid Res. Mol. Biol.* 21:1 (1978). QP551 P695.

26. Blobel G: Protein tightly bound to globin mRNA. *Biochem. Biophys. Res. Commun.* 47:88 (1972). QH 301 B358.

27. Schochetman G and Perry RP: Characterization of the messenger RNA released from L cell polyribosomes as a result of temperature shock. *J. Mol. Biol.* 63:577 (1972).

28. Civelli O, Vincent A, Buri JF, and Scherrer K: Evidence for a translational inhibitor linked to globin mRNA in untranslated free cytoplasmic messenger ribonucleoprotein complexes. *FEBS Lett.* 72:71 (1976).

29. Jain SK and Sarkar S: Poly (riboadenylate) containing messenger ribonucleoprotein particles of chick embryonic muscles. *Biochemistry* 18:745 (1979).

30. Jenkins NA, Kaumeyer JF, Young EM, and Raff RA: A test for masked message: the template activity of messenger ribonucleoprotein particles isolated from sea urchin eggs. *Dev. Biol.* 63:279 (1978).

31. Bag J and Sells BH: Heterogeneity of the non-polysomal cytoplasmic (free) mRNA-protein complexes of embryonic chicken muscle. *Eur. J. Biochem.* 99:507 (1979).

32. Moon RT, Danilchik MV, and Hille MB: An assessment of the masked message hypothesis: sea urchin egg messenger ribonucleoprotein complexes are efficient templates for *in vitro* protein synthesis. *Dev. Biol.* 93:389 (1982).

33. Buckingham ME, Cohen A, and Gros F: Cytoplasmic distribution of pulse-labelled poly(A)-containing RNA, particularly 26S RNA, during myoblast growth and differentiation. *J. Mol. Biol.* 103:611 (1976).

34. Dworkin MB and Infante AA: Relationship between the mRNA of polysomes and free ribonucleo-protein particles in the early sea urchin embryo. *Dev. Biol.* 53:73 (1976).

35. Saidapet CR, Munro HN, Valgeirsdottir K, and Sarkar S: Quantitation of muscle-specific mRNAs by using cDNA probes during chicken embryonic muscle development *in ovo*. *Proc. Natl. Acad. Sci. U.S.A.* 79:3087 (1982).

36. Croall DE and Morrison MR: Polysomal and non-polysomal messenger RNA in neuroblastoma cells. *J. Mol. Biol.* 140:549 (1980).

37. Baker EJ and Infante AA: Nonrandom distribution of histone mRNAs into polysomes and nonpolysomal ribonucleoprotein particles in sea urchin embryos. *Proc. Natl. Acad. Sci. U.S.A.* 79:2455 (1982).

38. Pramanik SK and Bag J: Translation of an mRNA in rat L6 muscle cells is regulated within the cell cycle. *Eur. J. Biochem.* 170:59 (1987).

39. Bag J and Sells BH: Presence of cyclic AMP independent protein kinase activity in RNA binding proteins of embryonic chicken muscle. *Eur. J. Biochem.* 106:411 (1980).

40. Dreyfuss G: Structure and function of nuclear and cytoplasmic ribonucleoprotein particles. *Annu. Rev. Cell Biol.* 2:459 (1986).

41. Kawn S and Brawerman G: A particle associated with the polyadenylate segment in mammalian messenger RNA. *Proc. Natl. Acad. Sci. U.S.A.* 69:3247 (1972).

42. Blobel G: A protein of molecular weight 78,000 bound to the polyadenylate region of eukaryotic messenger RNAs. *Proc. Natl. Acad. Sci. U.S.A.* 70:924 (1973).

43. Mueller RU, Chow V, and Gander ES: Characterization of the protein moiety of messenger ribonucleoprotein complexes from duck reticulocytes by two-dimensional polyacrylamide gel electrophoresis. *Eur. J. Biochem.* 77:287 (1977).

44. Bag J and Sells BH: The presence of protein kinase activity and acceptors of phosphate groups in the non-polysomal cytoplasmic messenger ribonucleoprotein complexes of embryonic chicken muscle. *J. Biol. Chem.* 254:3137 (1979).

45. Kumar A and Pederson T: Comparison of proteins bound to heterogeneous nuclear RNA and messenger RNA in HeLa cells. *J. Mol. Biol.* 96:353 (1975).

46. Jain SK, Pluskal MG, and Sarkar S: Thermal chromatography of eukaryotic messenger ribonucleoprotein particles on oligo(dT)-cellulose. *FEBS Lett.* 97:84 (1979).

47. Greenberg JR: Proteins cross-linked to messenger RNA by irradiating polyribosomes with ultraviolet light. *Nucleic Acids Res.* 8:5685 (1980).

48. Greenberg JR: The polyribosomal mRNA-protein complex is a dynamic structure. *Proc. Natl. Acad. Sci. U.S.A.* 78:2923 (1981).

49. Wagenmaker AJM, Reinders RJ, and van Venrooij WJ: Cross-linking of mRNA proteins by irradiation of intact cells with ultraviolet light. *Eur. J. Biochem.* 112:323 (1980).

50. Bag J: Cytoplasmic mRNA protein complexes of chicken muscle cells and their role in protein synthesis. *Eur. J. Biochem.* 141:247, (1984).

51. Spirin AS and Ajtkhozhin MA: Informosomes and polyribosome-associated proteins in eukaryotes. *TIBS* April:162 (1985).

52. Pramanik S, Walsh RW, and Bag J: Association of messenger RNA with the cytoskeletal framework in rat L6 myogenic cells. *Eur. J. Biochem.* 160:221 (1986).

53. Dworkin MB and Hershey JWB: Cellular titers and subcellular distributions of abundant polyadenylate-containing ribonucleic acid species during early development in the frog *Xenopus laevis. Mol. Cell Biol.* 1:983 (1981).

54. Yenofsky R, Bergmann I, and Brawerman G: Cloned complementary deoxyribonucleic acid probes for untranslated messenger ribonucleic acid components of mouse sarcoma ascites cells. *Biochemistry* 21:3909 (1982).

55. Schmid HP, Kohler K, and Setyono B: Possible involvement of messenger RNA-associated proteins in protein synthesis. *J. Cell Biol.* 93:893 (1982).

56. Setyono B and Greenberg JR: Protein associated with poly(A) and other regions of mRNA and hnRNA molecules as investigated by cross-linking. *Cell* 24:775 (1981).

57. Adam SA, Nakagawa T, Swanson MA, Woodruff TK, and Dreyfuss G: mRNA polyadenylate-binding protein: gene isolation and sequencing and identification of a ribonucleoprotein consensus sequence. *Mol. Cell Biol.* 6:2932 (1986).

58. Cobianchi F, Sengupta DN, Zmudka BZ, and Wilson SH: Structure of rodent helix-destabilizing protein revealed by cDNA cloning. *J. Biol. Chem.* 261:3536 (1986).

59. Williams KR, Stone KI, LoPresti MB, Merrill M, and Planck SR: Amino acid sequence of the UPI calf thymus helix-destabilizing protein and its homology to an analogous protein from mouse myeloma. *Proc. Natl. Acad. Sci. U.S.A.* 82:5666 (1985).
60. Sonenberg N: ATP/Mg^{+2}-dependent cross-linking of cap binding proteins to the 5' end of eukaryotic mRNA. *Nucleic Acids Res.* 9:1643 (1981).
61. Herman RC: Alternatives for the initiation of translation. *TIBS* 14:219 (1989).
62. Slobin LI and Greenberg JR: Purification and properties of a protein component of messenger ribonucleoprotein particles that shares a common epitope with eukaryotic elongation factor Tu. *Eur. J. Biochem.* 173:305 (1988).
63. Vincent A, Goldenberg S, and Scherrer K: Comparisons of proteins associated with duck-globin mRNA and its polyadenylated segment in polyribosomal and repressed free messenger ribonucleoprotein complexes. *Eur. J. Biochem.* 114:179 (1981).
64. Gedamu L, Dixon GH, and Davies PL: Identification and isolation of protamine messenger ribonucleoprotein particles from rainbow trout testis. *Biochemistry* 16:1383 (1977).
65. van Venrooij WJ, Riemen T, Van Eekelen CAG: Host proteins are associated with adenovirus specific mRNA in the cytoplasm. *FEBS Lett.* 145:62 (1982).
66. Ruzdijic S, Bag J, and Sells BH: Cross-linked proteins associated with a specific mRNA in the cytoplasm of HeLa cells. *Eur. J. Biochem.* 142:339 (1984).
67. Ruzdijic S, Bird RC, Jacobs FA, and Sells BH: Specific mRNP complexes characterization of the proteins bound to histone H4 mRNAs isolated from L6 myoblasts. *Eur. J. Biochem.* 153:587 (1985).
68. Pramanik S and Bag J: Expression of muscle-specific proteins is necessary to regulate translation of the mRNA for a 40 kDa housekeeping polypeptide in rat L6 cells. *Eur. J. Biochem.* 182:687 (1989).
69. Adam SA, Choi YD, and Dreyfuss G: Interaction of mRNA with proteins in vesicular stomatitis virus-infected cells. *J. Virol.* 57:614 (1986).
70. Greenberg JR and Carroll E III: Reconstitution of functional mRNA-protein complexes in a rabbit reticulocyte cell-free translation system. *Mol. Cell Biol.* 5:342 (1985).
71. Johnson TR and Ilan J: Hybrid selection of messenger ribonucleoprotein for serum albumin: analysis of specific message-bound proteins. *Proc. Natl. Acad. Sci. U.S.A.* 82:7327 (1985).
72. Chitpatima T and Brawerman G: Shifts in configuration of the 5'-noncoding region of a mouse messenger RNA under translational control. *J. Biol. Chem.* 263:7164 (1988).
73. Pramanik SK, Meadus WJ, and Bag J: Regulation of translation and stability of messenger RNA coding for a 40 kDa polypeptide in rat L6 muscle cells. *Eur. J. Biochem.* 172:355 (1988).
74. Richter JD: Information relay from gene to protein: the mRNP connection. *TIBS* 13:483 (1988).
75. Rouault T, Matthias A, Hentze W, Haile DJ, Harford JB, and Klausner RD: The iron-responsive element binding protein: a method for the affinity purification of a regulatory RNA-binding protein. *Proc. Natl. Acad. Sci. U.S.A.* 86:5768 (1989).

76. Walden WE, McQueen SD, Brown PH, Gaffield L, Russell DA, Bielser D, Bailey LC, and Thach RE: Translational repression in eukaryotes: partial purification and characterization of a repressor of ferritin mRNA translation. *Proc. Natl. Acad. Sci. U.S.A.* 85:9503 (1988).

77. Lenk R, Ransom L, Kaufmann Y, and Penmann S: A cytoskeletal structure with associated polyribosomes obtained from HeLa cells. *Cell* 10:67 (1977).

78. Nielsen P, Goelz S, and Trachsel H: The role of the cytoskeleton in eukaryotic protein synthesis. *Cell Biol. Int. Rep.* 7:245 (1983).

79. Lenk R and Penman S: The cytoskeletal framework and poliovirus metabolism. *Cell* 16:289 (1979).

80. Howe JG and Hershey JWB: Translational initiation factor and ribosome association with the cytoskeletal framework fraction from HeLa cells. *Cell* 37:85 (1984).

81. Cervera M, Dreyfuss G, and Penman S: Messenger RNA is translated when associated with the cytoskeletal framework in normal and VSV-infected HeLa cells. *Cell* 23:113 (1981).

82. Moon RT, Nicosia RF, Olsen C, Hille MB, and Jeffrey WR: The cytoskeletal framework of sea urchin eggs and embryos: developmental changes in the association of messenger RNA. *Dev. Biol.* 95:447 (1983).

83. Bonneau A, Darveau A, and Sonenberg N: Effect of viral infection on host protein synthesis and mRNA association with the cytoplasmic cytoskeletal structure. *J. Cell Biol.* 100:1209 (1985). PER QH 301.

84. Meadus WJ, Pramanik S, and Bag J: Cytoskeleton bound mRNA for a 40 kDa polypeptide in rat L6 cells is not always translated. *Exp. cell. Res.* (in press).

85. Zumbe A, Stahli C, and Trachsel H: Association of a Mr 50,000 cap-binding protein with the cytoskeleton in baby hamster kidney cells. *Proc. Natl. Acad. Sci. U.S.A.* 79:2927 (1982).

86. Brunel C, Sri-Widada J, and Jeanteur P: snRNPs and scRNPs in eukaryotic cells. *Prog. Mol. Sub. Cell. Biol.* 9:1 (1985).

87. Bag J, Hubley M, and Sells BH: A cytoplasmic ribonucleoprotein complex containing a small RNA inhibitor of protein synthesis. *J. Biol. Chem.* 255:7055 (1980).

88. Bag J, Maundrell K, and Sells BH: Low MrRNAs of chicken muscle cells and their interaction with messenger and ribosomal RNAs. *FEBS Lett.* 143:163 (1982).

89. Sarkar S: Translational control involving a novel cytoplasmic RNA and ribonucleoprotein. *Prog. Nucleic Acid Res. Mol. Biol.* 31:267 (1984).

90. van der Krol AR, Mol JNM, and Stuitje AR: Modulation of eukaryotic gene expression by complementary RNA or DNA sequences. *Biotechniques* 6:958 (1988).

91. Heywood SM: tcRNA as naturally occurring antisense RNA in eukaryotes. *Nucleic Acid. Res.* 14:6771 (1986). (Corrigendum, 15:384, 1987).

92. Schmid HP, Okhayat O, Desa CM, Puvion F, Koehler K, and Scherrer K: The prosome: an ubiquitous morphologically distinct RNP particle associated with repressed mRNPs and containing specific scRNA and a characteristic set of proteins. *EMBO J.* 3:29 (1984).

93. Arrigo AP, Darlix JL, Khandjian EW, Simon M, and Spahr PF: Characterization of the prosome from drosophila and its similarity to the cytoplasmic structures formed by the low molecular weight heat-shock proteins. *EMBO J.* 4:399 (1985).

Chapter

4

Initiation: Met-tRNA Binding

H.O. Voorma
Department of Molecular Cell Biology
University of Utrecht
Utrecht, The Netherlands

4.1. INTRODUCTION

Initiation of protein synthesis in eukaryotic cells is a complex process in which many protein and RNA molecules interact with each other giving rise to an initiation complex which consists of an 80S ribosome, messenger RNA, and Met-tRNA$_i^{met}$. The Met-tRNA$_i^{met}$ binds directly into the peptidyl (P) site of the ribosome, thus leaving the acceptor (A) site available for the binding of a second aminoacyl-tRNA, whereafter the elongation cycle of protein synthesis starts. To achieve the formation of an 80S initiation complex a number of events has to take place successively. First, the 80S ribosome is dissociated into the 40S and 60S ribosomal subunits by the concerted action of eIF-3 and eIF-4C (eIF-1A in new nomenclature, see Part III, Chapter 16 of this book). Then Met-tRNA$_i^{met}$ binds to the native 43S ribosomal subunit by means of eIF-2 and GTP, whereafter the messenger RNA interacts with eIF-4A, -4B, -4E, and -4F. These factors display an RNA-dependent ATPase activity that unwinds the secondary structure of the 5′ untranslated leader region of the messenger RNA and allows the 43S preinitiation complex to scan along the single-stranded RNA towards the initiation codon AUG. The selection of the AUG codon is made possible by the presence of the Met-tRNA$_i^{met}$ on the 43S ribosomal subunit. Whether unwinding and scanning are two distinct processes or are combined in the movement of the 43S particle along the messenger needs further research. The mechanism underlying unwinding and scanning and the initiation factors involved will be dealt with in

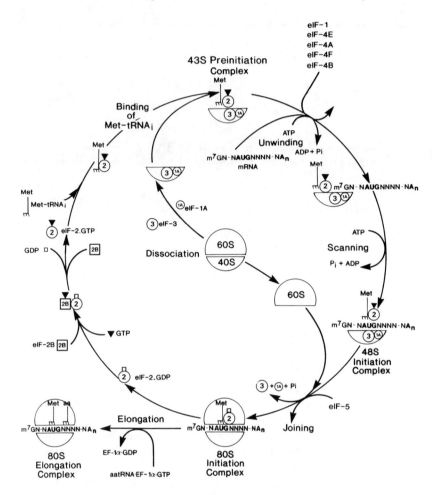

Figure 1. Initiation cycle of protein synthesis in eukaryotes.

detail in Part I, Chapter 5 of this book. After the 48S initiation complex has been formed, the joining of the 60S subunit occurs with the aid of eIF-5.

At this step all initiation factors which have not been released in preceding steps leave the ribosome with the possible exception of eIF-2, that in first instance may remain bound as an eIF-2.GDP complex. GDP is so tightly bound to eIF-2 that an eIF-2B-dependent GDP-GTP exchange reaction has to occur before eIF-2 complexed to GTP can enter a new round of initiation. The chain of initiation events is depicted in Figure 1. Recently the initiation of protein synthesis in mammalian cells has been reviewed quite adequately by Pain.[1]

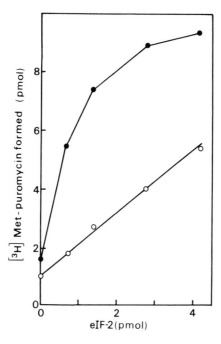

Figure 2. Effect of eIF-2B on methionyl-puromycin formation (○): no eIF-2B; (●): 2μg eIF-2B added.

4.2. FRACTIONATED CELL-FREE SYSTEMS

Since regulation of protein synthesis is mainly at the initiation step of protein synthesis a great deal of interest has been focused on initiation. One of the reasons that such studies could be carried out was the availability of a cell-free translation system derived from rabbit reticulocytes that synthesized α- and β-globin at a similar rate as occurs in reticulocytes, whereas regulation of initiation could be studied as well by deprivation of heme (see Part II, Chapter 8 of this book).

The reticulocyte lysate has also been a source for initiation factors and most of the activities of initiation factors have been purified and characterized from the ribosomal wash obtained by washing the reticulocyte ribosomes with high salt. To establish all the separate steps in initiation one needs a cell-free system in which all components can be added in a pure form. Another way to study overall translation is to destroy all the endogenous globin messenger RNA from reticulocytes by degradation with a Ca^{2+}-dependent micrococcal nuclease, followed by inactivation of this enzyme, whereafter the lysate can be programmed with every mRNA at will.

Let us first turn to the purified fractionated systems, in which the initiation factors have been used.

4.2.1. Dissociation of 80S Ribosomes

The eukaryotic 80S ribosome consists of two ribosomal subunits, 40S and 60S. A dynamic equilibrium ensures a very rapid exchange between the dissociated ribosomal subunits and those associated into 80S ribosomes (Equation 1).

$$80S \rightleftharpoons 60S + 40S \tag{1}$$

The dissociation reaction can best be described as follows: eIF-3 acts as an anti-association factor which prevents that the 40S.eIF-3 complex associates with 60S, whereas eIF-4C (eIF-1A) enhances the exchange rate in both directions which possibly means that this factor binds to both 80S and 40S. eIF-3 is the true anti-association factor. The presence of both factors pulls the equilibrium far to the right, resulting in the formation of native 40S ribosomal subunits, designated $43S_n$ (Equation 2).

$$80S.eIF-4C(eIF-1A) \rightleftharpoons 60S + 40S.eIF-4C.eIF-3 \tag{2}$$

Such a pool of free native 43S particles is required in order to facilitate the next steps in initiation, i.e., the binding of Met-tRNA$_i^{met}$ to the $43S_n$ ribosome, which is mediated by eIF-2 and GTP.

4.2.2. Eukaryotic Initiation Factor 2, eIF-2

eIF-2 consists of three nonidentical subunits; α (36 kDa), β (38 kDa), and γ (52 kDa). Hershey and co-workers[2, 3] have elucidated the nucleotide sequence of cDNAs encoding the α- and β-subunit comprising 315 and 333 amino acids, respectively. They established that the phosphorylation site for the heme-controlled repressor is Ser51 on the α-subunit;[4] whereas the β-subunit possesses putative GTP-binding sites, a zinc finger motif and three basic polylysine domains in the N-terminal part of the protein. The GTP-binding domain may be shared with the γ-subunit, since cross-linking experiments reveal contact sites with the phosphate group in the γ-subunit and a guanine binding pocket in the β-subunit, as has been demonstrated by Bommer and co-workers.[5] The role of the β-subunit appears to be dualistic in the sense that it is not required in the ternary complex formation per se, since factor preparations devoid of this subunit are still capable to bind Met-tRNA$_i^{met}$ and GTP.

On the other hand excellent genetic experiments with yeast *S. cerevisiae* carried out by Donahue and co-workers[6] reveal that the nucleic acid binding domain of the β-subunit may play a role in the scanning process of the 43S preinitiation complex and selects the ribosomal start site. A mutant β-subunit allows initiation at an UUG when AUG is absent in the HIS4 translational initiator region. These data may fill a gap in our understanding of the scanning process because the unwinding of the secondary structure is taken care of by eIF-4A, eIF-4B, and eIF-4F, whereafter eIF-2 with its β-subunit takes over the scanning during the movement of the 43S preinitiation complex.

4.2.3. Ternary Complex Formation

Binding of Met-tRNA$_i^{met}$ to eIF-2 can best be performed *in vitro* with purified eIF-2 and labeled Met-tRNA$_i^{met}$. The ionic conditions for such an assay are 20 mM Tris-HCl, pH 7.6, 80 mM KAc, 1 mM Mg(Ac)$_2$, and 7 mM β-mercaptoethanol, whereas 10 pmol of [^{3}H]Met-tRNA$_i^{met}$ and 1 to 3 μg eIF-2 are used. Incubation in 25 μl is for 10 min at 37°C in the presence of an energy regenerating system, containing 1 mM ATP, 0.4 mM GTP, 2 mM DTT, 5 mM creatine phosphate, and two units of creatine phosphokinase because traces of GDP impair the Met-tRNA$_i^{met}$ binding activity of eIF-2 to a great extent. After incubation the sample is diluted with the same buffer and filtered over cellulose nitrate filter. The complex of Met-tRNA$_i^{met}$, eIF-2 and GTP, in a ratio 1:1:1, is retained on the filter, whereas the unbound Met-tRNA$_i^{met}$ flows through. By means of the radioactive label in the Met-tRNA the extent of complex formation can be quantified (Equation 3).

$$\text{eIF-2} + \text{GTP} + \text{Met-tRNA}_i^{met} \rightleftharpoons \text{eIF-2.GTP.Met-tRNA}_i^{met} \quad (3)$$

4.2.4. Formation of the 43S Preinitiation Complex

The incubation conditions for the formation of this complex are as follows: 20 mM Tris-HCl, pH 7.6, 120 mM KAc, 0.6 Mg(Ac)$_2$, 0.1 mM spermine and 7 mM β-mercaptoethanol, 2 mM AUG, 0.5 μg eIF-1, 1 μg eIF-2, 0.16 μg eIF-4C, 7 μg eIF-3, 5 to 10 pmol [^{3}H]Met-tRNA$_i^{met}$, an energy-re-generating system, and 0.25 A$_{260}$ unit of 40S ribosomes. The incubation in 25 μl is for 20 min at 37°C, whereafter the preinitiation complexes are isolated by means of 15 to 32% sucrose gradients, centrifuged for 2.5 h at 50,000 rpm, or by Sepharose-6B column chromatography (see Equation 4).

$$\text{eIF-2.GTP.Met-tRNA}_i^{met} + \text{40S.eIF-4C.eIF-3} \rightleftharpoons$$
$$\text{40S.eIF-2.GTP.Met-tRNA}_i^{met}.\text{eIF-4C.eIF-3} \quad (4)$$
$$\text{ternary complex} + 43 \text{ Sn} \rightleftharpoons 43\text{S preinitiation complex}$$

80S-initiation complex formation is carried out under the same conditions as for the 43S complex, with the addition of 0.5 μg eIF-5 and 0.25 A_{260} unit of 60S ribosomes (Equation 5).

43S preinitiation complex + AUG + 60S + eIF-5 $\rightleftharpoons$ (5)
80S initiation complex + eIF-4C + eIF-3 + eIF-2.GDP + Pi

When natural mRNAs are used instead of AUG triplet, 2.5 μg eIF-4B, 0.5 μg pure eIF-4E, and 2.5 μg eIF-4A should be added. Incubation is for 20 min at 37°C, whereafter the extent of complex formation is measured on sucrose gradient, 15 to 32% sucrose in the same buffer as used in the assay, and centrifuged for 90 min at 50,000 rpm.

In Figure 2 one sees the effect of eIF-2B on the level of Met-tRNA$_i^{met}$ binding to 80S. Only Met-tRNA$_i^{met}$ that is bound into the P-site of the 80S ribosome reacts with puromycin and results in the synthesis of methionyl-puromycin. By this method one can discriminate between P-site-bound Met-tRNA$_i^{met}$ and Met-tRNA$_i^{met}$ either bound to eIF-2 or into 43S preinitiation complexes. In the absence of eIF-2B the efficiency of Met-tRNA$_i^{met}$ binding to 80S is astonishing low and a mere stoichiometric relationship exists between the amount of eIF-2 used and Met-tRNA$_i^{met}$ bound.

Addition of eIF-2B results in a four- to fivefold stimulation, especially with low eIF-2 input. Under these conditions there is a catalytic use of eIF-2, i.e., eIF-2 enters successive rounds of recycling provided that eIF-2B is present.

At this stage an explanation for the stimulatory activity of eIF-2B has to be found. From Figure 1 it is obvious that a roadblock for eIF-2 exists between the first initiation cycle and the second one. eIF-2 that enters the initiation cycle binds GTP and Met-tRNA$_i^{met}$, both remain bound to eIF-2 up to the joining reaction. Then the eIF-5-mediated joining gives rise to hydrolysis of GTP. However, GDP remains firmly bound to eIF-2. A dissociation constant K_D of 5×10^{-8} M means that the affinity of eIF-2 for GDP is about 100 times higher than for GTP. It is not completely clear as to whether the eIF-2.GDP complex remains bound and is translocated to the 60S ribosomal particle or is released directly into the supernatant. It may prove a strong point that phosphorylation of the complex and sequestration of eIF-2B with the phosphorylated eIF-2-αP.GDP both appear to occur on the 60S. In analogy with the prokaryotic initiation scheme, in which the presence of initiation factor IF-2, complexed with GDPCP, blocks the peptidyl transferase on the 50S particle, such a prediction cannot be firmly waved aside.

Experiments by Gross[7, 8] and Levin[9] strongly favor such a scheme. Although the GDP-GTP exchange reaction has not been elucidated in all its biochemical details, it appears that the guanine nucleotide exchange reaction follows rather a sequential than a substituted enzyme mechanism as has been recently proposed by Dholakia and Wahba.[10, 11] They presented evidence that eIF-2B interacts both with GTP and with eIF-2.GDP with no apparent preference for the order of events, which results in the formation of a quaternary complex, as is illustrated in Equation 6.

$$eIF\text{-}2.GDP + eIF\text{-}2B \rightleftharpoons eIF\text{-}2.GDP.eIF\text{-}2B + GTP \rightleftharpoons$$
$$eIF\text{-}2B + GTP \rightleftharpoons eIF\text{-}2B.GTP + eIF\text{-}2.GDP \rightleftharpoons$$
$$\rightleftharpoons eIF\text{-}2.GDP.eIF\text{-}2B.GTP \rightleftharpoons eIF\text{-}2.GTP + GDP + eIF\text{-}2B$$

(6)

The same authors report the important observation that the activity of eIF-2B in the GDP-GTP exchange reaction is enhanced by phosphorylation of the 82 kDa subunit of eIF-2B.

An opposite view on the mechanism of action of the guanine nucleotide exchange factor, designated as eIF-2B, has been expressed by Rowlands and co-workers.[12] Using kinetic enzyme methods their data are consistent with a substituted enzyme, a so-called ping-pong mechanism, which is comparable with the one depicted in Figure 5. From a number of reviews by Proud,[13] Manchester,[14] and Gupta[15] one may learn a lot more about the complex biochemical basis of the GDP/GTP exchange reaction and deduct that only rather complicated and certainly sophisticated experiments will really solve the mechanism of this reaction.

4.2.5. GDP:GTP Exchange Reaction

The exchange reaction can best be performed in two steps, in the preincubation step a complex is formed between [³H]GDP and eIF-2, which is followed by a second incubation of the samples with increasing amounts of GTP in the presence and absence of eIF-2B. As is shown in Figure 3 eIF-2B strongly enhances the GDP/GTP exchange reaction, it results in a decline of bound GDP which is exchanged by nonlabeled GTP. The reaction is severely impaired when phosphorylated eIF-2 is used, due to the formation of a complex of eIF-2 P.GDP.eIF-2B, see Part II, Chapter 8 of this book.

In the exchange experiments Salimans and co-workers[16] also noticed that the mere presence of native 40S particles, 43 Sn, enhances the GDP/GTP exchange which implies that the equilibrium reactions depicted in Equations 4 through 6 can be written even more complicated by including the binding and release of eIF-2B. It may also mean that the recycling of eIF-2B and eIF-2 takes place at different steps during the initiation event. To solve that part all the subunits of an eIF-2.eIF-2B complex have been radiolabeled by reductive alkylation using [¹⁴C]-formaldehyde, whereafter we could study the fate of the separate factors in the initiation assay. After formation of the 43S preinitiation complex by incubating 8 pmol of [¹⁴C]-eIF-2.eIF-2B with 40S, [³H]Met-tRNA$_i^{met}$, eIF-3, eIF-4C, and GTP, the sample was layered on a 15 to 32% isokinetic sucrose gradient and centrifuged for 3 h at 50,000 × g. Fractions were collected after the run and counted, the fractions comprising the 43S preinitiation complex and the top of the gradient were combined in order to facilitate the comparison of the subunit composition of these two fractions of the gradient with the original radiolabeled eIF-2.eIF-2B complex. The data in Figure 4A show that for every picomole of eIF-2 also 1 pmol Met-tRNA$_i^{met}$ is bound into the 43S initiation complex, whereas from Figure 4B it may be concluded that on the 43S complex only the three subunits of eIF-2 are present. On the other hand, the five subunits of eIF-2B are all present on the top of the gradient, together with nonparticipating eIF-2 subunits. It means that recycling of eIF-2B takes place at an earlier step, i.e., 43S preinitiation complex formation and eIF-2 recycles after the 80S initiation complex has been formed, starting with the release of eIF-2.GDP.

The recycling with emphasis on eIF-2 and eIF-2B has been depicted in Figure 5. The next step in initiation is the binding of messenger RNA, which is covered in the following chapter.

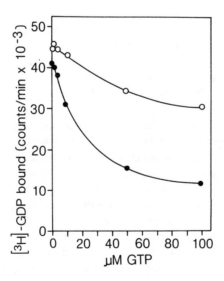

Figure 3. Effect of eIF-2B on GDP/GTP exchange (○): no eIF-2B, (●): 10μg eIF-2.eIF-2B added. (From Salimans et al: *Eur. J. Biochem.* 145:91 (1984). With permission.)

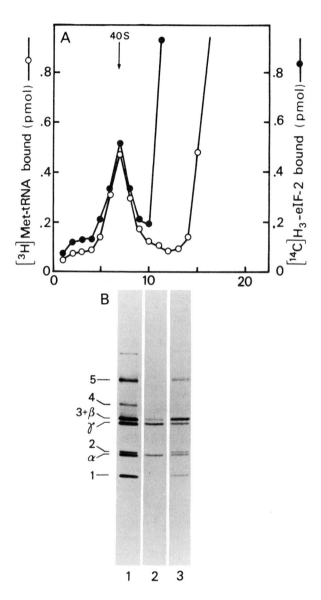

Figure 4. Analysis of subunit content on 43S preinitiation complexes. Panel A: stoichiometry of [³H]Met-tRNA$_i^{met}$ and [¹⁴C]-eIF-2 in 43S preinitiation complex, panel B: gel analysis of the protein composition in different fractions, lane 1, starting material; eIF-2.eIF-2B; lane 2, 43S preinitiation complex; lane 3, top of the gradient. (From Salimans et al: *Eur. J. Biochem.* 145:91 (1984). With permission.)

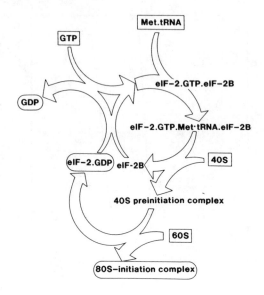

Figure 5. General scheme for the eIF-2 and eIF-2B cycle. (From Salimans et al: *Eur. J. Biochem.* 145:91 (1984). With permission.)

REFERENCES

1. Pain VM: Initiation of protein synthesis in mammalian cells. *Biochem. J.* 235:625 (1986).
2. Ernst H, Duncan RF, and Hershey JWB: Cloning and sequencing of complementary DNAs encoding the α-subunit of translational initiation factor eIF-2. Characterization of the protein and its messenger RNA. *J. Biol. Chem.* 262:1206 (1987).
3. Pathak VK, Nielsen PJ, Trachsel H, and Hershey JWB: Structure of the β subunit of translational initiation factor eIF-2. *Cell* 54:633 (1988).
4. Pathak VK, Schindler D, and Hershey JWB: Generation of a mutant form of protein synthesis initiation factor eIF-2 lacking the site of phosphorylation by eIF-2 kinases. *Mol. Cell Biol.* 8:933 (1988).
5. Bommer U-A and Kurzchalia TV: GTP interacts through its ribose and phosphate moieties with different subunits of the eukaryotic initiation factor eIF-2. *FEBS Lett.* 244:323 (1989).
6. Donahue TF, Cigan AM, Pabich EK, and Valavicius BC: Mutations at a Zn(11) finger motif in the yeast eIF-2β gene alter ribosomal start site selection during the scanning process. *Cell* 54:621 (1988).
7. Gross M, Redman R, and Kaplansky DA: Evidence that the primary effect of phosphorylation of eukaryotic initiation factor 2(α) in rabbit reticulocyte lysate is inhibition of the release of eukaryotic initiation factor-2.GDP from 60S ribosomal subunit. *J. Biol. Chem.* 260:9491 (1985).

8. Gross M, Wing M, Rundquist C, and Rubino MS: Evidence that phosphorylation of eIF-2(α) prevents the eIF-2B-mediated dissociation of eIF-2-GDP from the 60S subunit of complete initiation complexes. *J. Biol. Chem.* 262:6899 (1987).

9. Thomas NSB, Matts RL, Levin DH, and London IM: The 60S ribosomal subunit as a carrier of eukaryotic initiation factor 2 and the site of reversing factor activity during protein synthesis. *J. Biol. Chem.* 260:9860 (1985).

10. Dholakia JN and Wahba AJ: Phosphorylation of the guanine nucleotide exchange factor from rabbit reticulocytes regulates its activity in polypeptide chain initiation. *Proc. Natl. Acad. Sci. U.S.A.* 85:51 (1988).

11. Dholakia JN and Wahba AJ: Mechanism of the nucleotide exchange reaction in eukaryotic polypeptide chain initiation. Characterization of the guanine nucleotide exchange factor as a GTP-binding protein. *J. Biol. Chem.* 264:546 (1989).

12. Rowlands AG, Panniers R, and Henshaw EC: The catalytic mechanism of guanine nucleotide exchange factor action and competitive inhibition by phosphorylated eukaryotic initiation factor 2. *J. Biol. Chem.* 263:5526 (1988).

13. Proud CG: Guanine nucleotides, protein phosphorylation and the control of translation. *TIBS* 11:73 (1986).

14. Manchester KL: eIF-2B and the exchange of guanine nucleotides bound to eIF-2. *Int. J. Biochem.* 19:245 (1987).

15. Gupta NK: Regulation of eIF-2 activity and peptide chain initiation in animal cells. *TIBS* 12:15 (1987).

16. Salimans M, Goumans H, Amesz H, Benne R, and Voorma HO: Regulation of protein synthesis in eukaryotes. Mode of action of eRF, an eIF-2-recycling factor from rabbit reticulocytes involved in GDT/GTP exchange. *Eur. J. Biochem.* 145:91 (1984).

Chapter

5

Initiation: mRNA and 60S Subunit Binding

Robert E. Rhoads
Department of Biochemistry
College of Medicine
University of Kentucky
Lexington, Kentucky

5.1. INTRODUCTION

Initiation of protein synthesis may be divided into stages defined by the intermediates which are formed, termed initiation complexes, or in some literature, preinitiation complexes. The preceding chapter describes the formation of the 43S initiation complex, which consists of the 40S ribosomal subunit, met-tRNA$_i^{met}$, GTP, eIF-4C, and all of the polypeptides of eIF-2 and eIF-3. As indicated in Figure 1, the next two complexes are distinguishable by sedimentation value, contain mRNA, and can be designated 48S and 80S, although the precise S values will depend on the mRNA species present. The purpose of this chapter will be to summarize what is known about the mechanism by which these mRNA-containing complexes are formed.

The evidence supporting the model that met-tRNA$_i^{met}$ binds before mRNA is that a 40S:met-tRNA$_i^{met}$ complex, not containing mRNA, can be detected, and that upon addition of mRNA it is converted to an 80S complex, whereas the reverse is not true.[1, 2] The 43S complex has a buoyant density of 1.40 g/cc and contains both met-tRNA$_i^{met}$ and eIF-2.[3] With fractionated systems and purified initiation factors, it can be shown that met-tRNA$_i^{met}$ binding is a stringent prerequisite for mRNA binding, and that binding of mRNA requires specific factors and ATP.[4, 5] Nonetheless, Nasrin et al.[6] and Roy et al.[7] have performed experiments with purified factors, some of which are different than

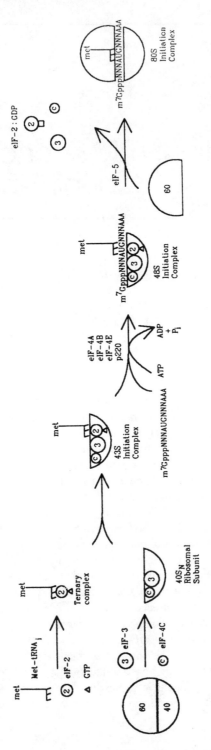

Figure 1. Formation of 48S and 80S initiation complexes.

those used in previous studies, and obtained evidence that mRNA is required for maximal binding of met-tRNA$_i^{met}$ to the 40S subunit, suggesting the opposite order of addition.

Interest in the mRNA-binding step of protein synthesis stems from the fact that under normal conditions, this step appears to be rate-limiting. This contention is supported by the findings that (1) 48S initiation complexes are far less abundant than 43S complexes,[1] so much so that they can usually be detected only in the presence of inhibitors,[8] and (2) the ATP-dependent step of initiation is rate limiting.[9] An exception to this is the situation in which production of active 43S initiation complexes becomes rate limiting, such as during virus infection or heme deprivation (see Part II, Chapters 8 and 11 of this book). There is an additional consequence of the fact that 48S complex production is ordinarily rate limiting: mRNAs are inherently different in their efficiency of initiation, and, according to a model developed by Lodish,[10] translation of mRNAs with low rate constants of initiation will be *preferentially* inhibited as the rate of 48S complex formation decreases. Thus, understanding the mechanism of the mRNA-binding step is important for both the quantitative and qualitative aspects of protein synthesis and has a bearing on mRNA competition and translational control.

The methodology used to study the mRNA binding step is generally similar to that described in the preceding chapter. However, in order to study messenger recruitment, it has been important to utilize messenger-dependent cell-free translational systems, the most popular being the nuclease-treated rabbit reticulocyte lysate[11] and the naturally mRNA-deficient wheat germ extract.[12] Cell-free translational systems have also been developed from a variety of other cell types (e.g., References 13 and 14). High fidelity of translation can be achieved in cell-free systems provided certain precautions are taken.[15] However, despite the advances which have come from a study of these systems and their fractionated components, there is always a possibility of *in vitro* artifacts; for many questions it has been necessary to use cultured cells transfected with plasmids which are then transcribed into specific mRNAs (e.g., References 16 and 17). The results of *in vivo* and *in vitro* translation are not always the same (e.g., reference 18 and references therein). Kozak[19] has recently reviewed other artifacts of cell-free systems.

5.2. FORMATION OF THE 48S INITIATION COMPLEX

Even though the 48S initiation complex is rapidly converted to an 80S complex under normal circumstances, its existence can be demonstrated under a variety of experimental conditions. In a fractionated translation system, omission of 60S subunits or the initiation factor eIF-5 causes build-up of the 48S complex, whereas omission of the eIF-3 or any of the eIF-4 group of initiation factors prevents it.[4, 5] In an unfractionated system it is necessary to use inhibitors. The peptide antibiotic edeine interferes with AUG recognition,[21] thus preventing 60S subunit joining and causing the accumulation of 48S complexes.[8, 22, 23] Fluoride also causes accumulation of mRNA-40S subunit complexes, but the met-tRNA$_i^{met}$ content is anomolously low.[24] Nonhydrolyzable GTP analogs prevent the eIF-5-catalyzed hydrolysis step and likewise cause build-up of the 48S complex.[2, 25, 26] More recently, an inhibitory activity found in sea urchin eggs has also been shown to cause accumulation of 48S complexes.[27]

As described in detail below, formation of the mature 48S initiation complex, as defined by the action of GMPPNP or edeine, occurs in several steps. Messenger first interacts with initiation factors. Next, the 43S initiation complex binds to the mRNA near the 5'-terminus. Unwinding of mRNA secondary structure by initiation factors occurs prior to or coincident with ribosome binding. The 40S subunit with associated factors migrates in the 3' direction and further unwinds the mRNA, generally stopping at the first initiation codon.

A major difference between prokaryotic and eukaryotic mechanisms for initiation is the requirement for ATP in the latter case. Marcus[9, 28] demonstrated that ATP is required for formation of an 80S initiation complex with natural (tobacco mosaic virus) but not synthetic [poly(U)] mRNA, and this role of ATP is independent of its function in charging tRNA. Additional experiments indicated that ATP is hydrolyzed, and that the step involved is formation of the 48S complex.[4, 5] Kozak[29] presented evidence that there are actually two ATP-requiring steps, the binding of mRNA to the 40S ribosomal subunit and the migration of the 40S subunit to the AUG, and that the first requirement is abolished when mRNA secondary structure is diminished. Tahara et al.,[30] on the other hand, found that ATP is required only after mRNA attachment to the 43S initiation complex. The initiation factors responsible for the ATP requirement are discussed below.

5.2.1. Structural Features in mRNA Which Affect Formation of the 48S Initiation Complex

A number of structural features in mRNA have a direct effect on the rate and extent of 48S initiation complex formation: (1) the presence of a cap, (2) the existence of secondary structure in the mRNA, and (3) the position of the initiation codon in the nucleotide sequence and the local sequence context surrounding it. In this section, these features will be considered in the presumed order of their interaction with the initiation machinery.

Presence of a cap—A prominent structure in eukaryotic mRNA which sets it apart from its prokaryotic counterpart is the presence of a 5'-terminal 7-methylguanosine-containing cap (reviewed in References 31 through 33).

The presence of this structure has a strong stimulatory effect on the translation of an mRNA bearing it. This can be demonstrated in a cell-free translational system using various preparations of mRNA: (1) viral mRNA synthesized *in vitro* in the absence of a methyl donor,[34] (2) chemically decapped mRNA obtained by periodate oxidation of the 2',2'-hydroxyls of m⁷Guo followed by β-elimination in the presence of aniline,[35] and (3) enzymatically decapped mRNA obtained with either potato nucleotide pyrophosphatase[36] or tobacco acid pyrophosphatase.[37] Additional evidence for the dependence of mRNA translation on the cap has come from the introduction of a reagent which has subsequently proved invaluable for the study of the mRNA-binding step: the cap analog. Including m⁷GMP, m⁷GTP or m⁷GpppGm in a cell-free translation mixture was shown to inhibit translation of capped but not naturally uncapped mRNAs such as encephalomyocarditis or satellite tobacco necrosis virus RNAs.[38–40]

The conclusion that cap recognition occurs during formation of the 48S initiation complex is based on two lines of reasoning: (1) uncapped (or unmethylated) mRNAs fail to become associated with 48S and 80S initiation complexes[38, 41] and (2) cap analogs do not inhibit formation of the 43S complex under conditions where 80S initiation complex formation is inhibited.[42]

The ability of cap analogs to inhibit translation of capped mRNAs has permitted delineation of the structural elements necessary for a cap to function. At least one phosphate group at the 5' position is necessary, since neither m⁷Guo nor 2'(3')m⁷GMP are inhibitors.[39, 40] Adding the β and γ phosphates improves the analog as an inhibitor,[43–45] but reducing the charge by methyl esterification of the α phosphate of m⁷GMP inactivates the analog.[46] Thus there appears to be a requirement for a negative charge center at the 5' position. Similarly, there is a requirement for a positive charge at the N-7 position of guanine; a variety of 7-alkyl substituents can replace the methyl group without diminishing cap analog activity, and in fact, 7-benzyl GMP is even more inhibitory than m⁷GMP.[47, 48] This, together with NMR data, has led to the hypothesis that an electrostatic interaction between the negative α phosphate and positive N-7 position locks the cap into an *anti* conformation in which the m⁷G is stacked on the adjacent base.[49, 50] Fluorescence and circular dichroism measurements also suggest a stacked configuration.[51, 52]

The behavior of other cap analogs gives further insight into the cap recognition process. The 2-amino group is important for cap activity; elimination of it reduces activity[47] whereas addition of one or two methyl groups to N-2 increases and decreases activity, respectively.[53] Methyl substitution at the N-1 position and chlorine substitution at C-6 both abolish cap analog activity.[47] This, together with information on the pH dependence of cap recognition,[48, 54, 55] has led to the hypothesis that only the enolate form of the cap, which is produced upon loss of the N-1 proton above pH 7.5, is active in translation.

To what degree does the presence of a cap on an mRNA stimulate its translation? Initial studies suggested that a cap was a strict requirement for translation, producing at least a five- to tenfold enhancement,[34, 35] but others favored a mediating role only, no more than two- to fourfold.[56] Subsequent studies have established that a wide range of variables determine the cap dependence of an mRNA. These may be divided into conditions of translation and the nature of the mRNA itself. When translation is carried out at increasingly higher ionic strength, dependence on the cap is increased, as measured either by the difference in translation of capped and uncapped forms of the same mRNAs or by the degree of inhibition by cap analogs.[45, 57, 58] The most likely interpretation is that increased mRNA secondary structure at higher ionic strength causes the 5'-terminus of the mRNA to be progressively less accessible; if the cap is missing, or if the protein which recognizes it is blocked with a cap analog, the mRNA is increasingly difficult to translate. Conversely, cap dependence is decreased at higher temperatures,[59] reflecting either a decrease in mRNA secondary structure or unstacking of the favorable conformation of m^7G with the adjacent base.[52] Cap dependence is greater at higher pH,[54] a likely reason being the enolate hypothesis discussed above. A final condition of cell-free translation which affects cap dependence is the nature of the extract. Whole reticulocyte lysate is relatively insensitive to cap analogs whereas wheat germ and nuclease-treated reticulocyte lysate are sensitive. Various interpretations have been put forth to explain this: the presence of a tightly bound initiation factor on endogenous mRNA, causing its insensitivity to cap analogs,[60] the sequestering of initiation factors by nuclease-treated mRNA fragments,[61] and reversible competition between mRNA and cap analog for a limiting initiation factor.[62] Regardless of the precise explanation, it seems clear that the overall concentration of mRNA directly affects the degree to which translation is dependent on the presence of a cap.

The examples of mRNA-specific differences in cap dependence are numerous (see Reference 33 for a partial list). These differences most likely result from the fact that different mRNAs have different amounts of secondary structure at their 5'-termini; those with high secondary structure have a greater need for factors which unwind secondary structure, a process which is initiated with cap recognition. Absence of a cap or inhibition of the cap-recognizing factor discriminates against those mRNAs with high secondary structure. This can be illustrated in several ways. Substitution of inosine for guanosine decreases the secondary structure of reovirus mRNA and makes its translation less cap dependent.[30, 63] The translation of alfalfa mosaic virus (AMV) 4 mRNA, even though it is a capped RNA, is almost cap independent, and this can be correlated with the fact that the 5'-terminus of the RNA is predominantly single stranded (Reference 64 and references therein). A direct measure of cap accessibility is the rate at which tobacco acid pyrophosphatase removes the cap. This has been used to demonstrate directly that cap accessibility is correlated with efficiency of initiation using the various AMV[65] and reovirus[66] mRNAs.

Secondary structure—The second variable in mRNA affecting 48S complex formation is secondary structure. Hairpins of 40 to 60 nucleotides occur upstream of the AUG more often in natural mRNA than by chance, suggesting a role in translational control.[67] Sequences from the 3'-terminus may also participate in the formation of secondary structure upstream of the AUG (see, for example, Reference 68). Pelletier and Sonenberg[69] have reviewed secondary structure in mRNA, including both prokaryotic and eukaryotic systems. Essentially all steps in the formation of 48S initiation complexes can be affected by secondary structure in mRNA: cap recognition (as mentioned above), mRNA binding to the 40S subunit, 40S subunit movement on the mRNA, and AUG recognition. Most assays do not distinguish between these steps. In this section, effects of mRNA secondary structure which cannot be assigned to individual steps will be considered.

Correlations have been made with naturally occurring mRNAs between low secondary structure and high translational efficiency. The 5'-terminus of AMV-4 mRNA, as noted above, contains little secondary structure, and this mRNA is translated more efficiently than other RNAs of that virus.[64] The 68-nucleotide leader of tobacco mosaic virus RNA is similarly low in secondary structure and serves as a translational enhancer (see Reference 70 and references therein). Another example (of many) from naturally occurring RNAs is the tenfold higher translation rate of rat aldehyde dehydrogenase mRNA compared to its bovine counterpart, which has high 5'-terminal secondary structure.[71]

The influence of secondary structure on efficiency of initiation has been examined more directly by altering the degree of secondary structure in a given mRNA. One way to increase secondary structure is to raise the ionic strength of a translation system. This causes inhibition of translation of most (e.g., globin, reovirus, VSV) mRNAs, whereas that of the low secondary structure AMV-4 mRNA is relatively resistant.[72]

A more specific way to alter secondary structure is modification or substitution of bases. Baim et al.[73] observed that a yeast mutant, in which several base substitutions increased secondary structure in the yeast iso-1-cytochrome c mRNA, exhibited decreased expression of the protein and produced iso-1-cytochrome c polysomes containing fewer ribosomes (an indicator of decreased initiation relative to elongation). Secondary structure is decreased by substitution of inosine (IMP) for guanosine and by bisulfite treatment to convert cytidine to uridine, whereas the strength of base-pairing is increased by substitution of bromouridine (BrUMP) for uridine. IMP-substituted reovirus mRNA promotes freer movement of wheat germ ribosomes, culminating in formation of 80S initiation complexes and larger structures.[26] This phenomenon can be further defined as an effect on 40S ribosomal subunit interactions with mRNA; wheat germ 40S subunits bind more readily to IMP-substituted reovirus mRNA than to unsubstituted mRNA.[63] Furthermore, IMP-substituted mRNA "outcompetes" unsubstituted mRNA for initiation complex formation while the opposite effect is observed with BrUMP-substituted mRNA.[63] Similar results are obtained with reticulocyte ribosomes.[30, 63] The combination of changing both ionic strength and base composition produces the expected result: initiation complex formation with BrUMP-substituted mRNA is severely inhibited at high ionic strength whereas that with its IMP-substituted counterpart is relatively resistant.[74]

Secondary structure of mRNA can also be altered by inserting or deleting nucleotide sequences. Thymidine kinase mRNAs were prepared with varying amounts of additional nucleotide sequence inserted into the 5' untranslated region in order to increase the secondary structure.[75] Translational efficiency was decreased, as measured by *in vitro* translation, ribosome binding, and expression in transfected cells. IMP-substitution reversed the inhibition in ribosome binding, supporting the view that these effects were due to increased secondary structure. The 5'-untranslated region of the c-*myc* mRNA appears to cause a similar *cis*-acting translational repression;[76] the fact that the sequences responsible are 240 nucleotides from the cap suggest that this is not due to an inhibition of cap recognition.[77] Insertion of hairpins of $\Delta G = -30$ kcal/mol into regions far from either cap or AUG in preproinsulin mRNA have no effect on *in vivo* expression, but more stable structures ($\Delta G = -50$ kcal/mol) strongly inhibit expression.[78] The opposite effect, translational enhancement, occurs when increasing lengths of unstructured leader are inserted upstream of a region of secondary structure in preproinsulin mRNA and the mRNA is expressed in cultured cells; the effect does not occur when the leader is downstream of the secondary structure region.[79] Other examples of translational repression due to secondary structure in the 5' untranslated region include pro-opiomelanocortin mRNA,[80] human immunodeficiency virus type 1,[81] and Yellow Fever virus RNA.[82]

A final illustration of the effect of secondary structure on mRNA translation comes from studies on short DNA oligonucleotides which are complementary to the 5'-terminus of mRNA. Whereas several investigators have demonstrated that cDNA can arrest translation (see references in Reference 69), Perdue et al.[83] investigated the effect of position of the hybrid on Rous sarcoma virus RNA translation. When the 5'-terminal 100 nucleotides of the 400-nucleotide untranslated leader were present in hybrid, translation was arrested, but when truncated cDNA allowed approximately 20 nucleotides at the 5'-terminus to be free, this effect was abolished. Lawson et al.[84] also observed that DNA complementary to the first 15 nucleotides of globin or reovirus mRNA inhibited translation while hybrid structures downstream did not. Both studies indicate that a free 5'-terminus is necessary for efficient initiation. This principle has also been demonstrated *in vivo* (frog oocytes) with antisense RNA complementary to the 5'-terminus of globin mRNA.[85]

The initiation codon—The third structural feature of mRNA which affects formation of the 48S initiation complex is the initiation codon itself. Studies in which the gene for yeast iso-1-cytochrome c was mutated to change the location of the AUG led Sherman and Stewart[86] to propose that translation is initiated at the most 5'-proximal AUG. The first AUG of reovirus mRNAs is also protected from nuclease by 48S and 80S initiation complexes.[20, 21] These observations were explained in a model for initiation codon selection called the scanning hypothesis,[87] in which the 43S initiation complex binds initially at or near the cap and then migrates until it encounters the first AUG, where it stops. Kozak[19] has recently reviewed the evidence supporting the scanning hypothesis. Briefly, translation of most vertebrate[88] and yeast[89] mRNAs begins at the first AUG. Adding new AUGs upstream of the natural one can cause initiation to begin there provided the context of the AUG is favorable, as discussed below.[16, 17] Eukaryotic ribosomes, unlike their prokaryotic counterparts, are unable to attach to circular mRNA.[90] The inhibitor sparsomycin prevents initiation from proceeding beyond the 80S complex; in its presence, RNAs with unusually long 5'-untranslated regions, such as tobacco mosaic virus and turnip yellow mosaic virus RNAs, bind two ribosomes (presumably one 80S ribosome at the initiation codon and one 40S subunit upstream[91]). Finally, as discussed above, secondary structure or hybrids with cDNA in regions which are between the cap and the AUG inhibit translation of mRNA.

What is known about the process of 40S subunit migration? The phenomenon has been directly demonstrated in unfractionated cell-free translation systems,[87, 92] but lack of a test system with purified components has hindered the acquisition of detailed information. Some of the partial activities of individual initiation factors detailed in the following section appear to be related to 40S subunit movement, but the picture is far from complete. One fact that is clear about 40S migration is a requirement for ATP hydrolysis.[29] This is distinct from the ATP requirement for mRNA binding to 40S subunits; the latter requirement can be eliminated by IMP substitution in reovirus mRNA or construction of a synthetic polynucleotide as template. In low secondary structure mRNAs thus obtained, the 40S subunit remains bound to the 5'-terminus in the absence of ATP and fails to migrate toward the AUG. GTP hydrolysis, by comparison, is not required for 48S complex formation, as this proceeds normally in the presence of GMPPCP.[63] Migration of 40S subunits per se does not appear to be rate limiting for initiation since the insertion of unstructured 5'-untranslated sequences does not decrease mRNA efficiency.[79] Insertion of secondary structure-conferring elements, however, does have an inhibitory effect on translation as noted above.

What is known about AUG recognition? Various agents or modifications of mRNA or cell-free translation conditions can affect this process specifically. Several inhibitors of initiation, edeine, pactamycin, and fluoride, cause 40S subunits to continue past the AUG and form "40S polysomes", in which internal sequences in the mRNA are protected from ribonuclease digestion.[87, 92] Secondary structure in mRNA, on the other hand, enhances AUG recognition. IMP-substituted mRNA forms "40S polysomes" whereas unsubstituted mRNA does not.[63] This may be an effect of generally reducing mRNA secondary structure or may be due to changing the initiation codon from AUG to AUI. A way to reduce secondary structure without altering the initiation codon is to treat mRNA with bisulfite; reovirus mRNA so treated exhibits the same property of "40S polysome" formation. However, this can be demonstrated only in wheat germ extracts which are deficient in 60S subunits; in their presence, "40S polysomes" are not formed. Recognition of the initiation codon is also impaired by low magnesium ion concentration,[92] which may be affecting either overall mRNA secondary structure or codon-anticodon interaction. Kozak[63] has speculated that mRNA secondary structure slows the advancing 40S subunit and allows more effective codon-anticodon interaction.

A direct demonstration that the anticodon of the initiator tRNA participates in AUG recognition comes from genetic studies. Cigan et al.[93] changed the anticodon site of one of the yeast tRNA$_i^{met}$ genes to 3'-UCC-5'. Strains in which the initiation codon of the *his4* mRNA had been mutated were able to grow in the absence of histidine only when the initiation codon was 5'-AGG-3'. This system provided further evidence for the scanning model with the observation that introduction of an upstream AGG caused the new initiation codon to be used in preference to the original AGG.

A major determinant in AUG recognition is the nucleotide sequence context in which the AUG appears (reviewed in Reference 19). Whereas eukaryotic mRNAs were initially believed to contain no initiation signals such as the Shine-Delgarno sequence of prokaryotic mRNAs,[94] examination of additional mRNA sequences revealed that the overwhelming majority contain a purine at the −3 position or a G at +4 or both.[95] A consensus sequence for 699 vertebrate mRNAs is found to be 5'-GCC(A/G)CCAUGG-3', where A/G indicates either A or G at position −3.[88] Of these nucleotides, the presence of a purine, generally A, at position −3 is the most conserved feature and occur in 97% of the mRNAs. A similar compilation from 131 yeast mRNAs yields a quite different consensus, 5'-(A/Y)A(A/U)AAUGUCU-3', but one which contains the A at −3.[89] A consensus for 23 plant and plant viral mRNAs is 5'-(U/A)ANNNNANNAUGGCU-3',[96] which, considering the small sample size, is not noticeably different from the vertebrate consensus. In the case of vertebrate mRNAs, the importance of the consensus nucleotides from −1 to −6 and at +4 has been demonstrated by site-directed mutagenesis;[97, 98] initiation codons can be designated "strong" or "weak" primarily on the basis of the -3 and $+4$ nucleotides. In the yeast case, genetic selection has failed to indicate a role for flanking nucleotides.[99]

Initiation codon selection by the 40S ribosomal subunit almost always follows the rules described above. Instances in which the first AUG is not used are generally attributable to a weak context.[19] Other unusual initiation events include reinitiation, i.e., bifunctional mRNAs (reviewed in Reference 100), use of non-AUG initiation codons,[101, 102] and internal initiation (see Part II, Chapter 11 of this book).

5.2.2. Proteins Involved in the Formation of the 48S Initiation Complex

The discussion up to this point has centered on the phenomenology of 48S initiation complex formation rather than biochemical mechanisms and has not brought in the polypeptides thought to be responsible. Research on the mRNA-binding initiation factors has proceeded in parallel with, but curiously, often without direct correspondence to, the phenomenological studies. This has produced the current situation in which many activities are known for individual initiation factors and complexes of factors, but the location of these polypeptides with respect to ribosomes and the mechanisms by which they facilitate the overall process are not yet known.

This section reviews information on the structure and activities of these factors. A somewhat arbitrary division is made on the basis of factors which have been *demonstrated* to be present on the 43S initiation complex prior to mRNA binding and those which have not. The intention is to separate out those factors whose primary role is mRNA binding to the 43S complex (eIF-4E, eIF-4A, eIF-4B, p220 and their complexes) from those which have other roles (eIF-3, eIF-2, eIF-4C). Unfortunately, the distinction is not clear cut; members of the latter group also appear to have roles in mRNA binding, as reviewed below, whereas some members of the former groups may well prove to be located on the 43S initiation complex. After considering the individual factors, the discussion will turn to the structures and activities of multifactor complexes.

Factors present on the 43S initiation complex: eIF-3, eIF-2, eIF-4C— The factors present on the 43S initiation complex prior to mRNA binding are discussed in detail in the previous chapter. However, in addition to their roles in formation of the 43S initiation complex, these factors participate in various steps leading to the 48S initiation complex.

The largest of the initiation factors, eIF-3, is required for maximal binding of mRNA to the 43S complex.[4, 5, 103] Although there is some uncertainty as to the number of subunits and whether some are derived from proteolysis,[103–106] most[5] or all[25] subunits appear to be present on the 43S initiation complex in amounts stoichiometric with met-tRNA$_i^{met}$.

Maximal binding of mRNA also requires the presence of all three subunits of eIF-2 on the 43S initiation complex.[4, 5, 25] The role of eIF-2 in mRNA recognition and/or AUG selection is not well defined, but Kaempfer et al.[107] have provided evidence that eIF-2 interacts directly with the ribosome binding site of mRNA. A remarkable finding which further implicates eIF-2 in the initiation codon selection process is that a mutation in yeast eIF-2β permits initiation at a UUG codon in place of AUG in *his4* mRNA.[108] The mutation occurs in a zinc-finger domain which is conserved in yeast and human eIF-2β and which is believed to participate in RNA binding.

The third factor on the 43S initiation complex prior to mRNA binding is eIF-4C, although there is some confusion on this point; Thomas et al.[109] demonstrated stoichiometric binding of eIF-4C to 40S subunits, but Peterson et al.[25] were unable to demonstrate binding. The role of eIF-4C is also not clear. Although it is required for maximal binding of met-tRNA$_i^{met}$, it also plays a role in maximal binding of mRNA.[4, 5] Wheat germ eIF-4C was shown to increase 40S subunit-bound met-tRNA$_i^{met}$ in the presence of GMPPNP, suggesting that it participates in 48S initiation complex formation.[103]

Factors not (yet) demonstrated to be on 43S initiation complexes: eIF-4E, eIF-4A, eIF-4B, p220—The cap-binding protein, or eIF-4E, is a 25 kDa polypeptide which has a specific affinity for the mRNA cap and is likely to be the first protein to come in contact with mRNA during the initiation cycle (reviewed in Reference 110). Early indications that eIF-4E was required for translation of capped mRNAs came from studies in a reconstituted cell-free translation system using fractionated initiation factors[111] and studies in an unfractionated translation system in which cap analog inhibition was reversed with eIF-4E.[112] More recently, development of a temperature-sensitive variant of eIF-4E and an eIF-4E-dependent cell-free translation system in yeast has provided the clearest demonstration.[113] In this system, translation of some mRNAs (total yeast mRNA, chloramphenicol acetyltransferase mRNA) was stimulated five- to sevenfold by exogenous eIF-4E, whereas that of AMV-4 mRNA was only marginally stimulated, in keeping with its low cap requirement and secondary structure. The importance of eIF-4E for translation is also shown by the fact that disruption of its gene prevents the growth of yeast.[114] Surprisingly, the eIF-4E gene of yeast is identical to a previously identified gene involved in controlling the cell cycle, *CDC33*.[115] Mutations in this gene cause a phenotype similar to nutritional arrest and also produce an eIF-4E protein which has reduced cap-binding activity.[116]

The primary structure from human[117] and mouse[118] eIF-4E are nearly identical but share only 33% identity with the yeast protein.[116] Nonetheless, the mouse protein can substitute for its yeast counterpart *in vivo* without major effects on cell viability, growth, and mating.[119] The protein which appears to be the *Drosophilia* equivalent of eIF-4E is, by contrast, very different in molecular weight (35 kDa) and does not cross-react immunologically with the mammalian factor.[120] Cap-binding proteins have been isolated from wheat germ extracts as well.[121, 122] Curiously, two cap-binding proteins of 26 and 28 kDa are found, but they are not immunologically cross-reactive.[123] Circular dichroism studies of the yeast protein indicate that the major element of secondary structure is β-sheets, with relatively little α-helix.[118, 124] Upon cap binding, there is a detectable change in secondary structure. Human eIF-4E appears to contain disulfide linkages, based on the inability of unreduced protein to react with iodoacetic acid, and is phosphorylated at Ser-53.[125, 126] The effects of phosphorylation of this and other initiation factors are discussed in Part II, Chapter 9 of this book.

Several activities can be demonstrated for eIF-4E: binding to caps, modulating the activities of other initiation-related polypeptides, and transfer to the 43S initiation complex along with mRNA, the latter two of which are considered in subsequent sections. Although a variety of cap-dependent activities can be demonstrated with combinations and complexes of initiation factors, it is only eIF-4E that, as an isolated polypeptide, exhibits specific binding to caps. Binding to caps was first demonstrated with the use of periodate-oxidized, cap-labeled reovirus mRNA.[127] Upon addition of Na-BH$_3$CN, it formed stable cross-links with various polypeptides including eIF-4E in unfractionated cell-free translational systems or initiation factor preparations. Photochemical cross-linking of eIF-4E to unmodified mRNA[128] and photoactive cap analogs[129, 130] can also be demonstrated. The affinity of eIF-4E for caps is great enough that reversible interactions can be exploited as well. The reversible binding of eIF-4E to cap analogs provides an assay for eIF-4E activity in unfractionated cell-free translational systems.[112] The binding of eIF-4E to immobilized cap analogs permits rapid purification of the protein.[131, 132] Capped oligonucleotides are retained on nitrocellulose filters by eIF-4E, providing a simple assay.[112, 132] Finally, quenching of tryptophan fluorescence in eIF-4E by cap analogs enables one to study variations in eIF-4E primary structure,[118, 124] cap analog structure,[48] and binding conditions.[55] These studies have shown that the affinity of eIF-4E for the cap structure ranges from 10^5 to 10^6 M^{-1}, depending on these variables.

The second factor to come in contact with mRNA during initiation is probably eIF-4A. This 46 kDa polypeptide is required for the mRNA binding step in mammalian[4, 5] and wheat germ.[133] In yeast it is essential for growth, as shown by gene disruption, and lysates from these cells are inactive in protein synthesis.[134] Cloning of mammalian eIF-4A cDNA reveals two mRNAs[135] and genomic cloning has led to the discovery of two functional genes and multiple pseudogenes.[136, 137] The eIF-4A genes of yeast have also been isolated, based on their ability to suppress a mitochondrial missense mutation[138], and a homologous gene has been found to be expressed only in the male germ line in mice.[139] The mammalian eIF-4A protein has been extensively characterized;[140] it is very similar in rabbit reticulocyte and HeLa cells[141] and is a glycoprotein.[142]

eIF-4A is the protein responsible for the ATP requirement of eukaryotic initiation. Although a variety of ATP-dependent activities can be demonstrated with combinations and complexes of initiation factors, only eIF-4A exhibits ATP-dependent activities as an isolated polypeptide. eIF-4A can be labeled with the ATP affinity analog 5'-*p*-fluorosulfonylbenzoyladenosine[143] or by direct UV cross-linking with ATP.[144] Mammalian eIF-4A hydrolyzes ATP in the presence of single- but not double-stranded RNA.[145, 146] Wheat germ eIF-4A was initially reported to hydrolyze ATP, but in a non-RNA-dependent manner.[147] Subsequent work, however, has indicated that factors from both reticulocyte and wheat germ have RNA-dependent ATPase activity and behave interchangeably.[148] eIF-4A also binds RNA in a manner which is enhanced in the presence of ATP.[140, 146, 149] Despite the fact that eIF-4A can be cross-linked to mRNA caps when in the presence of other initiation factors, the polypeptide alone has no cap-binding activity.[140, 150]

The next factor to be considered, eIF-4B, is less well understood. In mammalian systems it appears to be a dimer of identical 80 kDa polypeptides which is required for the mRNA binding step.[4, 5] The cDNA for human eIF-4B has been cloned and sequenced.[151] While rabbit and human eIF-4B are similar,[141] identifying the wheat germ equivalent has proved problematical. Seal et al.[133] originally considered their factor D2a, a polypeptide of 80 kDa, to be the same as eIF-4B. Subsequently, Lax et al.[121] obtained a different preparation, a complex composed of 82 and 28 kDa polypeptides which stimulated amino acid incorporation in an eIF-4B-deficient system; these authors presented evidence that this complex was the wheat germ equivalent of mammalian eIF-4B. The 28 kDa polypeptide is apparently a cap-binding protein since this complex binds to m⁷GTP-Sepharose. Seal et al.[122] obtained a similar preparation but termed it "cap-site factor". These authors also obtained a different factor from their fraction C1 (not D2a mentioned above) which did *not* bind to a cap affinity column and called this eIF-4B. Browning et al.[152] apparently isolated the same factor, which was not retained on m⁷GTP-Sepharose but did stimulate mRNA binding to ribosomes, determined its molecular weight to be 59 kDa and termed it eIF-4G. Finally, based on functional similarities to mammalian eIF-4B, the 59 kDa factor was renamed eIF-4B and the 82 kDa:28 kDa complex was renamed eIF-(iso)4F.[153] In reading the published literature from this period, it is therefore advisable to interpret activities on the basis of the polypeptide composition of a factor rather than its name. For the remainder of this chapter, the revised nomenclature will be used.

Unlike eIF-4E and eIF-4A, it is difficult to assign an activity to the eIF-4B polypeptide itself. eIF-4B stimulates incorporation of mRNA into initiation complexes in an eIF-4B-deficient translation system and participates in a variety of partial reactions in combination with other factors, to be considered below. Despite its ability to be cross-linked to mRNA caps in the presence of other polypeptides, eIF-4B alone does not have affinity for the cap.[150] The ability of wheat germ "eIF-4B" to bind RNA with high affinity was demonstrated by Butler and Clark,[154] but based on the polypeptide composition of their preparation, they were probably dealing with eIF-(iso)4F. RNA binding has also been demonstrated with rabbit eIF-4B.[149] In keeping with this, the cDNA-derived sequence of human eIF-4B[151] contains a consensus RNA-binding domain.[155]

The last polypeptide to be considered in this section is p220, first identified in high molecular weight complexes with other eIF-4 group factors based on the activity of restoring protein synthesis to lysates of poliovirus-infected HeLa cells.[156] Subsequently it was shown that p220 is proteolytically cleaved during poliovirus infection, causing inactivation of the machinery for translating capped mRNAs[157] (reviewed in Part II, Chapter 11 of this book). Virtually nothing is known about the protein chemistry of mammalian p220, other than that it appears as several polypeptides of 220 to 240 kDa on denaturing gel electrophoresis; all forms react with a monoclonal antibody against p220.[158] The wheat germ equivalent of p220 may be either the 220 kDa subunit of eIF-4F or the 82 kDa subunit of eIF-(iso)4F.[123] In yeast, a complex of polypeptides with apparent molecular masses of 24 and 150 kDa can be cross-linked to mRNA caps and behaves during purification like eIF-4F, suggesting that the 150 kDa polypeptide is the yeast equivalent of mammalian p220.[159] At this time, no biochemical activities can be assigned to the p220 polypeptide per se, although it probably mediates most of the cooperative activities of eIF-4 group initiation factor complexes discussed below.

Complexes of initiation factors—One of the most important characteristics of the initiation factors which participate in the incorporation of mRNA into the 48S initiation complex is their ability to form complexes. The nature of these complexes gives hints as to the concerted mechanism of action of the component factors. eIF-4E, eIF-4A, and eIF-4B are isolated predominantly as free polypeptides, but when ribosomes are extracted with 0.5 M KCl, and the extract sedimented ultracentrifugally in 0.5 M KCl, a high molecular weight complex, termed CBPII is obtained.[156] By functional and polypeptide pattern analysis, this complex contains eIF-4E, eIF-4A, eIF-4B and p220. Further purification yields a complex containing eIF-4E, eIF-4A, and p220 which was given factor status and named eIF-4F[160] or CBP complex.[150] Nonetheless, eIF-4B is only separated with some difficulty in high salt from eIF-4F and thus can be said to have an affinity for the eIF-4F complex.[160] The point has been raised that the eIF-4A component of eIF-4F may not be identical to free eIF-4A, since two-dimensional peptide maps[150] and HPLC patterns of CNBr-digested proteins[161] indicate slight differences. However, the size and isoelectric points of the two proteins are the same.[160] Alternative purification schemes to that used for eIF-4F yield a complex consisting of only p220 and eIF-4E, which has been termed CBPC.[162] and CBC-PC.[163] Passage of eIF-4F over phosphocellulose also removes eIF-4A to produce the same complex.[164]

Even larger complexes of the initiation factors involved in mRNA recruitment have been demonstrated. Trachsel et al.[165] originally detected poliovirus "restoring activity", which in retrospect was p220, in the eIF-3 region of a low salt sucrose gradient (15 to 18S) but found that it was released from eIF-3 in high salt (0.5 M KCl). In fact, the first clear identification of p220 was made using an antibody against eIF-3; p220 was present as a contaminant of the eIF-3 preparation used as immunogen.[157] eIF-4E, whether assayed using cap analog reversing activity[112] or cross-linking to mRNA caps,[166] is also isolated in a complex with eIF-3 which can be dissociated on high salt sucrose gradients. Such gradients are necessary to separate eIF-4F from eIF-3 as well.[150, 160] Etchison and Smith[167] have shown that the eIF-3:eIF-4F complex can be further purified by cap affinity chromatography.

Summing up this section, there appear to be a series of complexes of increasing size and decreasing stability to salt: p220 binds eIF-4E, followed by eIF-4A, eIF-4B and eIF-3. Given that the 43S and 48S initiation complexes contain a stoichometric complement of eIF-3 (see above), it seems likely that all of these initiation factors become bound to the 40S ribosomal subunit during some stage of initiation.

The biochemical activities which can be demonstrated for complexes and combinations of these initiation factors are both revealing in what they say about their concerted mechanism and bewildering in their sheer volume and complexity. In an attempt to deal with this complexity, the following discussion is divided into four areas: (1) the interaction of factors with caps on mRNA, which is primarily studied by cross-linking to cap-labeled, periodate-oxidized mRNA; (2) the interaction of factors with ATP, as studied by both photoaffinity labeling and RNA-stimulated ATPase activity; (3) the interaction of factors with mRNA per se (i.e., other than at the cap), as studied by RNA binding and RNA unwinding assays; and (4) the ability of factors to discriminate between mRNAs.

When purified eIF-4A, eIF-4B and eIF-4F from mammalian sources are tested for cross-linking to caps in the absence of ATP, only the eIF-4E component of eIF-4F reacts.[146, 150, 160] In the presence of ATP, however, Sonenberg[168] showed that cross-linking to a partially purified preparation of initiation factors yields labeled eIF-4A and eIF-4B. This reaction does not occur with nonadenosine nucleotides, with ADP, or with nonhydrolyzable ATP analogs, indicating that ATP hydrolysis is required to bring eIF-4A and eIF-4B into close proximity with the cap. Positive identification of the 80 kDa polypeptide as eIF-4B was made using an anti-eIF-4B antibody.[169] The labeling of eIF-4A is also demonstrated in highly purified eIF-4F,[150, 160] and this reaction is substantially enhanced by the presence of eIF-4B.[146] eIF-4A alone is very weakly cross-linked to caps (in an ATP-dependent fashion), and while this is slightly stimulated by eIF-4B, adding eIF-4F greatly enhances the labeling.[146] With wheat germ factors, the cap-site factor (probably eIF-

(iso)4F) similarly stimulates cross-linking of eIF-4A.[122] Cross-linking of purified wheat germ eIF-4A to mRNA caps is very weak and is not stimulated appreciably by eIF-4B (formerly eIF-4G), but eIF-4F stimulates it greatly, exceeded only by the combination of eIF-4B and eIF-4F.[153] Turning to eIF-4B, Abramson et al.[146] have demonstrated weak (ATP-independent) cross-linking of mammalian eIF-4B alone to mRNA caps, slight stimulation by either eIF-4A or eIF-4F, but large stimulation by eIF-4A *and* eIF-4F. Similarly, wheat germ eIF-4B cross links slightly as a single polypeptide; this is not stimulated by eIF-4A or eIF-4F alone but is greatly stimulated by a combination of the two.[153] Photochemical cross-linking of mRNA to factors yields, for reasons that are not understood, somewhat different results; the labeling of eIF-4B is considerably greater than that of eIF-4E, whereas eIF-4A is not labeled at all.[128] Taken together, all of these results indicate that eIF-4A and eIF-4B are cooperatively drawn close to the cap when eIF-4E and p220 are present, and that hydrolysis of ATP is required.

The presence of secondary structure in the region of mRNA near the cap has interesting effects on the cross-linking of caps to factors. Lee et al.[74] reported that with IMP-substituted mRNA, cross-linking to eIF-4A and eIF-4B was not dependent on ATP, but Tahara et al.[30] found no difference in ATP dependence between control and either IMP- or BrUMP-substituted mRNAs. Lawson et al.[84] approached this problem with cDNAs complementary to the extreme 5'-terminus (nucleotides 1 to 15) or more interior regions of globin or reovirus mRNAs. They found that cross-linking to the eIF-4E component of eIF-4F was not affected by any of the oligonucleotides, whereas that of the eIF-4A component was inhibited by 5'-proximal oligonucleotides only. This led them to propose a two-step mechanism: Step 1, in which mRNA first binds to the eIF-4E component in a secondary structure-independent, ATP-independent fashion, followed by Step 2, binding to the eIF-4A component attended by ATP hydrolysis, a step which can be inhibited by high secondary structure. Similar results were obtained by Pelletier and Sonenberg[128] who inserted secondary structure elements near the cap in thymidine kinase mRNA and observed that cross-linking to eIF-4A and eIF-4B in a total initiation factor preparation was inhibited while that of eIF-4E was unaffected; more cap-distal secondary structure did not have this effect. Another demonstration of this principle, that cross-linking to eIF-4E is independent of mRNA secondary structure while that of eIF-4A is dependent, was achieved with reovirus mRNAs under conditions of varying ionic strength.[66] These authors speculated that the rate-limiting step may vary between Steps 1 and 2 depending on secondary structure.

Another type of interaction involving the cap is the effect of eIF-4B on the ability of the eIF-4E component of eIF-4F to "recycle". In the presence of eIF-4B, but not eIF-4A, eIF-4E becomes free to bind to a second mRNA cap.[170]

The second activity of initiation factor complexes and combinations to be considered is interaction with ATP. Grifo et al.[145] showed that eIF-4F, like eIF-4A, has RNA-dependent ATPase activity and that both are stimulated by eIF-4B. The binding of ATP to the eIF-4A component of eIF-4F, as measured by photochemical labeling, is 60-fold greater than that to eIF-4A alone, and eIF-4B slightly stimulates the labeling of eIF-4A in either form.[144] Abramson et al.,[171] however, failed to find an effect of eIF-4B on the affinity of eIF-4A for ATP. Finally, both the wheat germ eIF-4F and eIF-(iso)4F complexes have RNA-dependent ATPase activity, and both are stimulated by eIF-4A.[147]

The interaction of combinations of factors with mRNA per se (as opposed to the cap) has been studied in a variety of ways. Grifo et al.[140] demonstrated that the ability of eIF-4A to cause retention of mRNA on nitrocellulose filters in an ATP-dependent fashion is stimulated by eIF-4B. The requirement for ATP involves hydrolysis but is not due to protein phosphorylation.[145] Using a different assay, RNA-dependent ATP hydrolysis, Abramson et al.[171] showed that the effect of eIF-4B was to cause an enormous decrease in the activation constant of eIF-4A for RNA, indicating an increase in affinity. The technique of fluorescence enhancement of polyethenoadenylic acid was also used to demonstrate the ATP-dependence of eIF-4F binding to RNA.[149]

Additional insight concerning the interaction of initiation factors with mRNA has come from assays which measure the unwinding of mRNA secondary structure near the cap. By measuring the nuclease sensitivity of cap-labeled mRNA, Ray et al.[164] were able to show that eIF-4A, in the presence of ATP, destabilized secondary structure, an activity which was stimulated by eIF-4B. However, eIF-4F was 20-fold more effective on a molar basis than free eIF-4A. By removing eIF-4A from eIF-4F, these authors showed that the activity resides in the eIF-4A component. Disruption of 5'-terminal secondary structure has also been demonstrated by hybridizing short labeled DNA oligonucleotides to mRNA and following their factor-mediated dissociation from mRNA.[172] In agreement with the nuclease-sensitivity technique, either eIF-4F alone or the combination of eIF-4A and eIF-4B was able to unwind the hybrids. Unwinding by eIF-4F was inhibited by cap analogs while unwinding by eIF-4A was not. More recently, Rozen et al.[173] have demonstrated, using short base-paired oligoribonucleotides, that eIF-4A and eIF-4F exhibit a bidirectional helicase activity. All of these findings are consistent with models in which the role of p220 is to juxtapose eIF-4E and eIF-4A in a complex, so that eIF-4A, with the cooperation of eIF-4B, can begin to unwind the secondary structure from the extreme 5'-terminus of the mRNA at the expense of ATP.

Another combined action of initiation factors is to discriminate between different mRNAs and affect their relative rates of initiation. This is presumably a reflection of the various partial activities discussed above working in concert on mRNAs which differ in cap accessibility or 5'-terminal secondary structure. An example is the relatively poor translation of α compared with β globin mRNA, a situation which is exaggerated at high ("competitive") mRNA concentrations.[10] Earlier thinking was in terms of a discriminatory factor which bound specifically to "strong" β globin mRNA,[174] but in light of the foregoing discussion, it is probably more useful to speak in terms of a greater requirement of "weak" mRNAs for the eIF-4 group factors, either to find a buried cap or to unwind secondary structure. In any case, Kabat and Chappell[174] found preferential stimulation of α globin mRNA with partially purified initiation factors, and Sarkar et al.[175] narrowed this down to eIF-4F. The reovirus mRNAs also provide an opportunity to test for mRNA discrimination. Ray et al.[176] found that of all the initiation factors tested, only eIF-4A and a preparation which apparently contained eIF-4F and eIF-4B (CBPII) were able to differentially stimulate translation of the various reovirus mRNAs. This type of activity is also illustrated by the differential effects of eIF-4F on various mRNAs as ionic strength is increased; the translation of those with high secondary structure is preferentially inhibited in high salt, and eIF-4F has a greater stimulatory effect under these circumstances.[72] In a wheat germ system, the unfavorable translation conferred by the 5'-leader of α-amylase mRNA can be overcome with eIF-4F, eIF-4B, and eIF-3, regardless of the coding region to which the leader is attached.[177]

Abundance and subcellular distribution of eIF-4 group initiation factors—Any model which describes how the initiation factors integrate mRNA into 48S complexes and which step is rate limiting must take into account the abundance of these initiation factors relative to each other, to ribosomes and to mRNA. This is particularly relevant now that accurate association constants for various binary interactions and rate constants for partial reactions are beginning to appear (e.g., References 55, 118, and 171). Additionally, it is essential to know the subcellular localization of factors if one is to evaluate the significance of complexes and deduce the correct sequence of events.

A fruitful approach to determining the total cellular content of several initiation factors simultaneously has been the use of a two-dimensional polyacrylamide gel electrophoresis on HeLa cell lysates.[178] eIF-4A was found to be the most abundant at 10^7 molecules per cell, whereas eIF-4B was at 1.4×10^6 molecules per cell, corresponding to 3 and 0.43 molecules per ribosome, respectively. In a subsequent study, Duncan and Hershey[179] determined that p220 was about as abundant as eIF-2, making it 1×10^5 molecules per cell, whereas eIF-4E was so low as to be not easily quantitated, an estimate being 0.01% of total cellular protein (roughly 2×10^5 molecules per cell). Improved methodology allowed Duncan et al.[180] to revise both of these estimates upwards, the final result being 2, 10, and 0.8×10^6 molecules per cell for p220, eIF-4A and eIF-4E, respectively. Hiremath et al.[181] immunologically estimated the concentration of eIF-4E in reticulocyte lysate to be 8 nM; this corresponds to an eIF-4E/ribosome ratio of 0.02, which is tenfold lower than that determined by Duncan et al.[180] for HeLa lysate. Whether this difference is due to cell type or method of measurement has not been determined.

In discussing subcellular distribution, one can consider both cytoskeleton vs. soluble fraction and ribosomal vs. supernatant fraction. A number of observations have suggested that protein synthesis may take place on the cytoskeleton (see references in Reference 182 and Part I, Chapter 3 of this book). Fractionation of HeLa cells revealed that eIF-4A and eIF-4B were enriched in the cytoskeletal fraction, and this correlated with the presence of ribosomes rather than mRNA.[182] Etchison and Etchison[183] concluded from immunofluorescence experiments that p220, although mostly cytoplasmic (as opposed to nuclear), was not affixed to the cytoskeleton.

The distribution of eIF-4 group factors in relation to ribosomes is particularly revealing. Nevertheless, one must always interpret sedimentation results carefully because of the possibility of artifactual aggregation on the one hand and artifactual stripping of factors from ribosomes by hydrodynamic shear forces on the other.

With these caveats in mind, let us first consider eIF-4E distribution. Although initial isolation of eIF-4E was from the ribosomal salt wash,[112, 131] the factor has also been found in postribosomal supernatant.[132, 184] The ribosome-bound form is mostly complexed with eIF-3[112, 166] (and presumably p220), whereas the supernatant form is a free polypeptide. Webb et al.[132] isolated more eIF-4E from the postribosomal supernatant than from the ribosomal salt wash. Using an antibody to follow eIF-4E during sucrose gradient sedimentation of whole reticulocyte lysate, Hiremath et al.[181] found all of the eIF-4E to be in the region of soluble proteins and ribosomal subunits with none on 80S monosomes or polysomes. This result was confirmed with radioactive eIF-4E in reticulocyte translation reaction mixtures; the majority of the eIF-4E sedimented at 3 to 4S.[185, 186] Westerman and Nygard[187] reported that eIF-4E was present on *in vitro* constituted 48S inititation complexes, but their evidence consisted only of a weakly staining protein band of 24 kDa among a large number of other bands; eIF-3, which was also present in their preparations, contains a similarly sized polypeptide which is not eIF-4E.[104] The question whether eIF-4E is present on initiation complexes was also investigated by Hiremath et al.,[185] who used radioactive eIF-4E and initiation inhibitors in an unfractionated reticulocyte lysate translation system. The results indicated that eIF-4E is present on 48S complexes but not 43S or 80S. A different kind of study was carried out by Görlach and Hilse,[188] who isolated polysomal mRNPs and found eIF-4E to be present in a 1:1 molar ratio; presumably these represent reinitiating polysomes, i.e., containing a 48S complex.

Less is known about the precise location of eIF-4A and eIF-4B. It is clear from the stoichiometries determined by Duncan et al.[180] that most of the eIF-4A cannot be present in an eIF-4F complex. Furthermore, Schreier et al.[189] isolated eIF-4A from a different ammonium sulfate cut of ribosomal salt wash than that containing eIF-4F. Thomas et al.[190] determined that most of eIF-4A was in postribosomal supernatant, whereas eIF-4B was almost exclusively in the ribosomal salt wash. The question of initiation complex localization of these factors is not definitively answered, but Safer et al.[191] deduced from the electrophoretic patterns of polypeptides that eIF-4A and eIF-4B were present in 48S but not 43S complexes.

In contrast to eIF-4E, which is mostly in the postribosomal supernatant, p220 is mostly in the ribosomal pellet, with little or none in the postribosomal supernatant (see Reference 167 and references therein). Significantly, m⁷GTP-Sepharose retains essentially all of the eIF-4E from HeLa cell lysates (not treated with high salt) but none of the p220.[162] These authors and others before them (e.g., Reference 150) have attributed this phenomenon to interference of eIF-4F binding to the affinity matrix by eIF-3, which must first be removed by sedimentation in high salt. A more likely hypothesis, however, given the relative abundances and subcellular distributions discussed above, is that the majority of eIF-4E and p220 molecules are not in a complex unless cell lysates are perturbed by high salt. In such a scenario, the effect of high salt is to strip p220 off ribosomes and allow it to form complexes with eIF-4E, eIF-4A, and eIF-4B, proteins for which it has high affinity but with which it normally forms transient complexes only at specific stages of initiation.

5.2.3. Models for 48S Initiation Complex Formation

Putting together the information reviewed in the first two sections, it is possible to construct models for 48S complex formation which, if not completely correct, are at least consistent with the available information. Two such models are presented in Figure 2, the main difference between them being the nature of the entity which first encounters mRNA. In Model 1, eIF-4F is that entity. As detailed above, eIF-4F possesses the combined activities of binding to the cap (Step 1) and, in an ATP-dependent step, bringing both eIF-4A (Step 2) and eIF-4B (Step 3) close enough to the cap to be cross-linked to it. eIF-4F also possesses unwinding activity which is stimulated by eIF-4B and which presumably accompanies this ATP-dependent binding to mRNA. However, the fact that these activities can be demonstrate with isolated factors does not prove they function this way in a complete system containing ribosomes. In Model 2 all of the same events occur, but on the surface of the 40S ribosome. The factor which first encounters the mRNA is eIF-4E. The appeal of this model is that a single entity, the 40S subunit with associated factors, carries out both unwinding and scanning. In Model 1, by contrast, some unwinding occurs in Steps 2 and 3, without a ribosome, and

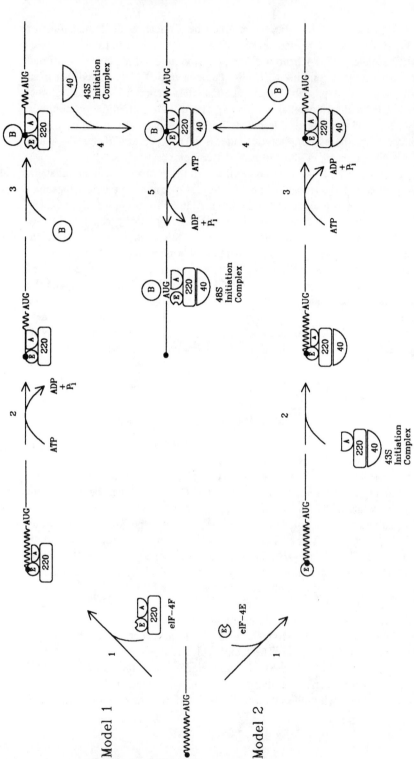

Figure 2. Two possible models for the participation of eIF-4 group initiation factors and ATP in 48S initiation complex formation. A, eIF-4A; B, eIF-4B; E,eIF-4E; 220, p220; 40, the 40S ribosomal subunit. The wavy line indicates secondary structure in mRNA.

some occurs in Step 5 after the ribosome joins. Since AUG codons quite near the cap can serve as initiation codons, it would be necessary to postulate a mechanism to prevent the eIF-4F:eIF-4B complex from continuing down the mRNA. Some models actually favor this notion and depict the factors unwinding the mRNA all the way to the initiation codon without participation of the 40S ribosome.[192] Such models cannot be ruled out with the available data but add another layer of complexity, e.g., what is the motive force for the 40S ribosome if ATP has already been expended in the binding of eIF-4A during factor-mediated unwinding? What prevents RNA secondary structure from reforming after the wave of unwinding but before the ribosome-joining step?

An argument for Model 1 is the simple fact that a complex of eIF-4E:eIF-4A:p220 can be isolated. However, the subcellular distribution studies reviewed above indicate that, in cell lysates which have not been subjected to high salt, most of the eIF-4E is in the $100,000 \times g$ soluble fraction whereas most of the p220 is present on ribosomes. Thus, the majority of molecules of both types are not in a complex with each other. If most of eIF-4E molecules in the cell are free and if eIF-4E is the least abundant of the initiation factors, as reviewed above, then mRNA is likely to encounter eIF-4E first as a free rather than complexed species. This places eIF-4E in the role of finding a cap which may sometimes be buried in secondary structure, a task more suited to a small (25 kDa) protein than a large (10S) complex.

An argument which might be raised in favor of Model 1 over Model 2 is that eIF-4F has a higher binding affinity to mRNA than eIF-4E. Lawson et al.[66] reported that the eIF-4F complex can be cross-linked to oxidized mRNA 15-fold more efficiently than can eIF-4E alone. Yet making inferences about cap:eIF-4E affinity from cross-linking data is not straightforward; first of all, Darzynkiewiwicz et al.[50] have found ribose ring-opened cap analogs to be inactive; second, affinity labeling does not represent an equilibrium situation; third, other proteins in eIF-4F (e.g., the RNA-binding protein eIF-4A) may produce a tighter binding of the factor to mRNA without affecting cap affinity. In fact, recent equilibrium studies have shown that the cap affinity of eIF-4E and eIF-4F is indistinguishable.[193] Thus, the ATP-independent Step 1, the "weak" binding step of Lawson et al.,[84] may occur with either the free or complexed form of eIF-4E with equal affinity.

The available data do not rule out a number of variations concerning the order of addition of eIF-4A and eIF-4B. Abramson et al.[171] favor a model in which additional eIF-4A molecules bind following the eIF-4B joining step. eIF-4A and eIF-4B may *both* join only after the ribosome binding step, in keeping with the evidence of Safer et al.[191] that eIF-4A and eIF-4B are on the 48S but not 43S initiation complexes. Conversely, eIF-4B may be complexed with eIF-4F (whether or not it is on the ribosome) prior to mRNA binding. Models 1 and 2 as well as those mentioned above are testable, given sufficiently discriminating tools for the detection of the various proteins and complexes. For example, a key test for Model 1 would be the detection of an eIF-4F:mRNA complex in cell lysates not subjected to high salt, perhaps freezing the complex by depleting the ATP. Support for Model 2 would be provided by the demonstration that both 43S and 48S initiation complexes contain p220.

5.3. FORMATION OF THE 80S INITIATION COMPLEX

The final stage of initiation is joining of the 60S ribosomal subunit, producing the 80S initiation complex. As summarized in Figure 3, this stage of initiation is thought to proceed in the order of (1) GTP hydrolysis, (2) release of initiation factors, and (3) joining of the 60S subunit. Elongation inhibitors like sparsomycin which inhibit the aminoacyl-tRNA transferase step cause protein synthesis to be arrested at the 80S stage before formation of the first peptide bond.[194] A more specific inhibitor which leads to accumulation of 80S complexes is the initiation inhibitor pactamycin, which causes polysomes to run off and form 80S ribosomes.[195] This compound is thought to interfere with the interaction of the 3′ end of the met-tRNA$_i^{met}$ with the P site in the peptidyl transferase center and thus freeze the process after 60S joining but before the first peptide bond is formed.[194]

The central initiation factor in 80S initiation complex formation is eIF-5. The protein was first purified from rabbit reticulocytes as one of the components necessary for the overall initiation process and was variously estimated to have a molecular weight in the range of 125 to 160 kDa.[189, 196, 197] Subsequently a wheat germ equivalent of 100 kDa was isolated.[103] Meyer et al.[104] found that the mammalian protein could be partially proteolyzed to yield active fragments ranging from 130 to 170 kDa. The factor in HeLa cells was shown to be similar to that of rabbit reticulocyte and have the same high molecular weight.[141] More recently, however, evidence that eIF-5 is considerably smaller has been presented. Raychaudhuri et al.[198, 199] found that a protein of 58 to 62 kDa purified from eIF-5 preparations could be shown to have all of the activities eIF-5 (see below). Experiments in which various tissues (reticulocyte, HeLa, L-cell) were lysed in sodium dodecylsulfate and probed with an antibody argued against proteolysis as an explanation for the smaller polypeptide; an antibody made against the 58 to 62 kDa protein reacted with only that size protein on immunoblots. This controversy has not yet been resolved.

Unlike the eIF-4 group factors, the role of eIF-5 has been rather well defined from the beginning. eIF-5 was shown to be a GTPase which required ribosomes for maximal activity and was proposed to be the subunit joining factor.[196] It was subsequently shown to carry out the conversion of met-

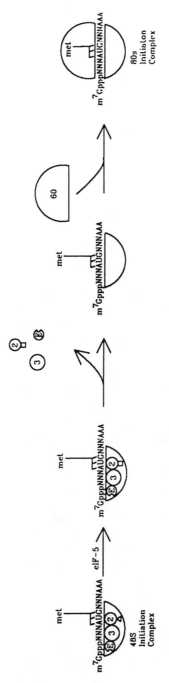

Figure 3. Formation of the 80S initiation complex.

tRNA$_i^{met}$ from a 40S to an 80S form in a reaction requiring hydrolysis of GTP.[4, 5] In the process, eIF-2 and eIF-3 are released from the 40S subunit.[5, 25] Further refinements of the model have established that the met-tRNA$_i^{met}$ is converted into a form which can react with puromycin (i.e., it is in the P site), that the loss of factors occurs in the absence of 60S subunits, and that 40S complexes treated with eIF-5 and GTP can subsequently react with the 60S subunit in the absence of GTP; thus, GTP hydrolysis and release of initiation factors occurs before and not coincident with 60S joining.[200] Interestingly, a form of eIF-2 which contains two rather than three subunits participates in all of these reactions.[201] Similar activities were demonstrated for the wheat germ eIF-5.[103]

More recent studies have established that isolated 40S initiation complexes (actually 43S complexes to which an AUG codon was added) will react with 60S subunits in a reaction catalyzed by eIF-5 and in which a stoichiometric amount of GTP is hydrolyzed.[202] The same authors showed that GDP is released as part of an eIF-2:GDP complex rather than free.

Another factor implicated in 80S complex formation is eIF-4C. It has been isolated from both reticulocyte[189, 197] and wheat germ[103] and is small (18 to 22 kDa) and heat stable. It is primarily located in the postribosomal supernatant,[190] but a controversy exists as to whether it becomes attached to initiation complexes (see above). The role of eIF-4C in 48S initiation complex formation, though not clear, was reviewed in the previous section. Unfortunately, its role in 80S complex formation is even less clear. Thomas et al.[109] reported that it stabilizes 48S initiation complexes, and Goumens et al.[203] found that it has a stimulatory effect on initiation complex formation when native 80S ribosomes are the only source of ribosomal subunits. These authors have speculated that the presence of eIF-4C on 40S subunits reduces their affinity for 60S subunits and other 40S subunits, thus ensuring a proper sequence of initiation events.

The last factor to be considered is eIF-4D. This has been isolated from reticulocytes as a 17 kDa polypeptide[197] required for optimal reaction with puromycin of met-tRNA$_i^{met}$ in the 80S initiation complex.[5] More is known about the structure of this protein than its function because of the intriguing discovery that it contains a unique amino acid residue, hypusine [*N*-(4-amino-2-hydroxybutyl)lysine], which is formed posttranslationally.[204] It is apparently the only protein in the cell to be so modified. The complete amino acid sequence and location of the hypusine residue are known.[205, 206] Changes in the rate of protein synthesis in HeLa cells do not correlate with the amount or posttranslational modification of eIF-4D,[207] but the hypusine modification is essential for eIF-4D function both *in vitro*[208] and *in vivo*.[209]

Despite the uncertainties in the roles of eIF-4C and eIF-4D, it is possible to formulate a model for the conversion of the 48S to the 80S initiation complex (Figure 3). Release of eIF-4E is depicted because of the study of

Hiremath et al.,[185] but it is also possible (and likely) that eIF-4A, eIF-4B, eIF-4C, and p220 are released as well. This only underscores the fact that the precise roles of factors in the mechanism of both 48S and 80S initiation complex formation are yet to be elucidated.

ACKNOWLEDGMENTS

The author gratefully acknowledges the expert assistance of Ms. Karen Fortenberry for library services, Mr. Kevin Burton for graphics, and Ms. Ginger Hodges and Ms. Jana De Benedetti for word processing.

REFERENCES

1. Darnbrough C, Legon S, Hunt T, and Jackson RJ: Initiation of protein synthesis: evidence for messenger RNA-independent binding of methionyl-transfer RNA to the 40S ribosomal subunit. *J. Mol. Biol.* 76:379 (1973).
2. Levin DH, Kyner D, and Acs G: Protein synthesis initiation in eukaryotes. Characterization of ribosomal factors from mouse fibroblasts. *J. Biol. Chem.* 248:6416 (1973).
3. Smith KE and Henshaw EC: Binding of Met-tRNA$_f$ to native and derived 40S ribosomal subunits. *Biochemistry* 14:1060 (1975).
4. Trachsel H, Erni B, Schrier MH, and Staehelin T: Initiation of mammalian protein synthesis. II. The assembly of the initiation complex with purified initiation factors. *J. Mol. Biol.* 116:755 (1977).
5. Benne R and Hershey JWB: The mechanism of action of protein synthesis initiation factors from rabbit reticulocytes. *J. Biol. Chem.* 253:3078 (1978).
6. Nasrin N, Ahmad MF, Nag MK, Tarburton P, and Gupta NK: Protein synthesis in yeast *Saccharomyces cerevisiae*. Purification of Co-eIF-2A and 'mRNA-binding factors(s)' and studies of their role in met-tRNA$_f$·40S·mRNA complex formation. *Eur. J. Biochem.* 161:1 (1986).
7. Roy AL, Chakrabarti D, Datta B, Hileman RE, and Gupta NK: Natural mRNA is required for directing Met-tRNA$_f$ binding to 40S ribosomal subunits in animal cells: involvement of Co-eIF-2A in natural mRNA-directed initiation complex formation. *Biochemistry* 27:8203 (1988).
8. Safer B, Kemper W, and Jagus R: Identification of a 48S preinitiation complex in reticulocyte lysate. *J. Biol. Chem.* 253:3384 (1978).
9. Marcus A: Tobacco mosaic virus ribonucleic acid-dependent amino acid incorporation in a wheat embryo system *in vitro*. Analysis of the rate-limiting reaction. *J. Biol. Chem.* 245:955 (1970).
10. Lodish HF: Model for the regulation of mRNA translation applied to haemoglobin synthesis. *Nature* 251:385 (1974).
11. Pelham HRB and Jackson RJ: An efficient mRNA-dependent translation system from reticulocyte lysates. *Eur. J. Biochem.* 67:247 (1976).

12. Roberts BE and Paterson BM: Efficient translation of tobacco mosaic virus RNA and rabbit globin 9S RNA in a cell-free system from commercial wheat germ. *Proc. Natl. Acad. Sci. U.S.A.* 70:2330 (1973).
13. Brown GD, Peluso RW, Moyer SA, and Moyer RW: A simple method for the preparation of extracts from animals cells which catalyze efficient *in vitro* protein synthesis. *J. Biol. Chem.* 258:14309 (1983).
14. Bader M and Sarre TF: A (re)initiation-dependent cell-free protein-synthesis system from mouse erythroleukemia cells. *Eur. J. Biochem.* 161:103 (1986).
15. Dasso MC and Jackson RJ: On the fidelity of mRNA translation in the nuclease-treated rabbit reticulocyte lysate system. *Nucleic Acids Res.* 17:3129 (1989).
16. Lomedico PT and McAndrew SJ: Eukaryotic ribosomes can recognize preproinsulin initiation codons irrespective of their position relative to the 5' end of mRNA. *Nature* 299:221 (1982).
17. Kozak M: Translation of insulin-related polypeptides from messenger RNAs with tandemly reiterated copies of the ribosome binding site. *Cell* 34:971 (1983).
18. Van Duijn LP, Holsappel S, Kasperaitis M, Bunschoten H, Konings D, and Voorma HO: Secondary structure and expression *in vivo* and *in vitro* of messenger RNAs into which upstream AUG codons have been inserted. *Eur. J. Biochem.* 172:59 (1988).
19. Kozak M: The scanning model for translation: an update. *J. Cell. Biol.* 108:229 (1989).
20. Kozak M and Shatkin AJ: Characterization of ribosome protected fragments from reovirus messenger RNA. *J. Biol. Chem.* 251:4259 (1976).
21. Kozak M and Shatkin AJ: Identification of features in 5'-terminal fragments from reovirus mRNA which are important for ribosome binding. *Cell* 13:201 (1978).
22. Legon S, Robertson HD, and Prensky W: The binding of ^{125}I-labeled rabbit globin messenger RNA to reticulocyte ribosomes. *J. Mol. Biol.* 106:23 (1976).
23. Hunter AR, Jackson RJ, and Hunt T: The role of complexes between the 40-S ribosomal subunit and met-tRNA$_f^{met}$ in the initiation of protein synthesis in the wheat-germ system. *Eur. J. Biochem.* 75:159 (1977).
24. Jagus R and Safer B: Quantitation and localization of globin messenger RNA in rabbit reticulocyte lysate. *J. Biol. Chem.* 254:6865 (1979).
25. Peterson DT, Merrick WC, and Safer B: Binding and release of radiolabeled eukaryotic initiation factors 2 and 3 during 80S initiation complex formation. *J. Biol. Chem.* 254:2509 (1979).
26. Morgan MA and Shatkin AJ: Initiation of reovirus transcription by inosine 5'-triphosphate and properties of 7-methylinosine-capped, inosine-substituted messenger RNA. *Biochemistry* 19:5960 (1980).
27. Hansen LJ, Huang W-I, and Jagus R: Inhibitor of translational initiation in sea urchin eggs prevents mRNA utilization. *J. Biol. Chem.* 262:6114 (1987).
28. Marcus A: Tobacco mosaic virus ribonucleic acid-dependent amino acid incorporation in a wheat embryo system *in vitro*. Formation of a ribosome-messenger "initiation" complex. *J. Biol. Chem.* 245:962 (1970).
29. Kozak M: Role of ATP in binding and migration of 40S ribosomal subunits. *Cell* 22:459 (1980).

30. Tahara SM, Morgan MA, and Shatkin AJ: Binding of inosine-substituted mRNA to reticulocyte ribosomes and eukaryotic initiation factors 4A and 4B requires ATP. *J. Biol. Chem.* 258:11350 (1983).
31. Banerjee AK: 5'-Terminal cap structure in eucaryotic messenger ribonucleic acids. *Microbiol. Rev.* 44:175 (1980).
32. Shatkin AJ: mRNA cap binding proteins: essential factors for initiating translation. *Cell* 40:223 (1985).
33. Rhoads RE: The cap structure of eukaryotic messenger RNA and its interaction with cap-binding protein, in Hahn FE, Kopecko DJ, and Muller WEG (Eds): *Progress in Molecular and Subcellular Biology,* vol. 9, Berlin, Springer Verlag, 104 (1985).
34. Both GW, Banerjee AK, and Shatkin AJ: Methylation-dependent translation of viral messenger RNAs *in vitro. Proc. Natl. Acad. Sci. U.S.A.* 72:1189 (1975).
35. Muthukrishnan S, Both GW, Furuichi Y, and Shatkin AJ: 5'-Terminal 7-methylguanosine in eukaryotic mRNA is required for translation. *Nature* 255:33 (1975).
36. Zan-Kowalczewska M, Bretner M, Sierakowska H, Szczesna E, Filipowicz W, and Shatkin AJ: Removal of 5'-terminal m7G from eukaryotic mRNAs by potato nucleotide pyrophosphatase and its effect on translation. *Nucleic Acid Res.* 4:3065 (1977).
37. Shimotohno K, Kodama Y, Hashimoto J, and Miura K: Importance of 5'-terminal blocking structure to stabilize mRNA in eukaryotic protein synthesis. *Proc. Natl. Acad. Sci. U.S.A.* 74:2734 (1977).
38. Both GW, Furuichi Y, Muthukrishnan S, and Shatkin AJ: Effect of 5'-terminal structure and base composition on polyribonucleotide binding to ribosomes. *J. Mol. Biol.* 104:637 (1976).
39. Hickey ED, Weber LA, and Baglioni C: Inhibition of initiation of protein synthesis by 7-methylguanosine-5'-monophosphate. *Proc. Natl. Acad. Sci. U.S.A.* 73:19 (1976).
40. Canaani D, Revel M, and Groner Y: Translational discrimination of "capped" and "non-capped" mRNAs: inhibition of a series of chemical analogs of m7GpppX. *FEBS Lett.* 64:326 (1976).
41. Both GW, Furuichi Y, Muthukrishnan S, and Shatkin AJ: Ribosome binding to reovirus mRNA in protein synthesis requires 5' terminal 7-methylguanosine. *Cell* 6:185 (1975).
42. Roman R, Brooker JD, Seal SN, and Marcus A: Inhibition of the transition of a 40S ribosome-met-tRNA$_i^{met}$ complex to an 80S ribosome-met-tRNA$_i^{met}$ complex by 7-methylguanosine-5'-phosphate. *Nature (London)* 260:359 (1976).
43. Hickey ED, Weber LA, Baglioni C, Kim CH, and Sarma RH: A relation between inhibition of protein synthesis and conformation of 5'-phosphorylated 7-methylguanosine derivatives. *J. Mol. Biol.* 109:173 (1977).
44. Suzuki H: Effect of m7G5'ppp5'Nm on the rabbit globin synthesis. *FEBS Lett.* 79:11 (1977).
45. Chu L-Y and Rhoads RE: Translational recognition of the 5'-terminal 7-methylguanosine of globin mRNA as a function of ionic strength. *Biochemistry* 17:2450 (1978).

46. Darzynkiewicz E, Antosiewicz J, Ekiel I, Morgan MA, Tahara SM, and Shatkin AJ: Methyl esterification of m⁷G⁵′p reversibly blocks its activity as an analog of eukaryotic mRNA 5′-caps. *J. Mol. Biol.* 153:451 (1981).
47. Adams BL, Morgan M, Muthukrishnan S, Hecht SM, and Shatkin AJ: The effect of "cap" analogs on reovirus mRNA binding to wheat germ ribosomes. *J. Biol. Chem.* 253:2589 (1978).
48. Carberry SE, Darzynkiewicz E, Stepinski J, Tahara SM, Rhoads RE, and Goss DJ: A spectroscopic study of the binding of N-7-substituted cap analogs to human protein synthesis initiation factor 4E. *Biochemistry* 29:3337 (1990).
49. Kim CH and Sarma RH: Spatial configuration of mRNA 5′-terminus. *Nature* 270:223 (1977).
50. Darzynkiewicz E, Ekiel I, Lassota P, and Tahara S: Inhibition of eukaryotic translation by analogs of mRNA 5′-cap: chemical and biological consequences of 5′-phosphate modifications of 7-methyl guanosine 5′-monophosphate. *Biochemistry* 26:4372 (1987).
51. Hattori M, Miura K, Yamaguchi K, Ohtani S, and Hata T: Interaction between bases involved in the 5′-terminal cap structure of eukaryotic mRNA. *Nucleic Acids Res.* Spec. publ. no. 5 (1978).
52. Nishimura Y, Takahashi S, Yamamoto T, Tsuboi M, Hattori M, Miura K, Yamaguchi K, Ohtani S, and Hata T: On the base-stacking in the 5′-terminal cap structure of mRNA: a fluorescence study. *Nucleic Acids Res.* 8:1107 (1980).
53. Darzynkiewicz E, Stepinski J, Ekiel I, Jin Y, Haber D, Sijuwade T, and Tahara SM: β-globin mRNAs capped with m⁷G, m₂²′⁷G or m₃²′²′⁷G differ in intrinsic translation efficiency. *Nucleic Acids Res.* 16:8953 (1988).
54. Rhoads RE, Hellmann GM, Remy P, and Ebel J-P: Translational recognition of messenger RNA caps as a function of pH. *Biochemistry* 22:6084 (1983).
55. Carberry SE, Rhoads RE, and Goss DJ: A spectroscopic study of the binding of m⁷GTP and m⁷GpppG to human protein synthesis initiation factor 4E. *Biochemistry* 28:8078 (1989).
56. Rose JK and Lodish HF: Translation *in vitro* of vesicular stomatitis virus mRNA lacking 5′-terminal 7-methylguanosine. *Nature* 262:32 (1976).
57. Weber LA, Hickey ED, Nuss DL, and Baglioni C: 5′-Terminal 7-methylguanosine and mRNA function: influence of potassium concentration on translation *in vitro*. *Proc. Natl. Acad. Sci. U.S.A.* 74:3254 (1977).
58. Kemper B and Stolarsky L: Dependence on potassium concentration of the inhibition of the translation of messenger ribonucleic acid by 7-methylguanosine 5′-phosphate. *Biochemistry* 16:5676 (1977).
59. Weber LA, Hickey ED, and Baglioni C: Influence of potassium salt concentration and temperature on inhibition of mRNA translation by 7-methylguanosine 5′-monophosphate. *Proc. Natl. Acad. Sci. U.S.A.* 253:178 (1978).
60. Asselbergs FAM, Peters W, van Venrooij WJ, and Bloemendal H: Diminished sensitivity of re-initiation of translation to inhibition by cap analogues in reticulocyte lysates. *Eur. J. Biochem.* 88:483 (1978).
61. Willems M, Wieringa B, Mulder J, Ab G, and Gruber M: Translation of vitellogenin mRNA in the presence of 7-methylguanosine 5′-triphosphate. Cap analogs compete with mRNAs on the basis of affinity for initiation complex formation. *Eur. J. Biochem.* 93:469 (1979).

62. Chu L-Y and Rhoads RE: Inhibition of cell-free messenger ribonucleic acid translation by 7-methylguanosine 5'-triphosphate: effect of messenger ribonucleic acid concentration. *Biochemistry* 19:184 (1980).

63. Kozak M: Influence of mRNA secondary structure on binding and migration of 40S ribosomal subunits. *Cell* 19:79 (1980).

64. Gehrke L, Auron PE, Quigley GJ, Rich A, and Sonenberg N: 5'-Conformation of capped alfalfa mosaic virus ribonucleic acid 4 may reflect its independence of the cap structure or of cap-binding protein for efficient translation. *Biochemistry* 22:5157 (1983).

65. Godefroy-Colburn T, Ravelonandro M, and Pinck L: Cap accessibility correlates with the initiation efficiency of alfalfa mosaic virus RNAs. *Eur. J. Biochem.* 147:549 (1985).

66. Lawson TG, Cladaras MH, Ray BK, Lee KA, Abramson RD, Merrick WC, and Thach RE: Discriminatory interaction of purified eukaryotic initiation factors 4F plus 4A with the 5' ends of reovirus messenger RNAs. *J. Biol. Chem.* 263:7266 (1988).

67. Konings DAM, van Duijn LP, Voorma HO, and Hogeweg P: Minimal energy foldings of eukaryotic mRNAs form a separate leader domain. *J. Theor. Biol.* 127:63 (1987).

68. Browning KS, Fletcher L, and Ravel JM: Evidence that the requirements for ATP and wheat germ initiation factors 4A and 4F are affected by a region of satellite tobacco necrosis virus RNA that is 3' to the ribosomal binding site. *J. Biol. Chem.* 263:8380 (1988).

69. Pelletier J and Sonenberg N: The involvement of mRNA secondary structure in protein synthesis. *Biochem. Cell Biol.* 65:576 (1987).

70. Gallie DR, Walbot V, and Hershey JWB: The ribosomal fraction mediates the translational enhancement associated with the 5'-leader of tobacco mosaic virus. *Nucleic Acids Res.* 16:8675 (1988).

71. Guan KL and Weiner H: Influence of the 5'-end region of aldehyde dehydrogenase mRNA on translational efficiency. *J. Biol. Chem.* 264:17764 (1989).

72. Edery I, Lee KAW, Sonenberg N: Functional characterization of eukaryotic mRNA cap binding protein complex: effects on translation of capped and naturally uncapped RNAs. *Biochemistry* 23:2456 (1984).

73. Baim SB, Pietras DF, Eustice DC, and Sherman F: A mutation allowing an mRNA secondary structure diminishes translation of *Saccharomyces cerevisiae* iso-l-cytochrome c. *Mol. Cell Biol.* 5:1839 (1985).

74. Lee KAW, Guertin D, and Sonenberg N: mRNA secondary structure as a determinant in cap recognition and initiation complex formation: ATP-Mg^{2+} independent cross-linking of cap binding proteins to m^{7}I-capped inosine-substituted reovirus mRNA. *J. Biol. Chem.* 258:707 (1983).

75. Pelletier J and Sonenberg N: Insertion mutagenesis to increase secondary structure within the 5' noncoding region of a eukaryotic mRNA reduces translational efficiency. *Cell* 40:515 (1985).

76. Darveau A, Pelletier J, and Sonenberg N: Differential efficiencies of in vitro translation of mouse *c-myc* transcripts differing in the 5'-untranslated region. *Proc. Natl. Acad. Sci. U.S.A.* 82:2315 (1985).

77. Parkin N, Darveau A, Nicholson R, and Sonenberg N: cis-Acting translational effects of the 5' noncoding region of *c-myc* mRNA. *Mol. Cell. Biol.* 8:2875 (1988).
78. Kozak M: Influences of mRNA secondary structure on initiation by eukaryotic ribosomes. *Proc. Natl. Acad. Sci. U.S.A.* 83:2850 (1986).
79. Kozak M: Leader length and secondary structure modulate mRNA function under conditions of stress. *Mol. Cell. Biol.* 8:2737 (1988).
80. Chevrier D, Vezina C, Bastille J, Linard C, Sonenberg N, and Boileau G: Higher order structures of the 5'-proximal region decrease the efficiency of translation of the porcine pro-opiomelanocortin mRNA. *J. Biol. Chem.* 263:902 (1988).
81. Parkin NT, Cohen EA, Darveau A, Rosen C, Haseltine W, and Sonenberg N: Mutational analysis of the 5' non-coding region of human immunodeficiency virus type 1: effects of secondary structure on translation. *EMBO J.* 7:2831 (1988).
82. Ruiz-Linares A, Bouloy M, Girard M, and Cahour A: Modulations of the *in vitro* translational efficiencies of yellow fever virus mRNAs: interactions between coding and noncoding regions. *Nucleic Acid Res.* 17:2463 (1989).
83. Perdue ML, Borchelt D, and Resnick RJ: Participation of 5'-terminal leader sequences in *in vitro* translation of Rous sarcoma virus RNA. *J. Biol. Chem.* 257:6551 (1982).
84. Lawson TG, Ray BK, Dodds JT, Grifo JA, Abramson RD, Merrick WC, Betsch DF, Weith HL, and Thach RE: Influence of 5' proximal secondary structure on the translational efficiency of eukaryotic mRNAs and on their interaction with initiation factors. *J. Biol. Chem.* 261:13979 (1986).
85. Melton DA: Injected anti-sense RNAs specifically block messenger RNA translation *in vivo*. *Proc. Natl. Acad. Sci. U.S.A.* 82:144 (1985).
86. Sherman F and Stewart JW: The use of iso-l-cytochrome c mutants of yeast for elucidating the nucleotide sequences that govern initiation of translation, in Bernardi G and Gros F (Eds): *Organization and Expression of the Eukaryotic Genome: Biochemical Mechanisms of Differentiation in Prokaryotes and Eukaryotes,* vol. 38 New York, Elsevier, 175 (1975).
87. Kozak M and Shatkin AJ: Migration of 40S ribosomal subunits on messenger RNA in the presence of edeine. *J. Biol. Chem.* 253:6568 (1978).
88. Kozak M: An analysis of 5'-noncoding sequences from 699 vertebrate messenger RNAs. *Nucleic Acids Res.* 15:8125 (1987).
89. Cigan AM and Donahue TF: Sequence and structural features associated with translational initiator regions in yeast - a review. *Gene* 59:1 (1987).
90. Kozak M: Inability of circular mRNA to attach to eukaryotic ribosomes. *Nature* 280:82 (1979).
91. Filipowicz W and Haenni A-L: Binding of ribosomes to 5'-terminal leader sequences of eukaryotic mRNAs. *Proc. Natl. Acad. Sci. U.S.A.* 76:3111 (1979).
92. Kozak M: Migration of 40S ribosomal subunits on messenger RNA when initiation is perturbed by lowering magnesium or adding drugs. *J. Biol. Chem.* 254:4731 (1979).
93. Cigan AM, Feng L, and Donahue TF: tRNA$_i^{met}$ functions in directing the scanning ribosome to the start site of translation. *Science* 242:93 (1988).

94. Baralle FE and Brownlee GG: AUG is the only recognisable signal sequence in the 5' non-coding regions of eukaryotic mRNA. *Nature* 274:84 (1978).
95. Kozak M: Possible role of flanking nucleotides in recognition of the AUG initiator codon by eukaryotic ribosomes. *Nucleic Acids Res.* 9:5233 (1981).
96. Heidecker G and Messing J: Structural analysis of plant genes. *Annu. Rev. Plant Physiol.* 37:439 (1986).
97. Kozak M: Point mutations define a sequence flanking the AUG initiator codon that modulates translation by eukaryotic ribosomes. *Cell* 44:283 (1986).
98. Kozak M: At least six nucleotides preceding the AUG initiator codon enhance translation in mammalian cells. *J. Mol. Biol.* 196:947 (1987).
99. Donahue TF and Cigan AM: Genetic selection for mutations that reduce or abolish ribosomal recognition of the *HIS4* translational initiator region. *Mol. Cell. Biol.* 8:2955 (1988).
100. Kozak M: Bifunctional messenger RNA's in eukaryotes. *Cell* 47:481 (1986).
101. Peabody DS: Translation initiation at Non-AUG triplets in mammalian cells. *J. Biol. Chem.* 264:5031 (1989).
102. Dasso MC and Jackson RJ: Efficient initiation of mammalian mRNA translation at a CUG codon. *Nucleic Acids Res.* 17:6485 (1989).
103. Seal SN, Schmidt A, and Marcus A: Fractionation and partial characterization of the protein synthesis of wheat germ. II. Initiation factors D1 (eukaryotic initiation factor 3), D2c (eukaryotic initiation factor 5), and D2d (eukaryotic initiation factor 4C). *J. Biol. Chem.* 258:866 (1983).
104. Meyer LJ, Brown-Luedi ML, Corbett S, Tolan DR, and Hershey JWB: The purification and characterization of multiple forms of protein synthesis eukaryotic initiation factors 2, 3 and 5 from rabbit reticulocyte. *J. Biol. Chem.* 256:351 (1981).
105. Nygard O and Westermann P: Purification of protein synthesis initiation factor eIF-3 from rat liver microsomes by affinity chromatography on rRNA-cellulose. *Biochim. Biophys. Acta* 697:263 (1982).
106. Lauer SJ, Burks EA, and Ravel JM: Characterization of initiation factor 3 from wheat germ. I. Effects of proteolysis on activity and subunit composition. *Biochemistry* 24:2924 (1985).
107. Kaempfer R, Emmelo JV, and Fiers W: Specific binding of eukaryotic initiation factor 2 to satellite tobacco necrosis virus RNA at a 5'-terminal sequence comprising the ribosome binding site. *Proc. Natl. Acad. Sci. U.S.A.* 78:1542 (1981).
108. Donahue TF, Cigan AM, Pabich EK, and Valavicius BC: Mutations at a Zn(II) finger motif in the yeast eIF-2β gene alter ribosomal start-site selection during the scanning process. *Cell* 54:621 (1988).
109. Thomas A, Goumans H, Voorma HO, and Benne R: The mechanism of action of eukaryotic initiation factor 4C in protein synthesis. *Eur. J. Biochem.* 107:39 (1980).
110. Rhoads RE: Cap recognition and the entry of mRNA into the protein synthesis initiation cycle. *Trends Biochem. Sci.* 13:52 (1988).
111. Sonenberg N, Trachsel H, Hecht S, and Shatkin AJ: Differential stimulation of capped mRNA translation *in vitro* by cap binding protein. *Nature* 285:331 (1980).
112. Hellmann GM, Chu L-Y, and Rhoads RE: A polypeptide which reverses cap analogue inhibition of cell-free protein synthesis. *J. Biol. Chem.* 257:4056 (1982).

113. Altmann M, Sonenberg N, and Trachsel H: Translation in *Saccharomyces cerevisiae*: initiation factor 4E-dependent cell-free system. *Mol. Cell. Biol.* 9:4467 (1989).
114. Altmann M, Handschin C, and Trachsel H: Yeast mRNA cap binding protein gene: cloning of the gene encoding protein synthesis initiation factor eIF-4E from the yeast *Saccharomyces cerevisiae*. *Mol. Cell. Biol.* 7:998 (1987).
115. Brenner C, Nakayama N, Goebl M, Tanaka K, Toh-E A, and Matsumoto K: *CDC33* encodes mRNA cap-binding protein eIF-4E of *Saccharomyces cerevisiae*. *Mol. Cell. Biol.* 8:3556 (1988).
116. Altmann M and Trachsel H: Altered mRNA cap recognition activity of initiation factor 4E in the yeast cell cycle division mutant cdc33. *Nucleic Acids Res.* 17:5923 (1989).
117. Rychlik W, Domier LL, Gardner PR, Hellmann GM, and Rhoads RE: Amino acid sequence of the mRNA cap-binding protein from human tissues. *Proc. Natl. Acad. Sci. U.S.A.* 84:945 (1987).
118. McCubbin WD, Edery I, Altmann M, Sonenberg N, and Kay CM: Circular dichroism and fluorescence studies on protein synthesis initiation factor eIF-4E and two mutant forms from the yeast *Saccharomyces cerevisiae*. *J. Biol. Chem.* 263:17663 (1988).
119. Altmann M, Müller PP, Pelletier J, Sonenberg N, and Trachsel H: A mammalian translation initiation factor can substitute for its yeast homologue *in vivo*. *J. Biol. Chem.* 264:12145 (1989).
120. Maroto FG and Sierra JM: Purification and characterization of mRNA cap-binding protein from *Drosophila melanogaster* embryos. *Mol. Cell. Biol.* 9:2181 (1989).
121. Lax S, Fritz W, Browning K, and Ravel J: Isolation and characterization of factors from wheat germ that exhibit eIF-4B activity and overcome m⁷GTP inhibition of polypeptide synthesis. *Proc. Natl. Acad. Sci. U.S.A.* 82:330 (1985).
122. Seal SN, Schmidt A, Marcus A, Edery I, and Sonenberg N: A wheat germ cap-site factor functional in protein chain initiation. *Arch. Biochem. Biophys.* 246:710 (1986).
123. Browning KS, Lax SR, and Ravel JM: Identification of two messenger RNA cap binding proteins in wheat. *J. Biol. Chem.* 262:11228 (1987).
124. McCubbin WD, Edery I, Altmann M, Sonenberg N, and Kay CM: Circular dichroism and fluorescence studies on five mutant forms of protein synthesis initiation factor eIF-4E from the yeast *Saccharomyces cerevisiae*. *FEBS Lett.* 245:261 (1989).
125. Rychlik W, Gardner PR, Vanaman TC, and Rhoads RE: Structural analysis of the messenger RNA cap-binding protein. Presence of phosphate, sulfhydryl and disulfide groups. *J. Biol. Chem.* 261:71 (1986).
126. Rychlik W, Russ MA, and Rhoads RE: Phosphorylation site of eukaryotic initiation factor 4E. *J. Biol. Chem.* 262:10434 (1987).
127. Sonenberg N and Shatkin AJ: Reovirus mRNA can be covalently crosslinked via the 5′ cap to proteins in initiation complexes. *Proc. Natl. Acad. Sci. U.S.A.* 74:4288 (1977).
128. Pelletier J and Sonenberg N: Photochemical cross-linking of cap binding proteins to eucaryotic mRNAs: effect of mRNA 5′ secondary structure. *Mol. Cell. Biol.* 5:3222 (1985).

129. Patzelt E, Blaas D, and Kuechler E: CAP binding proteins associated with the nucleus. *Nucleic Acids Res.* 11:5821 (1983).
130. Chavan AJ, Rychlik W, Blaas D, Kuechler E, Watt DS, and Rhoads RE: Phenyl azide- and benzophenone-substituted phosphonamides of 7-methyl guanosine 5'-triphosphate as photoaffinity probes for protein synthesis initiation factor eIF-4E. *Biochemistry* 29:5521 (1990).
131. Sonenberg N, Rupprecht KM, Hecht SM, and Shatkin AJ: Eukaryotic mRNA cap binding protein: purification by affinity chromatography on Sepharose-m⁷GDP. *Proc. Natl. Acad. Sci. U.S.A.* 76:4345 (1979).
132. Webb NR, Chari RVJ, DePillis G, Kozarich JW, Rhoads RE: Purification of the messenger RNA cap-binding protein using a new affinity medium. *Biochemistry* 23:177 (1984).
133. Seal SN, Schmidt A, and Marcus A: Fractionation and partial characterization of the protein synthesis of wheat germ. I. Resolution of two elongation factors and 5 initiation factors. *J. Biol. Chem.* 258:859 (1983).
134. Blum S, Mueller M, Schmid S, Linder P, and Trachsel H: Translation in *Saccharomyces cerevisiae:* Initiation factor 4A-dependent cell-free system. *Proc. Natl. Acad. Sci. U.S.A.* 86:6043 (1989).
135. Nielsen PJ, McMaster GK, and Trachsel H: Cloning of eukaryotic protein synthesis initiation factor genes: isolation and characterization of cDNA clones encoding factor eIF-4A. *Nucleic Acids Res.* 13:6867 (1985).
136. Nielsen PJ and Trachsel H: The mouse protein synthesis initiation factor 4A gene family includes two related functional genes which are differentially expressed. *EMBO J* 7:2097 (1988).
137. Reddy NS, Roth WW, Bragg PW, and Wahba AJ: Isolation and mapping of a gene for protein synthesis initiation factor 4A and its expression during differentiation of murine erythroleukemia cells. *Gene* 70:231 (1988).
138. Linder P and Slonimski PP: An essential yeast protein, encoded by duplicated genes TIF1 and TIF2 and homologous to the mammalian translation initiation factor eIF-4A, can suppress a mitochondrial missense mutation. *Proc. Natl. Acad. Sci. U.S.A* 86:2286 (1989).
139. Leroy P, Alzari P, Sassoon D, Wolgemuth D, and Fellows M: The protein encoded by a murine male germ cell-specific transcript is a putative ATP-dependent RNA helicase. *Cell* 57:549 (1989).
140. Grifo JA, Tahara SM, Leis JP, Morgan MA, Shatkin AJ, and Merrick WC: Characterization of eukaryotic initiation factor 4A, a protein involved in ATP-dependent binding of globin mRNA. *J. Biol. Chem.* 257:5246 (1982).
141. Brown-Luedi ML, Meyer LJ, Milburn SC, Yau PM-P, Corbett S, and Hershey JWB: Protein synthesis initiation factors from HeLa cells and rabbit reticulocytes are similar: comparison of protein structure, activities and immunochemical properties. *Biochemistry* 21:4202 (1982).
142. Van der Mast C and Voorma HO: Eukaryotic initiation factor eIF-4A from rabbit reticulocytes is a heterogeneous glycoprotein. *Biochim. Biophys. Acta* 739:141 (1983).
143. Seal SN, Schmidt A, and Marcus A: Eukaryotic initiation factor 4A is the component that interacts with ATP in protein chain initiation. *Proc. Natl. Acad. Sci. U.S.A.* 80:6562 (1983).

144. Sarkar G, Edery I, and Sonenberg N: Photoaffinity labeling of the cap-binding protein complex with ATP/dATP. *J. Biol. Chem.* 260:13831 (1985).

145. Grifo JA, Abramson RD, Satler CA, and Merrick WC: RNA-stimulated ATPase activity of eukaryotic initiation factors. *J. Biol. Chem.* 259:8648 (1984).

146. Abramson RD, Dever TE, Lawson TG, Ray BK, Thach RE, and Merrick WC: The ATP-dependent interaction of eukaryotic initiation factors with mRNA. *J. Biol. Chem.* 262:3826 (1987).

147. Lax SR, Browning KS, Maia DM, and Ravel JM: ATPase activities of wheat germ initiation factors 4A, 4B and 4F. *J. Biol. Chem.* 261:15632 (1986).

148. Abramson RD, Browning KS, Dever TE, Lawson TG, Thach RE, Ravel JM, and Merrick WC: Initiation factors that bind mRNA. *J. Biol. Chem.* 263:5462 (1988).

149. Goss DJ, Woodley CL, and Wahba AJ: A fluorescence study of the binding of eukaryotic initiation factors to messenger RNA and messenger RNA analogues. *Biochemistry* 26:1551 (1987).

150. Edery I, Hümbelin M, Darveau A, Lee KAW, Milburn S, Hershey JWB, Trachsel H, and Sonenberg N: Involvement of eukaryotic initiation factor 4A in the cap recognition process. *J. Biol. Chem.* 258:11398 (1983).

151. Milburn SC, Hershey JWB, Davies MV, Kelleher K, and Kaufman RJ: Cloning and expression of eukaryotic initiation factor 4B cDNA: Sequence determination identifies a common RNA recognition motif. *EMBO J.* 9:2783 (1990).

152. Browning KS, Maia DM, Lax SR, and Ravel JM: Identification of a new protein synthesis initiation factor from wheat germ. *J. Biol. Chem.* 262:538 (1987).

153. Browning KS, Fletcher L, Lax SR, and Ravel JM: Evidence that the 59-kDa protein synthesis initiation factor from wheat germ is functionally similar to the 80-kDa initiation factor 4B from mammalian cells. *J. Biol. Chem.* 264:8491 (1989).

154. Butler JS and Clark JM Jr: Eucaryotic initiation factor 4B of wheat germ binds to the translation initiation region of a messenger ribonucleic acid. *Biochemistry* 23:809 (1984).

155. Bandziulis RJ, Swanson MS, and Dreyfuss G: RNA-binding proteins as developmental regulators. *Genes Dev.* 3:431 (1989).

156. Tahara SM, Morgan MA, and Shatkin AJ: Two forms of purified m⁷G-cap binding protein with different effects on capped mRNA translation in extracts of uninfected and poliovirus-infected HeLa cells. *J. Biol. Chem.* 256:7691 (1981).

157. Etchison D, Milburn SC, Edery I, Sonenberg N, and Hershey JWB: Inhibition of HeLa cell protein synthesis following poliovirus infection correlates with the proteolysis of a 220,000-dalton polypeptide associated with eukaryotic initiation factor 3 and a cap binding protein complex. *J. Biol. Chem.* 257:14806 (1982).

158. Etchison D and Fout S: Human rhinovirus 14 infection of HeLa cells results in the proteolytic cleavage of the p220 cap-binding complex subunit and inactivates globin mRNA translation *in vitro. J. Virol.* 54:634 (1985).

159. Goyer C, Altmann M, Trachsel H, and Sonenberg N: Identification and characterization of cap-binding proteins from yeast. *J. Biol. Chem.* 264:7603 (1989).

160. Grifo JA, Tahara SM, Morgan MA, Shatkin AJ, and Merrick WC: New initiation factor activity required for globin mRNA translation. *J. Biol. Chem.* 258:5804 (1983).

161. Merrick WC, Abramson RD, Anthony DD, Dever TE, and Caliendo AM: Involvement of nucleotides in protein synthesis initiation, in Ilan J (Ed): *Translational Regulation of Gene Expression,* New York, Plenum Press, 265 (1987).
162. Buckley B and Ehrenfeld E: The cap-binding protein complex in uninfected and poliovirus-infected HeLa cells. *J. Biol. Chem.* 262:13599 (1987).
163. Etchison D and Milburn S: Separation of protein synthesis initiation factor eIF4A from a p220-associated cap binding complex activity. *Mol. Cell. Biochem.* 76:15 (1987).
164. Ray BK, Lawson TG, Kramer JC, Cladaras MH, Grifo JA, Abramson RD, Merrick WC, and Thach RE: ATP-dependent unwinding of messenger RNA structure by eukaryotic initiation factors. *J. Biol. Chem.* 260:7651 (1985).
165. Trachsel H, Sonenberg N, Shatkin AJ, Rose JK, Leong K, Bergmann JE, Gordon J, and Baltimore D: Purification of a factor that restores translation of vesicular stomatitis virus mRNA in extracts from poliovirus-infected HeLa cells. *Proc. Natl. Acad. Sci. U.S.A.* 77:770 (1980).
166. Hansen J, Etchison D, Hershey JWB, and Ehrenfeld E: Association of cap-binding protein with eucaryotic initiation factor 3 in initiation factor preparations from uninfected and poliovirus-infected HeLa cells. *J. Virol.* 42:200 (1982).
167. Etchison D and Smith K: Variations in cap-binding complexes from uninfected and poliovirus-infected HeLa cells. *J. Biol. Chem.* 265:7492 (1990).
168. Sonenberg N: ATP/Mg^{++}-dependent cross-linking of cap binding proteins to the 5' end of eukaryotic mRNA. *Nucleic Acids Res.* 9:1643 (1981).
169. Milburn SC, Pelletier J, Sonenberg N, and Hershey JWB: Identification of the 80-kDa protein that crosslinks to the cap structure of eukaryotic mRNA as initiation factor eIF-4B. *Arch. Biochem. Biophys.* 264:348 (1988).
170. Ray BK, Lawson TG, Abramson RD, Merrick WC, and Thach RE: Recycling of messenger RNA cap-binding proteins mediated by eukaryotic initiation factor 4B. *J. Biol. Chem.* 261:11466 (1986).
171. Abramson RD, Dever TE, and Merrick WC: Biochemical evidence supporting a mechanism for cap-independent and internal initiation of eukaryotic mRNA. *J. Biol. Chem.* 213:6016 (1988).
172. Lawson TG, Lee KA, Maimone MM, Abramson RD, Dever TE, Merrick WC, and Thach RE: Dissociation of double stranded polynucleotide helical structures by eukaryotic initiation factors as revealed by a novel assay. *Biochemistry* 28:4729 (1989).
173. Rozen F, Edery I, Meerovitch K, Dever TE, Merrick WC, and Sonenberg N: Bidirectional RNA helicase activity of eukaryotic translation initiation factors 4A and 4F. *Mol. Cell. Biol.* 10:1134 (1990).
174. Kabat D and Chappell MR: Competition between globin messenger ribonucleic acids for a discriminating initiation factor. *J. Biol. Chem.* 252:2684 (1977).
175. Sarkar G, Edery I, Gallo R, and Sonenberg N: Preferential stimulation of rabbit α globin mRNA translation by cap binding protein complex. *Biochim. Biophys. Acta* 783:122 (1984).
176. Ray BK, Brendler TG, Adya S, Daniels-McQueen S, Miller JK, Hershey JWB, Grifo JA, Merrick WC, and Thach RE: Role of mRNA competition in regulating translation: further characterization of mRNA discriminatory initiation factors. *Proc. Natl. Acad. Sci. U.S.A.* 80:663 (1983).

177. Browning KS, Lax SR, Humphreys J, Ravel JM, Jobling SA, and Gehrke L: Evidence that the 5'-untranslated leader of mRNA affects the requirement for wheat germ initiation factors 4A, 4F, and 4G. *J. Biol. Chem.* 263:9630 (1988).

178. Duncan R and Hershey JWB: Identification and quantitation of levels of protein synthesis initiation factors in crude HeLa cell lysates by two-dimensional polyacrylamide gel electrophoresis. *J. Biol. Chem.* 258:7228 (1983).

179. Duncan R and Hershey JWB: Regulation of initiation factors during translational repression caused by serum depletion. Abundance, synthesis and turnover rates. *J. Biol. Chem.* 260:5486 (1985).

180. Duncan R, Milburn SC, and Hershey JWB: Regulated phosphorylation and low abundance of HeLa cell initiation factor eIF-4F suggest a role in translational control. *J. Biol. Chem.* 262:380 (1987).

181. Hiremath LS, Webb NR, and Rhoads RE: Immunological detection of the messenger RNA cap-binding protein. *J. Biol. Chem.* 260:7843 (1985).

182. Howe JG and Hershey JWB: Translational initiation factor and ribosome association with the cytoskeletal framework fraction from HeLa cells. *Cell* 37:85 (1984).

183. Etchison D and Etchison JR: Monoclonal antibody-aided characterization of cellular p220 in uninfected and poliovirus-infected HeLa cells: subcellular distribution and identification of conformers. *J. Virol.* 61:2702 (1987).

184. Hansen JL, Etchison DO, Hershey JWB, and Ehrenfeld E: Localization of cap-binding protein in subcellular fractions of HeLa cells. *Mol. Cell. Biol.* 2:1639 (1982).

185. Hiremath LS, Hiremath ST, Rychlik W, Joshi S, Domier LL, and Rhoads RE: In vitro synthesis, phosphorylation, and localization on 48S initiation complexes of human protein synthesis initiation factor 4E. *J. Biol. Chem.* 264:1132 (1989).

186. Joshi-Barve S, Rychlik W, and Rhoads RE: Alteration of the major phosphorylation site of eukaryotic protein synthesis initiation factor 4E prevents its association with the 48S initiation complex. *J. Biol. Chem.* 265:2979 (1990).

187. Westerman P and Nygard O: Cross-linking of mRNA to initiation factor eIF-3, 24 kDa cap binding protein and ribosomal proteins S1, S3/3a and S11 within the 48S pre-initiation complex. *Nucleic Acids Res.* 12:8887 (1984).

188. Görlach M and Hilse K: Stoichiometric association of cap-binding protein I with translated polysomal globin mRNP. *EMBO J.* 5:2629 (1986).

189. Schreier MH, Erni B, and Staehelin T: Initiation of mammalian protein synthesis. I. Purification and characterization of seven initiation factors. *J. Mol. Biol.* 116:727 (1977).

190. Thomas A, Goumans H, Amesz H, Benne R, and Voorma HO: A comparison of initiation factors of eukaryotic protein synthesis from ribosomes and from the postribosomal supernatant. *Eur. J. Biochem.* 98:329 (1979).

191. Safer B, Jagus R, and Kemper WM: Analysis of initiation factor function in highly fractionated and unfractionated reticulocyte lysate systems. *Methods Enzymol.* 60:61 (1979).

192. Sonenberg N: Cap-binding proteins of eukaryotic messenger RNA: functions in initiation and control of translation. *Prog. Nucleic Acid Res. Mol. Biol.* 35:173 (1988).

193. Goss DJ, Carberry SE, Dever TE, Merrick WC, and Rhoads RE: A fluorescence study of the binding of m⁷GpppG and rabbit globin mRNA to protein synthesis initiation factors 4A, 4E and 4F. *Biochemistry* 29:5008 (1990).

194. Vasquez D, Dölz E, and Jimenez A: Action of inhibitors of protein biosynthesis, in Perez-Bercoff R (Ed): *Protein Biosynthesis in Eukaryotes,* New York, Plenum Press, 311 (1982).

195. Stewart-Blair ML, Yanowitz IS, and Goldberg IH: Inhibition of synthesis of new globin chains in reticulocyte lysates by pactamycin. *Biochemistry* 10:4198 (1971).

196. Merrick WC, Kemper WM, and Anderson WF: Purification and characterization of homogeneous initiation factor M2A from rabbit reticulocytes. *J. Biol. Chem.* 250:5556 (1975).

197. Benne R, Brown-Luedi ML, and Hershey JWB: Purification and characterization of protein synthesis initiation factors eIF-1, eIF-4C, eIF-4D, and eIF-5 from rabbit reticulocytes. *J. Biol. Chem.* 253:3070 (1978).

198. Raychaudhuri P, Chaudhuri A, and Maitra U: Eukaryotic initiation factor 5 from calf liver is a single polypeptide chain protein of Mr = 62,000. *J. Biol. Chem.* 260:2132 (1985).

199. Raychaudhuri P, Chevesich J, Ghosh S, and Maitra U: Characterization of eukaryotic initiation factor 5 from rabbit reticulocytes. *J. Biol. Chem.* 262:14222 (1987).

200. Peterson DT, Safer B, and Merrick WC: Role of eukaryotic initiation factor 5 in the formation of 80S initiation complexes. *J. Biol. Chem.* 254:7730 (1979).

201. Chaudhuri A, Stringer EA, Valenzuela D, and Maitra U: Characterization of eukaryotic initiation factor 2 containing two polypeptide chains of Mr = 48,000 and 38,000. *J. Biol. Chem.* 256:3988 (1981).

202. Raychaudhuri P, Chaudhuri A, and Maitra U: Formation and release of eukaryotic initiation factor 2-GDP complex during eukaryotic ribosomal polypeptide chain initiation complex formation. *J. Biol. Chem.* 260:2140 (1985).

203. Goumans H, Thomas A, Verhoeren A, Voorma HO, and Benne R: The role of eIF-4C in protein synthesis initiation complex formation. *Biochem. Biophys. Acta* 608:39 (1980).

204. Cooper HL, Park MH, Folk JE, Safer B, and Braverman R: Identification of the hypusine-containing protein Hy⁺ as translation initiation factor eIF-4D. *Proc. Natl. Acad. Sci. U.S.A.* 80:1854 (1983).

205. Park MH, Liu T-Y, Neece SH, and Swiggard WJ: Eukaryotic initiation factor 4D. Purification from human red blood cells and the sequence of amino acids around its single hypusine residue. *J. Biol. Chem.* 261:14515 (1986).

206. Smit-McBride Z, Dever TE, Hershey JWB, and Merrick WC: Sequence determination and cDNA cloning of eukaryotic initiation factor 4D, the hypusine-containing protein. *J. Biol. Chem.* 264:1578 (1989).

207. Duncan RF and Hershey JWB: Changes in eIF-4D hypusine modification or abundance are not correlated with translational repression in HeLa Cells. *J. Biol. Chem.* 261:12903 (1986).

208. Park MH: The essential role of hypusine in eukaryotic translation initiation factor 4D (eIF-4D). *J. Biol. Chem.* 264:18531 (1989).

209. Smit-McBride Z, Schnier J, Kaufman RJ, and Hershey JWB: Protein synthesis initiation factor eIF-4D. Functional comparison of native and unhypusinated forms of the protein. *J. Biol. Chem.* 264:18527 (1989).

Polypeptide Chain Elongation

Lawrence I. Slobin
Department of Biochemistry
University of Mississippi School of Medicine
Jackson, Mississippi

6.1. INTRODUCTION

The study of eukaryotic polypeptide chain elongation is about to enter its fourth decade. In 1961 Bishop and Schweet[1] reported the participation of two enzyme fractions in the transfer of amino acids from aminoacyl-tRNA to reticulocyte ribosomes. Since that historic communication hundreds of laboratories have produced thousands of publications which deal with the subject of this chapter. A number of excellent reviews on polypeptide chain elongation have appeared over the past two decades. Some of the more recent ones should be consulted to gain access to the earlier literature.[2–4] The first part of this review deals with process of peptide elongation while the later half discusses the biochemistry and molecular biology of the elongation factors. Neither subject could be adequately treated without occasional mention of work on that humble prokaryote, *Escherichia coli.*

6.2. ELONGATION RATE

Protein synthesis represents the final process in the direct transfer of genetic information within the cell. Inevitably, the rate of protein synthesis will limit how quickly genetic events can be transmuted into cell growth and division. In *Escherichia coli* growing at 37° the peptide chain elongation rate increases with growth rate from about 13 to 20 amino acid residues per second between 0.6 and 2.5 doublings per hour. It appears that the limitation during slow growth is not some structural component of the protein synthetic apparatus.[5, 6]

A number of procedures have been devised to measure peptide chain elongation rates in eukaryotic cells and organisms.[7-9] Haschemeyer derived the following relationship for the incorporation of radioactive amino acids into protein:

$$S/T = t/2t_c$$

In cells or organisms, the incorporation of radioactive amino acids into completed polypeptides (S) is equal to the total incorporation in the system (T) minus the incorporation into the growing peptide chain (R). In the above equation t equals the time of labeling and t_c equals the average polypeptide chain assembly time or transit time. The equation states that when the time of labeling equals the transit time, completed polypeptides will comprise $1/2$ the total incorporation into peptide bonds. Average elongation rates have been estimated for a number of different vertebrate systems, including mouse liver, rat liver, toadfish liver, HeLa cells, Chinese hamster ovary cells, etc. (see Reference 10 for a review). Most of the published data for *in vivo* systems give an elongation rate of about six amino acid residues per second per ribosome at 37°. Some *in vitro* studies with cultured cells have shown a lower elongation rate of about three residues per second. Studies on the elongation rate of individual proteins have given values which are in agreement with the above mentioned average rates.[11, 12] Thus the average rate of polymerization of amino acids into polypeptides in vertebrates is approximately threefold slower than the same process in rapidly growing *E. coli*. The rate limiting step in peptide elongation in eukaryotes is unknown. There is reasonable evidence that in certain circumstances (e.g., hormonal stimulation, cold adaptation in fish, stimulation of stationary mammalian cell cultures by addition of fresh medium, etc.) the rate of peptide elongation can be augmented.[11, 13-16] Changes in polypeptide elongation rates have been correlated with an increase in the methylation of EF-1α during morphogenesis of the fungus *Mucor racemosus*[17] (see below). There also is evidence that elongation rates are reduced during aging.[18-20]

Elongation rates may be linked to gene expression via control of mRNA stability. Work in the laboratory of Don Cleveland has shown that tubulin synthesis is controlled, in part, by an autoregulatory mechanism that affects the stability of ribosomal bound tubulin mRNAs.[21, 22] The same laboratory has demonstrated, using peptide elongation inhibitors, that ongoing ribosome translocation is required to destabilize tubulin mRNAs.[23] Furthermore, 95% inhibition rather than complete inhibition of translation elongation was actually found to stimulate tubulin mRNA degradation. A model of regulated tubulin instability was proposed, in which the binding of tubulin monomers to a nascent tubulin peptide emerging from ribosomes causes ribosome stalling on tubulin mRNAs. According to this model, ribosome stalling leads to a gap between ribosomes which could provide an effective target for a non-specific ribonuclease. Selective ribosome stalling would clearly be disrupted by complete inhibition of elongation (translocation) since complete translational arrest would prevent mRNA gaps from forming.

There are a number of reports which provide indications that peptide chain elongation plays a role in controlling mRNA stability.[24-26] Although regulation of tubulin mRNA stability is a highly specific process, it may well be that global changes in elongation rates in a cell can alter ribosome stalling[27] and thereby promote or retard the degradation of certain mRNAs. If so, such a mechanism might provide a direct linkage between the rate of translation (transit times) and gene expression.

6.3. THE ELONGATION CYCLE

Figure 1 shows what we know to be the general features of the eukaryotic elongation cycle. In the following I will attempt to summarize our current understanding of the individual steps in the cycle, leaving for later any discussion of the structures of individual factors. Emphasis throughout will be placed on recent findings.

6.3.1. EF-1α Cycle

EF-1α (also designated in the literature as eEF-Tu) is a basic polypeptide of approximately 50,000 MW which has been highly purified from at least a dozen different sources.[2-4] The *Artemia* factor binds both GDP and GTP with apparently almost equal avidity [K_d (EF-1α·GTP) = $8.5 \times 10^{-7} M$ and K_d (EF-1α·GDP) = 5.3×10^{-7} M];[28] essentially identical results have been reported with EF-1α from rabbit reticulocytes and calf brain.[29, 30] The binding of GTP to the factor is energetically favored by the subsequent attachment of aminoacyl tRNA (ternary complex formation) followed by binding of the ternary complex to the ribosome. Subsequent to ribosome binding, GTP is hydrolyzed to GDP and the factor is released as an EF-1α·GDP complex. Just as for ET-Tu from *E. coli*,[31] EF-1α possess an intrinsic GTPase activity[32, 33] which is stimulated upon interaction of ternary complexes with ribosomes.

Although EF-1α binds GTP and GDP with about equal affinity, the spontaneous rate of dissociation of GDP from the factor is slow. Janssen and Möller, working with *Artemia* EF-1α, report a dissociation rate of 14×10^{-3} s^{-1}, which they calculate to be too slow to explain observed *in vivo* peptide

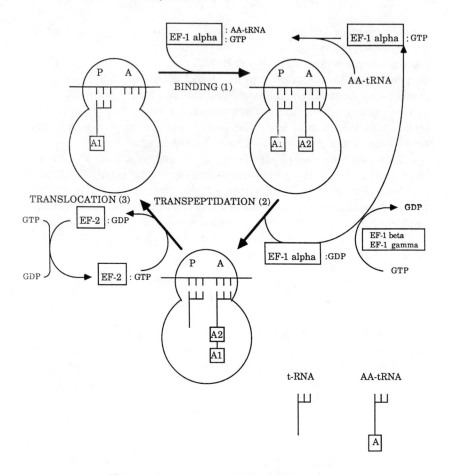

Figure 1. Eukaryotic polypeptide elongation cycle.

elongation rates.[34] To recycle, the EF-1α bound to GDP eukaryotic organisms depend on a factor, EF-1β (also designated as eEF-Ts) which performs a function identical to EF-Ts from *E. coli*. This factor (MW = 23,000) is usually found in association with a polypeptide of 49,000 MW designated as EF-1γ. Both EF-1β and EF-1γ have been purified and characterized from many sources (References 4 and 34 and references therein). EF-1β, by itself, has been found to catalyze a guanine nucleotide exchange reaction in the manner of EF-Ts. Although an early report indicated that EF-1γ is without effect on the exchange reaction,[35] more recent evidence suggests EF-1γ stimulates the exchange reaction slightly.[34] Janssen and Möller also studied the mechanism of the guanine nucleotide exchange reaction. They found the exchange involves the formation of a transient ternary complex of EF-1α·GDP·EF-1βγ. This ternary complex releases GDP with a catalytic rate constant of at least 14 s^{-1} at 27°, 10^3 times faster than the spontaneous rate of dissociation of the nucleotide from EF-1α (see above). Of interest is the observation that EF-1β can be phosphorylated at serine 89 by an endogenous kinase present in EF-1βγ preparations.[36] The phosphorylation site resembles the recognition sequence for casein kinase II. The rate of guanine nucleotide exchange on EF-1α was moderately affected by the phosphorylation state of EF-1β: the rate of exchange catalyzed by the phosphorylated factor was reported to be half that of the unmodified EF-1β. To pique interest in EF-1βγ phosphorylation, Bellé and co-workers have observed that EF-1γ is phosphorylated in *Xenopus laevis* oocytes by the cell division control protein kinase p34^{cdc2}.[37] Both p34^{cdc2} kinase and casein kinase 2 are activated during meiotic cell division; this activation is correlated with changes in protein synthesis.[38] However, correlation between the phosphorylation state of EF-1βγ and peptide elongation rates *in vivo* remains to be established.

6.3.2. Binding of Aminoacyl tRNA to Ribosomes

According to the canonical model for the EF-1α dependent binding of aminoacyl tRNA to ribosomes, the aminoacyl tRNA is deposited at the ribosomal acceptor or A site prior to the transpeptidation reaction. New findings with yeast suggest that this view of aminoacyl-tRNA binding may be too simple. It has been known for some time that yeast requires, in addition to the above mentioned elongation factors, another polypeptide factor designated as EF-3.[39–40] This factor is a single polypeptide chain of about 120 kDa and has been purified to homogeneity by several laboratories.[40–41] The gene for EF-3 has been isolated recently[42] and we shall undoubtedly know its primary structure in the near future. EF-3 was uniquely required by yeast ribosomes for the translation of poly U[39] and natural mRNA.[43, 44] An isolated temperature-sensitive mutant of yeast harboring an altered EF-3 was unable to grow at a nonpermissive temperature, indicating that the factor is essential for translation *in vivo*.[44–46] There is evidence that EF-3 plays an important role in correctly positioning aminoacyl tRNA on ribosomes[41, 47] and that a functionally similar factor may be an essential constituent of all eukaryotic ribosomes.

The EF-1α dependent binding of aminoacyl tRNA to yeast ribosomes·mRNA complexes is saturated at low concentrations of the factor. The presence of EF-3 markedly stimulates the binding reaction. Uritani and Miyazaki[41] have shown that the Phe-tRNA·poly(U)·ribosome complexes formed with EF-1α alone were less stable under certain ionic conditions than the complexes formed with the participation of both EF-1α and EF-3. Furthermore, while no enhancement of the poly(U)-directed Leu-tRNA binding (misreading) was observed, EF-3 in combination with ATP greatly stimulated Phe-tRNA binding in the same system (correct decoding). The effect of EF-

3 could readily be distinguished from that of EF-1βγ. Based on these observations, Uritani and Miyazaki[41] suggest that the binding of aminoacyl tRNA on ribosomes with EF-1α alone positions the aminoacyl tRNA at a site which is distinct from the site occupied when EF-1α and EF-3 act in concert. They propose that the EF-1α ternary complex first binds aminoacyl tRNA to an I- ("introducing") site. Following I-site binding, EF-3 acts by catalyzing an ATP dependent transfer of aminoacyl tRNA from the I-site to the canonical A site (see Figure 1).

EF-3 activity has not been detected in eukaryotic systems other than fungi.[40] It has been shown, however, that ribosomes from a number of non-fungal sources possess strong ATPase and GTPase activities whereas yeast ribosomes do not hydrolyze these nucleotides.[41] The claim was made (no data were given) that anti-yeast EF-3 cross-reacts with ribosomal protein components from *Artemia* and rat liver. These findings, if further corroboration is forthcoming, imply that the function of EF-3 in nonfungal eukaryotes is associated with a ribosomal protein(s). The relationship, if any, between the I-site in eukaryotic ribosomes and the recognition (R)-site, proposed by Lake[48] as the initial aminoacyl tRNA binding site on *E. coli* ribosomes, remains to be established.

The two-step model for aminoacyl tRNA binding to the ribosomal A site is consistent with the view that this reaction plays an important role in limiting the translational error frequency. Thompson has recently proposed that the rate constants governing ribosome interactions with EF-Tu act as an internal kinetic standard to measure ribosome interaction with aminoacyl tRNA, and thereby help "proofread" aminoacyl tRNA.[49] Support for this proposal comes from observations that an EF-Tu mutant, EF-TuAr[50] lowers the normal translational accuracy in *E. coli*.[51, 52] EF-1α is also a proofreading enzyme. Sandbaken and Culbertson have demonstrated that mutations in EF-1α affect the frequency of frameshifting and amino acid misincorporation in yeast.[53] These mutations were analyzed in terms of our current knowledge of the structural domains of EF-1α.

Similarities between prokaryotes and eukaryotes in the binding of aminoacyl tRNA to ribosomes are highlighted by recent observations by Moazed and co-workers.[54] These investigators used chemical probes to examine the "footprints" that resulted when prokaryotic elongation factors interact with *E. coli* ribosomes. EF-Tu gave an overlapping footprint *in vitro* at positions 2665 and 2661 in 23S rRNA. These positions are part of a universally conserved loop in rRNA. This loop is the locus of action of the RNA specific cytotoxins α-sarcin and ricin whose effects abolish translation by inhibiting EF-1α dependent binding of aminoacyl tRNA[55] and EF-2 dependent translocation,[56] respectively. Thus the evidence for evolutionary conservation of the ribosome binding site of the elongation factors is rather compelling. A 28S rRNA binding site for EF-1α is plausible, given the documented ability of the factor to bind RNA.[57, 58]

It also has been shown that EF-1α has a high binding specificity for synthetic poly (U) and poly (G) and that this RNA binding site is separate and distinct from the aminoacyl tRNA binding site of the factor.[32] Rat liver 28S rRNA has the sequence UUGGUGAUUGUUUGGUG starting 19 residues downstream from the site of α-sarcin cleavage. If attachment of EF-1α to ribosomes is mediated through its RNA binding site, it seems reasonable to suppose that just such a sequence in rRNA would be involved.

Chemical cross-linking has been employed to identify r-proteins which interact with EF-1α during aminoacyl tRNA binding. Complexes containing rat liver 80S ribosomes, poly(U), Phe-tRNA, EF-1α, and GMPPCP were prepared and treated with the bifunctional cross-linking reagent 2-iminothiolane. The following proteins of the 40S and 60S subunits were identified as cross-linked to EF-1α: L12, L23, L39, S23/24, and S26.[59] It is clear also that acidic r-proteins of the 60S subunit (L40, L41), the evolutionary descendants of *E. coli* r-protein L12, are important in EF-1α dependent reactions.[60] The homology between EF-Tu and EF-1α (see below) and the conservation of the ribosome A site structure lend additional support to the hypothesis that the fundamental mechanism of binding aminoacyl-tRNA to ribosomes has been conserved during the course of evolution. Some details have changed, however, since EF-Tu is not capable of supplying aminoacyl tRNA to the A site in a eukaryotic protein synthesis system.[61]

6.3.3. Peptide Bond Formation

As illustrated in Figure 1, when aminoacyl tRNA is correctly positioned at the ribosomal A site and the P site is occupied by peptidyl tRNA, the ribosome has the capacity to catalyze peptide bond formation. The exact nature of the catalytic agent, generally referred to as peptidyl transferase, has remained elusive despite several decades of intense investigation.[4] Relatively little work has been performed on the eukaryotic peptidyl transferase center during the past decade. It seems likely that we shall first understand the molecular details of peptide bond formation from work on *E. coli* ribosomes and that, whatever the mechanism, it will be rather faithfully adopted by eukaryotic ribosomes.

6.3.4. Translocation

Upon completion of the peptidyl transferase reaction the ribosomal P site is occupied with deacylated tRNA and the A site contains the newly formed peptidyl tRNA (Figure 1). The translocation reaction results in the movement of the ribosome and/or mRNA in relation to each other, by precisely one triplet, from the 5′- to the 3′-end of mRNA. Translocation is dependent upon EF-2 (EF-G in prokaryotes) and GTP. EF-2 is a single polypeptide of approximately 100 kDa and has been purified from many different sources.[3, 4] The translocation process commences when an EF-2·GTP complex binds to a pretranslocation complex (see above) containing peptidyl tRNA in the A site. At some point in the translocation reaction GTP is hydrolyzed to GDP and P_i and EF-2 is released from the posttranslocated ribosome. Just as for EF-1α (see above), the catalytic center for GTP hydrolysis appears to be located within EF-2.[62] However, the GTPase activity of EF-2 is very low in the absence of ribosomes.[63]

Complexes of EF-2 and ribosomes can be formed in the presence of nonhydrolyzable analogs of GTP, indicating that GTP hydrolysis is is not required to bind the factor to ribosomes. Furthermore, Tanaka and co-workers have demonstrated that EF-2 dependent translocation is supported by non-hydrolyzable analogs of GTP.[64] These same investigators also found that translocation in the presence of GMP-P(NH)P was reversible and stoichiometric with EF-2 whereas in the presence of GTP translocation was unidirectional and EF-2 acted catalytically. It is likely that the EF-2 catalyzed hydrolysis of GTP is required for the release and recycling of EF-2 rather than for the translocation reaction per se. In this regard, the GTP hydrolysis reaction plays a functionally identical role in aminoacyl tRNA binding (EF-1α recycling) and translocation (EF-2 recycling).[2] The laboratory of Nilsson and Nygård has provided additional evidence that GTP is hydrolyzed by EF-2 on posttranslocated ribosomes.[63, 65, 66] If GTP hydrolysis occurs subsequent to translocation of peptidyl tRNA and drives the recycling of EF-2 we are left with the question of the function of EF-2 in the translocation process. One attractive hypothesis which has been often suggested, but for which there is little direct support, is that EF-2 catalyzes the release of uncharged tRNA from the P site.

Modest progress has been made in defining the ribosome binding site for EF-2. There is evidence that EF-1α acts only on posttranslocational ribosomes whereas EF-2 appears to interact specifically with pretranslocational ribosomes.[67, 68] Footprinting experiments (see above) with *E. coli* elongation factors and ribosomes indicate that both EF-Tu and EF-G bind at or near the universally conserved loop in the 2600 region of 23S rRNA.[54] The notion that the two elongation factors bind at overlapping sites on the ribosome is also supported by recent chemical cross-linking studies. In separate experiments, the bifunctional cross-linking reagent 2-iminothiolane was found to cross-link L12, L23, and S23/24 to EF-1α and EF-2.[59, 69] A number of other r-proteins were identified as cross-linked to EF-2, including L9, LA33, P2 and S6. The cross-link between L23 and EF-2 has been confirmed using dimethylsuberimidate as a cross-linking reagent.[70] Additionally, a number of other cross-links between r-proteins and EF-2 have been found.[70] 5S rRNA was also reported to cross-link to EF-2 when EF-2·GuoPP[CH$_2$]P·ribosome

TABLE 6–1.
Elongation Factors and their Genes

Factor	Organism	Sequence	Gene Number	Ref.
EF-1α	Yeast	Genomic	2	93, 141–143
EF-1α	*Mucor*	Genomic	3	144–146
EF-1α	*Artemia*	Genomic cDNA protein		86, 96
EF-1α	*Drosophila*	Genomic cDNA	2	97
EF-1α	*Xenopus*	cDNA	ND	137
EF-1α	Tomato	cDNA	ND	147
EF-1α	Mouse	cDNA	ND	148
EF-1α	Human	Genomic cDNA	ND	75, 98, 99
EF-1β	*Artemia*	cDNA	ND	101
EF-1γ	*Artemia*	cDNA	ND	102
EF-2	Hamster	Genomic cDNA	1	126
EF-2	Rat	cDNA	ND	117, 124
EF-2	Slime mold	cDNA	ND	118

ND = not determined.

complexes were treated with the bifunctional reagent, diepoxybutane.[71] Our current information on EF-2 GTPase indicates that both ribosomal subunits make a significant contribution to factor binding.

6.4. STRUCTURE OF ELONGATION FACTORS AND THEIR GENES

The past 6 years has seen an enormous increase in our knowledge of eukaryotic elongation factors and their genes. While much of this new knowledge has confirmed the conservative evolution of the process of peptide elongation, there have also been many surprises which point to elaborate differences between prokaryotic and eukaryotic elongation factor structures. From what we already know, it seems likely that much additional structural information is built into the eukaryotic factors: altered activities, stabilities, or capacities to target intracellular locations important for protein synthesis, are predicted consequences of these evolutionary changes. Table 6–1 gives a list of known complete elongation factor protein sequences (in almost all cases deduced from DNA sequences) and elongation factor genes.

6.4.1. EF-1α

EF-1α is perhaps the most well studied and is certainly the most abundant eukaryotic protein synthesis factor in mammalian cells.[72, 73] It comprises about 5% of the soluble protein in yeast cells.[74] The amino acid sequence of EF-1α is known for eight different organisms (Table 6–1) and the extremely conservative evolution of the protein has been well established. For example, human EF-1α is >80% identical in sequence to the yeast and *Artemia* factor and the yeast factor is 32% identical to EF-Tu from *E. coli*. The conservation in structure is particularly striking in the amino-terminal half of EF-1α, where the binding sites for guanine nucleotides and aminoacyl tRNA are thought to reside.[75–79] This same region also shows a clear structural similarity with a number of G proteins, such as IF-2, EF-G, EF-2, the α subunit of transducin, vRas, etc.[80, 81]

Despite its conservative evolution, the structure of EF-1α is not without novelty. Hiatt and co-workers have shown that 8 or 9 of the 44 lysine residues of EF-1α are modified from mycelia of *Mucor racemosus* and exist as mono-, di-, and trimethyllysine.[82] Further work demonstrated that EF-1α methylation is regulated during the growth of the organism: the hyphal form of the factor is methylated whereas EF-1α from sporangiophores is unmethylated.[83] The specific activity of EF-1α was reported to increase sixfold when crude extracts of *Mucor* were compared prior and subsequent to germination. It was suggested that lysine methylation might modulate EF-1α activity. More recently,[84] spore and mycelial forms of EF-1α were compared in a series of assays to determine their relative affinities for various substrates and cofactors known to interact with the factor during the elongation cycle. No significant difference in activities between the methylated and unmethylated forms of EF-1α were found. A difference in the ability of the two forms of EF-1α to bind to EF-1βγ was noted. However, it is not clear how this observation can account for the large increase in elongation rates which occur during the yeast-to-hyphae morphogenesis in the fungus (see above). Lysine methylation has been observed in EF-1α from *Artemia*[85, 86] and murine 3T3 cells.[87] EF-Tu is methylated at a single lysine residue in *E. coli*.[88] Analysis of the posttranslational modifications of EF-1α by Dever and co-workers has shown that five lysines are methylated: residues 55 and 165 are dimethyllysine and trimethyllysine is found at residues 36, 79, and 318.[89]

Two additional posttranslational modifications of rabbit EF-1α have been found. They involve the addition of ethanolamine to glutamic acid residues 301 and 374.[89–91] A glycerylphosphoryl moiety is attached to an ethanolamine-modified Glu residues of both murine[90] and rabbit factors.[89] This posttranslational modification has not yet been observed with any other protein. The glycerylphosphorylethanolamine structure appears to be synthesized directly on EF-1α; CDP-ethanolamine does not serve as a substrate for the modification reaction.[92] The Glu-301 is not highly conserved in eukaryotes whereas Glu-374 is present in all known EF-1α sequences as well imbedded in a highly conserved region of the factor. The function of the glycerylphosphorylethanolamine modification is unknown. It may be possible to determine the function of these posttranslational modifications by isolation of the unmodified factor from an *E. coli* expression system.

Studies on EF-1α genes are in their infancy. There are probably at least two genes that code for EF-1α in all organisms. Cottrelle and co-workers, using a gene interruption technique, have demonstrated in yeast that possession of one of the two EF-1α genes is sufficient for cell viability.[93] As might be expected, inactivation of both genes in yeast proved lethal. Further work has revealed a DNA footprinting activity that interacted specifically with the 5′ upstream regions of the two EF-1α genes.[94] The protected regions encompassed conserved sequences previously detected 200 to 400 bp upstream of most of the yeast r-protein genes examined. It was suggested that this DNA factor is required for activation and coordinated regulation of genes coding for components of the translational apparatus. The three closely related genes coding for EF-1α in *Mucor racemosus* have also been extensively investigated.[95] The three genes were found to be transcribed at different rates, and the expression of one of the genes (TEF-3) showed a morphology-specific pattern of transcript accumulation. *Artemia* probably contains three to four EF-1α genes, one of which was sequenced and found to contain five exons divided over 10^4 base pairs.[96] *Drosophila* was found to possess two EF-1α genes, one of which (F1) contains two exons; the other gene (F2) has five exons.[97] The nucleotide sequence differences between the two genes in the coding region were large enough to suggest the genetic separation of F1 and F2 is not a recent event. In contrast with the sequence of the *Artemia* and human EF-1α genes[98] Hovemann and co-workers did not recognize a TATA transcription motif in either of the two EF-1α genes. The *Drosophila* F1 gene contained a sequence 373 bases upstream of the transcription start site which was very similar to general promoter enhancer motif, the HOMOL box, which precedes the start of transcription of yeast EF-1α genes.

Little sequence similarity was observed in front of the transcription start sites of the two *Drosophila* EF-1α genes. However, additional data indicates that the expression of the two factor genes is highly regulated during the lifespan of *Drosophila* (see below).

Obtaining the structure of a human EF-1α gene must have required considerable diligence. As is the case for many gene families, the human genome contains at least 20 processed EF-1α pseudogenes.[98, 99] The active human gene which has been sequenced consists of eight exons and seven introns spanning about 3.5 kb. Four splice junctions are conserved between *Artemia* and human EF-1α genes. Using an *in vitro* transcription system, Uetsuki and co-workers were able to show that the EF-1α promoter could synthesize RNA at least twofold more effectively than the adenovirus major late promoter. This result as well as other recent findings (see below) indicate that the EF-1α promoter is among the strongest known. Like ribosomal protein genes[100] the major site of transcription initiation of the human EF-1α gene is located at a cytosine residue which is followed by a tract of consecutive pyrimidines. Given the large number of EF-1α pseudogenes, it seems likely that we will have to await work on the genome project to determine if there are additional functional EF-1α genes in humans.

6.4.2. EF-1βγ

Using an immunoscreening procedure, Maessen and co-workers[101, 102] have obtained cDNA clones for EF-1β and EF-1γ from *Artemia*. The derived amino acid sequence of EF-1β corresponds to 208 amino acids and a calculated M_r of 23,300. Although the size, isoelectric point, and hydrophobicity of EF-1β and EF-Ts were found to be comparable, little if any primary structural homology between the two factors could be detected. Since EF-1β has not yet been sequenced from any other organism it is not clear to what extent EF-1β structure has been conserved in eukaryotes. The derived amino acid sequence of *Artemia* EF-1γ corresponds to 429 amino acids with a M_r of 49,200. The protein is relatively rich in aromatic amino acids (13%) and possesses alternating hydrophobic and hydrophilic regions bordering a large hydrophilic region in the middle of the sequence. No significant homology of EF-1γ with any know protein was observed. As yet we do not know any EF-1γ structures from other organisms.

6.4.3. EF-1

We now know that the factor originally referred to as EF-1[103] consists of at least three different subunits which are functionally and structurally distinct: EF-1α, EF-1β, and EF-1γ (see above). The literature on this factor is often muddled by a nomenclature which arose prior to characterization of its individual subunits. Thus the terms $EF-1_H$, $EF-1_L$, and even eEF-1, are commonly encountered. There is no reason to employ the lower case e which is superfluous. As for $EF-1_L$, which was originally employed to designate a lower molecular weight form of an aggregate $EF-1_H$, its raison d'être was eliminated when it became clear that it corresponds to EF-1α. Presently, the designation $EF-1_H$ is also of dubious merit. Generally preparations designated $EF-1_H$ were so called because they contained EF-1α activity but were of greater size than $EF-1_L$. Bearing these considerations in mind, there is still reason to believe that the polypeptide components of EF-1αβγ are organized into higher order structures within eukaryotic cells. In addition, several reports indicate the association of EF-1α with EF-1βγ is altered during embryonic development,[104-106] although the significance of these changes in association of factor subunits remain to be clarified.

More recent work indicates that EF-1αβγ specifically associate with other polypeptides. There is evidence that val-tRNA synthetase is tightly associated with EF-1αβγ in rabbit reticulocytes and probably in other eukaryotic cells.[107] Janssen and Moller have found that about 5% of EF-1βγ in *Artemia* extracts is present in membrane fractions.[108] In the same work it was noted that EF-1βγ copurifies with tubulin on DEAE cellulose and that both EF-1βγ and tubulin can be simultaneously immunoprecipitated from mixtures of both proteins with either anti-EF-1βγ or anti-tubulin antibodies. A prudent speculator might suggest that the various posttranslational modifications of EF-1α, EF-1β, and EF-1γ serve the function of organizing the factors into complex intracellular structures that facilitate protein synthesis.

6.4.4. EF-1α-Like Polypeptides

There are some reasons to believe that EF-1α is part of a larger gene family whose gene products may serve functions other than those depicted in Figure 1. Several laboratories have found that the most abundant soluble protein of previtellogenic *Xenopus* oocytes is structurally and functionally related to EF-1α. The tRNA in these oocytes is found in a 42S ribonucleoprotein (RNP) which, in addition to tRNA, consists of two proteins.[109] One of these proteins, called either thesaurin or 42Sp48, is immunologically cross-reactive with anti EF-1α antibodies.[110] Furthermore, this cross-reactive protein can transfer charged amino acids to proteins. Thesaurin may have different physical properties than EF-1α and appears to have a different amino acid sequence than EF-1α of both embryonic and somatic tissues.[111] Consistent with a specialized role for thesaurin in the storage of tRNA in oocytes is the observed rapid degradation of the protein soon after the onset of vitellogenesis.

In order to confirm the association of EF-1α with mRNA in rabbit reticulocyte lysates,[112] an anti-peptide antibody was produced against KAERERGITIDISLW. This sequence is located near the amino-terminus of EF-1α, in a very highly conserved portion of the factor. It was found that these antibodies cross-react with a 62 kDa polypeptide present in rabbit reticulocytes.[113] Previous work has shown that this polypeptide is a constituent of mRNPs from a variety of mammalian cell types.[114] The purified 62 kDa polypeptide did not bind guanine nucleotides but did possess the same binding specificity toward synthetic polynucleotides as EF-1α. These observations hint that EF-1α may be part of a family of RNA binding proteins.

6.4.5. EF-2

Investigation of EF-2 has been dominated by the discovery that it is specifically ADP ribosylated and inactivated by fragment A of diphtheria toxin (DT) or by *Pseudomonas aeruginosa* extoxin A. The basis for these unique reactions was clarified by studies from James Bodley's laboratory[115] where it was shown that EF-2 contains an unusual amino acid derived from a histidine residue and dubbed diphthamide (see Chapter 13).

Recent studies have sought to understand the molecular basis of toxin resistance. A cDNA clone for toxin resistant EF-2 from a mutant hamster cell line was carefully examined.[116] Resistance to either DT or PA is conferred by a G-to-A transition in the first nucleotide of codon 717, which substitutes a glycine residue for an arginine. (Codon 715 encodes the histidine residue that is modified posttranslationally to diphthamide.) Transfection of mouse L cells with this mutant cDNA clone conferred resistance to PA toxin. The mutation in codon 717 created a new MboII restriction site. Kohno and Uchida observed that several independently isolated DT-resistant CHO cell lines showed the same alteration in the MBOII restriction pattern. Apparently, the G-to-A transition is a hot spot for toxin resistance.

The study of toxin resistance points to a central role of His 715 in the function of EF-2. The amino acid sequence (from Asp-712 to Arg-716) which surrounds diphthamide is probably invariant[117, 118] and EF-2 bound to ribosomes is resistant to DT action.[119] Further work using mutant EF-2 cDNAs has demonstrated that substitution of His 715 with Arg or Lys did not confer toxin resistance to the transfected cells.[120] These results imply the absence of His 715 results in inactivation of EF-2 *in vivo*. This inactivation is unlikely to be due to a toxin mediated ADP-ribosylation reaction (since the mutant EF-2 lacks His 715) but rather to the ineffectiveness of the mutant EF-2 in catalyzing translocation of peptidyl tRNA. The results of *in vivo* competition assays between toxin-resistant and functionless (minus His 715) EF-2 led Omura and co-workers to infer that His 715 is probably not essential for factor binding to ribosomes. The implication is that His 715 is indispensable for the translocase activity of EF-2. New impetus for work on EF-2 has been the discovery that the factor can be phosphorylated by a specific Ca^{2+}/calmodulin-dependent protein kinase.[121, 122] The phosphorylated factor is unable to catalyze translocation of peptidyl tRNA from the A to the P site[123] (see Chapter 13).

The cloning of EF-2 genes and cDNAs has revealed another elongation factor with a remarkably conservative evolution.[117] Out of 857 amino acid residues there was an observed 61% identity between hamster and *Dictyostelium discoideum* EF-2 and only two amino acid differences between the rat and hamster factor.[118, 124] The structural similarities between EF-2 and EF-1α are apparently confined to the amino-terminal regions (including the first

160 residues of EF-2) of both factors where the structures responsible for GTP binding and catalysis reside.[117] The structural similarities between EF-2 and EF-G are more extensive and include regions near the C-terminal end of both factors which are important in the interaction of EF-2 with toxins as well as ribosomes (see above). The sites of guanine nucleotide binding and ADP ribosylation predicted by DNA sequences are in agreement with the locations of sites suggested by Nilsson and Nygård on the basis of experiments using limited proteolysis of EF-2.[125]

The hamster EF-2 gene is unusual: many short introns and 13 exons are contained within a modest stretch 5.7 kb.[126] The 5'flanking region contains a TATA sequence and CACCC-like repeats characteristic of housekeeping genes.[127] The later sequence is contained within residues -31 to -92 which has been shown to contain the main promoter activity of the gene by deletion mapping. The $G + C$ content of this promoter region is very high (79%) and it contains many stretches of GC repeats. Unlike EF-1α which possesses many pseudogenes (see above), only a single copy of the EF-2 gene per haploid genome was observed by Southern blot analysis.[116] Another similarity with r-protein and EF-1α genes should also be noted: transcription of the EF-2 gene in hamsters begins with a polypyrimidine sequence (see above).

6.5. ELONGATION FACTOR SYNTHESIS

We know very little about the biosynthesis of elongation factors. EF-3, EF-1β, and EF-1γ synthesis have not been investigated. Gill and Dinus[128] have noted that the EF-2 content of different mammalian tissues is closely correlated with the number of ribosomes, with an average ratio of about 1.2 molecules of factor per ribosome. These observations suggest a coordination between the biosynthesis of EF-2 and ribosomes. The recent availability of EF-2 cDNAs has permitted a more detailed study of EF-2 expression in *Dictyostelium discoideum.*[118]

These studies have demonstrated developmental regulation of EF-2 mRNA and protein. The development of *D. discoideum* leads to the formation of prespore and prestalk cells. Toda and co-workers reported there is about 94 times as much EF-2 mRNA in prespore cells over prestalk cells. This despite finding no significant difference in the amount of total RNA from the two cell types. They suggest that either EF-2 mRNA is much more unstable and selectively degraded in prestalk cells or that transcription of the EF-2 gene is downregulated as prestalk cells differentiate. Of additional interest was the observation that, despite large differences in mRNA levels, the amount of EF-2 in prestalk and prespore cells was comparable. These initial studies on EF-2 synthesis hint that complex mechanisms may operate to achieve a nearly constant stoichiometry of EF-2 to ribosomes.

The mechanisms regulating EF-1α synthesis must also be complex. Over the past two decades there have been numerous reports that EF-1 activity is altered during development, senescence, by hormone treatment, in cultured cells which are growth arrested etc.[4, 19, 129–131] However, these investigations invariably report measurements of EF-1α activity in crude cells extracts. In most cases it is not clear whether the factor itself is actually changing in abundance when growth conditions are altered. Furthermore, these studies provide little indication as to whether changes in *in vivo* peptide elongation rates or ribosome abundance accompany changes in EF-1α activity. Finally, little effort has been made to distinguish EF-1α activity from the activity of the EF-1βγ. For these, as well as other reasons, it is difficult to assess the significance of most reports on changes in EF-1 activity in response to changing physiological circumstances.

There are, however, strong indications that EF-1α expression is subject to regulation at a number of different levels. Addition of serum or epidermal growth factor to quiescent Swiss mouse 3T3 cells in cultures results in the rapid augmentation of synthesis of a limited number of proteins,[132] among which is EF-1α.[133] After 60 min of serum stimulation of 3T3 cells it was reported that EF-1α synthesis, as measured by incorporation of [^{35}S]methionine, increased about sixfold. Experiments with actinomycin D indicated that the altered expression of EF-1α in 3T3 cells was controlled at the posttranscriptional level. Experiments with Friend erythroleukemia cells demonstrated that the expression of EF-1α is subject to translational control.[134] When these cells were grown to stationary phase, approximately 60% of the EF-1α mRNA in cell extracts sedimented at ≤80S in the form of nontranslating messenger ribonucleoprotein particles (mRNPs). The association of EF-1α mRNA with polysomes changed dramatically when stationary phase cells were treated with fresh medium. After 1 h in fresh medium, approximately 90% of the EF-1α mRNA in Friend cells was found in heavy polysomes. Treatment with fresh medium also resulted in 2.6-fold increase in EF-1α synthesis as well as a surprising 40% decrease in the total amount of EF-1α mRNA per cell. The data raised the possibility that the EF-1α accumulated in mRNPs in stationary-phase cells was degraded rather than reutilized for EF-1α synthesis. This suggestion was confirmed by subsequent studies which demonstrated that EF-1α mRNA stability varies with Friend cell growth rate.[135] Although EF-1α mRNA was found to be stable in growing cells ($t_{1/2}$ = 9 h) and stationary cells ($t_{1/2}$ = 24 h), rapid decay ($t_{1/2}$ = 1 h) of the mRNP form of EF-1α mRNA was observed when stationary cells were treated with fresh medium. The suggestion was made that reutilization of the nontranslating EF-1α mRNA by cells would result in overexpression of the factor, a circumstance that might prove inimical to cell growth.

Transcriptional processes also dictate EF-1α expression. Using differential hybridization, Lu and co-workers[136] have prepared cell cycle phase-specific cDNA libraries which reflect phase-specific gene expression of murine Ehrlich ascites cells growing *in vivo*. They identified EF-1α as an S-phase specific clone. The discovery that EF-1α gene transcription appears to be confined to S-phase is noteworthy and certainly demands further investigation. Hoverman and co-workers[97] found, using primer extension and analysis, that the F1 EF-1α gene (see above) corresponds to the housekeeping gene that gives rise to EF-1α present in all somatic cells. The F2 EF-1α gene, whose expression peaks in the pupal stage of *Drosophila* development, is specifically expressed during certain developmental stages, possibly in a tissue specific manner. The relationship of F2 EF-1α to thesaurin (see above) is uncertain at this time. The possibility that there are embryonic forms of EF-1α in other organisms or that there are embryonic forms of other protein synthesis factors needs to be investigated.

Kreig and co-workers have investigated the appearance of the EF-1α mRNA during early development of *Xenopus laevis*.[137] They report that the levels of EF-1α mRNA in the embryo increase about 50-fold in the period between midblastula and gastrula (called the midblastula transition or MBT). No evidence for nontranslating forms of EF-1α were observed, implying that the MBT is characterized by a substantial increase in EF-1α content of embryonic cells. However, earlier investigations have shown that total protein synthesis does not show any significant changes following the MBT.[138] Kreig and co-workers argue that EF-1α does not appear to be a limiting factor for protein synthesis in the early embryo nor does it seem likely that it regulates the expression of specific genes. Rather, they suggest that the production of large amounts of EF-1α at the MBT is part of a general commitment of the *Xenopus* embryo to accumulate large amounts of the translational machinery (large amounts of ribosomal RNA and protein are synthesized during or slightly after the MBT).

There appear to be two active EF-1α genes during the early development of *Xenopus*.[137] From the data on the appearance of EF-1α transcripts at the MBT, it was estimated that EF-1α mRNA is synthesized at a rate of three to five molecules per minute per gene. This rate of transcription is comparable to that of other very active genes, such as globin[139] and the estrogen induced ovalbumin gene.[140] It agrees with the *in vitro* data on the human EF-1α gene (see above) as well as data on F1 EF-1α gene in *Drosophila*.

6.6. CONCLUSIONS

It seems safe to suppose that the detailed molecular mechanisms that govern the elongation process will first be known for prokaryotic systems, viz., *E. coli*. However, there are many questions concerning elongation in eukaryotes that need to be addressed. The role of EF-3 in the elongation process still remains to be fully described. The polypeptide corresponding to EF-3 in nonfungal organisms remains to be discovered. The exact function of EF-1γ in elongation remains to be clarified. We now know that eukaryotic elongation factors, particularly EF-1α and EF-2, contain peculiar posttranslational modifications. The functions of these modifications in protein synthesis needs to be understood. There is little information on the tertiary and higher order structures of eukaryotic elongation factors. Understanding the interactions of the factors with ligands as well as the conformational alterations produced by posttranslational modifications will depend on further knowledge of the higher order structures of elongation factors. If the levels of elongation factor activity are altered during growth or senescence, it should be demonstrated whether these changes result from loss or inactivation of the factor. Do these changes in activity change elongation rates *in vivo*? The *cis* and *trans*-acting structures that regulate transcription of the elongation factors need to be further investigated. Ditto for the structures concerned with posttranscriptional control of factor expression. The mechanisms governing coordination of protein synthesis factors with ribosomes and mRNA are not yet understood. On this score we cannot appeal to *E. coli* for help. As the reader can see there is more than enough work to keep those with an interest in these problems busy for a while longer.

ACKNOWLEDGMENTS

This work was supported in part by a grant from the U.S. Public Health Service (GM 25434). Thanks are due Dr. A. Miseta for aid in the preparation of the figure and Ms. R. Brown for help in typing this manuscript.

REFERENCES

1. Bishop JO and Schweet RS: Participation of two enzymes in the transfer of amino-acyl RNA to reticulocyte ribosomes. *Biochim. Biophys. Acta* 54:617 (1961).
2. Kaziro Y: The role of guanosine 5'-triphosphate in polypeptide chain elongation. *Biochim. Biophys. Acta* 505:95 (1978).
3. Bermek E: Mechanisms in polypeptide chain elongation. *Prog. Nucl. Acid. Res. Mol. Biol.* 21:64 (1978).
4. Moldave K: Eucaryotic protein synthesis. *Annu. Rev. Biochem.* 54:1109 (1985).
5. Dennis P and Bremer H: Differential rate of ribosomal protein synthesis in *Escherichia coli* B/R. *J. Mol. Biol.* 84:407 (1974).
6. Shepherd NS, Churchward G, and Bremer H: Synthesis and function of ribonucleic acid polymerase and ribosomes in *Escherichia coli* B/R after a nutritional shift-up. *J. Bacteriol.* 143:1332 (1980).

7. Haschemeyer AEV: Rates of polypeptide chain assembly *in vivo. Proc. Natl. Acad. Sci. U.S.A.* 62:128 (1969).
8. Fan H and Penman S: Regulation of protein synthesis in mammalian cells. *J. Mol. Biol.* 50:655 (1970).
9. Ledford BE and Davis DF: Analysis of translational parameters in cultured cells. *Biochim. Biophys. Acta* 519:204 (1978).
10. Haschemeyer AEV: Kinetic of protein synthesis in higher organisms *in vivo. Trends Biochem. Sci.* 1:133 (1976).
11. Palmiter RD: Regulation of protein synthesis in the chick oviduct L1. Modulation of polypeptide elongation and initiation rates by estrogen and progesterone. *J. Biol. Chem.* 247:6770 (1972).
12. Goustin AS and Wilt FH: Direct measurement of histone peptide elongation rate in cleaving sea urchin embryos. *Biochim. Biophys. Acta* 699:22 (1982).
13. Nielsen JBK, Plant PW, and Haschemeyer AEV: Control of protein synthesis in temperature acclimation. 11. Correlation of elongation factor 1 activity with elongation rate *in vivo. Physiol. Zool.* 50:22 (1977).
14. Nielsen JBK, Plant PW, and Haschemeyer AEV: Polypeptide elongation factor 1 and control of elongation rate in rat liver *in vivo. Nature* 264:804 (1976).
15. Nielsen PJ and McConkey EH: Evidence for control of protein synthesis in HeLa cells via the elongation rate. *J. Cell Physiol.* 104:269 (1980).
16. Simon E: Effect of acclimation temperature on the elongation step of protein synthesis in different organs of the rainbow trout. *J. Comp. Physiol.* 157:201 (1987).
17. Orlowski M and Sypherd P: Regulation of translation rate during morphogenesis in the fungus *Mucor. Biochemistry* 17:569 (1978).
18. Khasigov PZ and Nikolaev AY: Age-related changes in the rate of peptide chain elongation. *Biochem. Int.* 15:1171 (1987).
19. Webster GC and Webster SL: Specific disappearance of translatable messenger RNA for elongation factor 1 in aging *Drosophila melanogaster. Mech. Ageing Dev.* 24:335 (1984).
20. Castañeda M, Vargas R, and Galvan SC: Stagewise decline in the activity of brain protein synthesis factors and relationship between this decline and longevity in two rodent species. *Mech. Ageing Dev.* 38:197 (1986).
21. Gay DA, Yen TJ, Lau JTY, and Cleveland DW: Sequences that confer β-tubulin autoregulation through modified mRNA stability reside within exon 1 of a β-tubulin mRNA. *Cell* 50:671 (1987).
22. Yen TJ, Machlin PS, and Cleveland DW: Autoregulated instability of β-tubulin mRNAs by recognition of the nascent amino terminus of β-tubulin. *Nature* 334:580 (1988).
23. Gay DA, Sisodia SS, and Cleveland DW: Autoregulatory control of β-tubulin mRNA stability is linked to translational elongation. *Proc. Natl. Acad. Sci. U.S.A.* 86:5763 (1989).
24. Muller EW and Kuhn LC: A stem-loop in the 3′ untranslated region mediates iron-dependent regulation of transferrin mRNA stability in the cytoplasm. *Cell* 53:815 (1988).
25. Sive HL, Heintz N, and Roeder RG: Regulation of human histone gene expression during the HeLa cell cycle requires protein synthesis. *Mol. Cell. Biol.* 4:2723 (1984).

26. Linial M, Gunderson N, and Groudine M: Enhanced transcription of c-myc in bursal lymphoma cells requires continuous protein synthesis. *Science* 230:1126 (1985).

27. Wolin SL and Walter P: Ribosome pausing and stacking during translation of a eukaryotic mRNA. *EMBO J.* 7:3559 (1988).

28. Roobol K and Möller W: The role of guanine nucleotides in the interaction between aminoacyl-tRNA and elongation factor 1 of *Artemia salina*. *Eur. J. Biochem.* 90:471 (1978).

29. Carvalho MG, Carvalho JF, and Merrick W: Biological characterization of the various forms of elongation factor 1 from rabbit reticulocytes. *Arch. Biochem. Biophys.* 234:603 (1984).

30. Crechet J-B and Parmeggiani A: Characterization of the elongation factors from calf brain. 2. Functional properties of EF-1α, the action of physiological ligands and kirromycin. *Eur. J. Biochem.* 161:647 (1986).

31. Wolf H, Chinali G, and Parmeggiani A: Kirromycin, an inhibitor of protein synthesis that acts on elongation factor Tu. *Proc. Natl. Acad. Sci. U.S.A.* 71:4910 (1974).

32. Slobin LI: The binding of eucaryotic elongation factor Tu to nucleic acids. *J. Biol. Chem.* 258:4895 (1983).

33. Crechet J-B and Parmeggiani A: Characterization of the elongation factors from calf brain. 3. Properties of the GTPase activity of EF-1α and mode of action of kirromycin. *Eur. J. Biochem.* 161:655 (1986).

34. Janssen GMC and Möller W: Kinetic studies on the role of elongation factors 1 beta and 1 gamma in protein synthesis. *J. Biol. Chem.* 263:1773 (1988).

35. Motoyoshi K and Iwasaki K: Resolution of the polypeptide chain elongation factor-1beta gamma into subunits and some properties of the subunits. *J. Biochem.* 82:703 (1977).

36. Janssen GMC, Maessen, GDF, Amons R, and Möller W: Phosphorylation of elongation factor 1 beta by an endogenous kinase affects its catalytic nucleotide exchange activity. *J. Biol. Chem.* 263:11063 (1988).

37. Bellé R, Derancourt J, Poulhe R, Capony J-P, Ozon R, and Mulner-Lorillon O: A purified complex from *Xenopus* oocytes contains a P47 protein, an *in vivo* substrate of MPF, and a P30 protein respectively homologous to elongation factors EF-1 gamma and EF-1β. *FEBS Lett.* 255:101 (1989).

38. Wasserman WJ, Richter JD, and Smith LD: Protein synthesis during maturation promoting factor-and progesterone-induced maturation in *Xenopus* oocytes. *Dev. Biol.* 89:152 (1982).

39. Skogerson L and Wakatama E: A ribosome-dependent GTPase from yeast distinct from elongation factor 2. *Proc. Natl. Acad. Sci. U.S.A.* 73:73 (1976).

40. Dasmahapatra B and Chakraburrty K: Protein synthesis in yeast. 1. Purification and properties of elongation factor 3 from *Saccharomyces cerevisiae*. *J. Biol. Chem.* 256:9999 (1981).

41. Uritani M and Miyazaki M: Characterization of the ATPase and GTPase activities of elongation factor 3 (EF-3) purified from yeasts. *J. Biochem.* 103:522 (1988).

42. Qin SL, Moldave K, and McLaughlin CS: Isolation of the yeast gene encoding elongation factor 3 for protein synthesis. *J. Biol. Chem.* 262:7802 (1987).

43. Hutchinson JS, Feinberg B, Rothwell TC, and Moldave K: Monoclonal antibody specific for yeast elongation factor 3. *Biochemistry* 23:3055 (1984).

44. Herrera F, Martinez JA, Moreno N, Sadnik I, McLaughlin CS, Feinberg B, and Moldave K: Identification of an altered elongation factor in temperature-sensitive mutant as 7-14 of *Saccharomyces cerevisiae. J. Biol. Chem.* 259:14347 (1984).

45. Kamath A and Chakraburrty K: Protein synthesis in yeast. Identification of an altered elongation factor in thermolabile mutants of *Saccharomyces cerevisiae. J. Biol. Chem.* 261:12593 (1986a).

46. Kamath A and Chakraburrty K: Protein synthesis in yeast. Purification of elongation factor 3 from temperature-sensitive mutant 13-06 of the yeast *Saccharomyces cerevisiae. J. Biol. Chem.* 261:12596 (1986b).

47. Kamath A and Chakraburrty K: Role of yeast elongation factor 3 in the elongation cycle. *J. Biol. Chem.* 264:15423 (1989).

48. Lake JK: Aminoacyl-tRNA binding at the recognition site is the first step of the elongation cycle of protein synthesis. *Proc. Natl. Acad. Sci. U.S.A.* 74:1903 (1977).

49. Thompson RC: EFTu provides an internal kinetic standard for translational control. *Trends Biochem. Sci.* 13:91 (1988).

50. van de Klundert JAM, de Turk E, Borman AH, van der Meide PH, and Bosch L: Isolation and characterization of a mocimycin resistant mutant of *Escherichia coli* with an altered elongation factor EF-Tu. *FEBS Lett.* 81:303 (1977).

51. Tapio S and Kurland CG: Mutant EF-Tu increases missense mutations *in vitro. Mol. Gen. Genet.* 205:186 (1986).

52. Vijgenboom E, Vink T, Kraal B, and Bosch L: Mutants of the elongation factor EF-Tu, a new class of nonsense suppressors. *EMBO J.* 4:1049 (1985).

53. Sandbaken MG and Culbertson MR: Mutations in elongation factor EF-1α affect the frequency of frameshifting and amino acid misincorporation in *Saccharomyces cerevisiae. Genetics* 120:923 (1988).

54. Moazed D, Robertson JM, and Noller HF: Interaction of elongation factors EG-G and EF-Tu with a conserved loop in 23S RNA. *Nature* 334:362 (1988).

55. Endo Y and Wool I: The site of action of alpha-sarcin on eucaryotic ribosomes. *J. Biol. Chem.* 257:9054 (1982).

56. Endo Y, Mitsui K, Motizuki M, and Tsurgi K: The mechanism of action of ricin and related toxin lectins on eucaryotic ribosomes. *J. Biol. Chem.* 262:5908 (1987).

57. Ovchinnikov LP, Spirin AS, Erni B, and Staehelin T: RNA-binding proteins of rabbit reticulocytes contain the two elongation factors and some initiation factors of translation. *FEBS Lett.* 88:21 (1978).

58. Kolb AJ, Redfield B, Twardowski T, and Weissbach H: Binding of reticulocyte elongation factor 1 to ribosomes and nucleic acids. *Biochim. Biophys. Acta* 519:398 (1978).

59. Uchiumi T and Ogata K: Cross-linking study on the localization of the binding site for elongation factor 1 alpha on rat liver ribosomes. *J. Biol. Chem.* 262:9668 (1986).

60. Möller W, Slobin LI, Amons R, and Richter D: Isolation and characterization of two acidic proteins of 60s ribosomes from *Artemia salina* cysts. *Proc. Natl. Acad. Sci. U.S.A.* 72:4744 (1975).

61. Slobin LI: The inhibition of eucaryotic protein synthesis by procaryotic elongation factor Tu. *Biochem. Biophys. Res. Commun.* 101:1388 (1981).
62. De Vendettis E, Masullo M, and Bocchini V: The elongation factor G carries a catalytic site for GTP hydrolysis, which is revealed by using 2-propanol in the absence of ribosomes. *J. Biol. Chem.* 261:4445 (1986).
63. Nygård O and Nilsson L: Characterization of the ribosomal properties required for formation of a GTPase active complex with eukaryotic elongation factor 2. *Eur. J. Biochem.* 179:603 (1989).
64. Tanaka M, Iwasaki K, and Kaziro Y: Translocation reaction promoted by polypeptide chain elongation factor-2 from pig liver. *J. Biochem.* 82:1035 (1977).
65. Nilsson L and Nygård O: The mechanism of the protein-synthesis elongation cycle in eukaryotes. *Eur. J. Biochem.* 161:111 (1986).
66. Nilsson L and Nygård O: Structural and functional studies on the interactions of eukaryotic elongation factor EF-2 with GTP and ribosomes. *Eur. J. Biochem.* 171:293 (1988).
67. Nombela C and Ochoa S: Conformational control of the interaction of eucaryotic elongation factors EF-1 and EF-2 with ribosomes. *Proc. Natl. Acad. Sci. U.S.A.* 70:3556 (1973).
68. Modellell J, Carber B, and Vazquez D: The interaction of elongation factor G with N-acetylphenylalanine transfer RNA·ribosome complexes. *Proc. Natl. Acad. Sci. U.S.A.* 70:3561 (1973).
69. Uchiumi T, Kikuchi M, Terao K, Iwasaki K, and Ogata K: Cross-linking of elongation factor 2 to rat liver ribosomal proteins by 2-iminothiolane. *J. Biol. Chem.* 261:9663 (1986).
70. Nygård O, Nilsson L, and Westerman P: Characterization of the ribosomal binding site for eukaryotic elongation factor 2 by chemical cross-linking. *Biochim. Biophys. Acta* 910:245 (1987).
71. Nygård O and Nilsson O: The ribosomal binding site for eukaryotic elongation factor EF-2 contains 5S ribosomal RNA. *Biochim. Biophys. Acta* 908:46 (1987).
72. Slobin LI: The role of eucaryotic elongation factor Tu in protein synthesis. *Eur. J. Biochem.* 110:555 (1980).
73. Lenstra JA and Bloemendal H: The major proteins from HeLa cells. *Eur. J. Biochem.* 130:419 (1983).
74. Thiele D, Cottrelle P, Iborra F, Buhler JM, Sentenac A, and Fromageot P: Elongation factor 1alpha from *Saccharomyces cerevisiae*. *J. Biol. Chem.* 260:3084 (1985).
75. Brands JHGM, Maassen JA, Van Hemert FJ, Amons R, and Möller W: The primary structure of the alpha subunit of human elongation factor 1. *Eur. J. Biochem.* 155:167 (1986).
76. Jurnak F: Structure of the GDP domain of EF-Tu and location of the amino acids homologous to ras oncogene proteins. *Science* 230:32 (1985).
77. Slobin LI, Clark RV, and Olson MOJ: Limited cleavage of eucaryotic elongation factor Tu by trypsin: alignment of the tryptic fragments and effect of nucleic acids on the enzymatic reaction. *Biochemistry* 22:1911 (1983).
78. Möller W, Schipper A, and Amons R: A conserved amino acid sequence around Arg[68] of *Artemia* elongation factor 1α is involved in the binding of guanine nucleotides and aminoacyl transfer RNAs. *Biochemie* 69:983 (1987).

79. Miyazaki M, Uritani M, Fujimura K, Yamakatsu H, Kageyama T, and Takahashi K: Peptide elongation factor 1 from yeasts: purification and biochemical characterization of peptide elongation factors 1alpha and 1beta(gamma) from *Saccharomyces carlsbergensis* and *Schizosaccharomyces pombe. J. Biochem.* 103:508 (1988).

80. Möller W and Amons R: Phosphate-binding sequences in nucleotide-binding proteins. *FEBS Lett.* 186:1 (1985).

81. Dever TE, Glynias MJ, and Merrick W: GTP-binding domain: three consensus sequence elements with distinct spacing. *Proc. Natl. Acad. Sci. U.S.A.* 84:1814 (1987).

82. Hiatt WR, Garcia R, Merrick WC, and Sypherd PS: Methylation of elongation factor 1alpha from the fungus *Mucor. Proc. Natl. Acad. Sci. U.S.A.* 79:3433 (1982).

83. Fonzi WA, Katayama C, Leathers T, and Sypherd PS: Regulation of protein synthesis factor EF-1α in *Mucor racemosus. Mol. Cell Biol.* 5:1100 (1985).

84. Sherman M and Sypherd P: Role of lysine methylation in the activities of EF-1α. *Arch. Biochem. Biophys.* 274:371 (1989).

85. Amons R, Pluijms W, Roobol K, and Möller W: Sequence homology between EF-1alpha, the alpha-chain of elongation factor 1 from *Artemia salina* and elongation factor EF-Tu from *Escherichia coli. FEBS Lett.* 153:37 (1983).

86. van Hemert FJ, Amons R, Pluijms WJM, van Ormondt H, and Möller W: The primary structure of elongation factor EF-1α from the brine shrimp *Artemia. EMBO J.* 3:1109 (1984).

87. Coppard NJ, Clark BFC, and Cramer F: Methylation of elongation factor 1alpha in mouse 3T3B and 3T3B/SV40 cells. *FEBS Lett.* 164:330 (1983).

88. Arai K, Clark BFC, Duffy L, Jones MD, Kaziro Y, Laursen RA, L'Italien J, Miller DL, Nagarkatti S, Nakamura S, Nielsen KM, Petersen TE, Takahashi K, and Wade M: Primary structure of elongation factor Tu from *Escherichia coli. Proc. Natl. Acad. Sci. U.S.A.* 77:1326 (1980).

89. Dever TE, Costello CE, Owens CL, and Merrick WC: Location of seven post-translational modifications in rabbit EF-1α including dimethyllysine, trimethyllysine and glycerylphosphorylethanolamine. *J. Biol. Chem.* 264:20518 (1989).

90. Whitehearst SW, Shenbagamurthi P, Chen L, Cotter RJ, and Hart GW: Murine EF-1α is posttranslationally modified by novel amide-linked ethanolamine-phosphoglycerol moieties. *J. Biol. Chem.* 264:14334 (1989).

91. Rosenberry TL, Krall JA, Dever TE, Hass R, Louvard D, and Merrick WC: Biosynthetic incorporation of [³H]ethanolamine in protein synthesis elongation factor 1α reveals a new post translational protein modification. *J. Biol. Chem.* 264:7096 (1989).

92. Whitehearst SW and Hart GW: Specific incorporation of [³H]ethanolamine into protein in an *in vitro* system. *J. Cell. Biol.* 107:854a (1988).

93. Cottrelle P, Cool M, Thuriaux P, Price VL, Thiele D, Buhler JM, and Fromageot P: Either one of the two yeast EF-1alpha genes is required for cell viability. *Curr. Gen.* 9:693 (1985).

94. Huet J, Cottrelle P, Cool M, Vignais ML, Thiele D, Marck C, Buhler JM, Sentenac A, and Fromageot P: A general upstream binding factor for genes of the yeast translational apparatus. *EMBO J.* 4:3539 (1985).

95. Linz JE and Sypherd PS: Expression of three genes for elongation factor 1α during morphogenesis of *Mucor racemosus*. *Mol. Cell. Biol.* 7:1925 (1987).

96. Lenstra JA, Vliet AV, Arnberg AC, Van Hemert FJ, and Möller W: Genes coding for the elongation factor EF-1alpha in *Artemia*. *Eur. J. Biochem.* 155:475 (1986).

97. Hovemann B, Richter S, Walldorf U, and Cziepluch C: Two genes encode related cytoplasmic elongation factors 1alpha (EF-1alpha) in *Drosophila melanogaster* with continuous and stage specific expression. *Nucleic Acids. Res.* 16:3175 (1988).

98. Uetsuki T, Naito A, Nagata S, and Kaziro Y: Isolation and characterization of the human chromosomal gene for polypeptide chain elongation factor-1α. *J. Biol. Chem.* 264:5791 (1989).

99. Opdenakker G, Cabeza-Arvelaiz Y, Fiten P, Dijkmans R, Van Damme J, Volckaert G, Billiau A, Van Elsen A, and Cassiman J-J: Human elongation factor 1α a polymorphic and conserved multigene family with multiple chromosomal localizations. *Hum. Genet.* 75:339 (1987).

100. Dudov KP and Perry RP: Properties of a mouse ribosomal protein promoter. *Proc. Natl. Acad. Sci. U.S.A.* 83:8545 (1986).

101. Maessen GDF, Amons R, Maassen JA, and Möller W: Primary structure of elongation factor 1beta from *Artemia*. *FEBS Lett.* 208:77 (1986).

102. Maessen GDF, Amons R, Zeelen JP, and Moller W: Primary structure of elongation factor 1gamma from *Artemia*. *FEBS Lett.* 223:181 (1987).

103. Caskey T, Leder P, Moldave K, and Schlessinger D: Translation: its mechanism and control. *Science* 176:195 (1972).

104. Slobin LI and Möller W: Changes in the form of elongation factor 1 during development of *Artemia salina*. *Nature* 258:452 (1975).

105. Murakami K and Miyamoto K: Polypeptide elongation factors of the developing chick brain. *J. Neurochem.* 38:1315 (1982).

106. Sacchi GA, Zocchi G, and Cocucci S: Changes in form of elongation factor 1 during germination of wheat seeds. *Eur. J. Biochem.* 139:1 (1984).

107. Motorin YA, Wolfson AD, Orlovsky AI, and Gladilin KL: Mammalian valyl-tRNA synthetase forms a complex with the first elongation factor. *FEBS Lett.* 238:262 (1988).

108. Janssen GMC and Möller W: Elongation factor 1 beta gamma from *Artemia*: purification and properties of its subunits. *Eur. J. Biochem.* 171:119 (1988).

109. Picard B, le Maire M, Wegnez M, and Denis H: Biochemical research on oogenesis. Composition of the 42S storage particles of *Xenopus laevis* oocytes. *Eur. J. Biochem.* 109:359 (1980).

110. Mattaj IW, Coppard NJ, Brown RS, Clark BFC, and De Robertis EM: 42S P48—the most abundant protein in previtellogenic *Xenopus* oocytes—resembles elongation factor-1α structurally and functionally. *EMBO J.* 6:2409 (1987).

111. Viel A, Dje MK, Mazabrand A, Denis H, and le Maire M: Thesaurin A, the major protein of previtellogenic *Xenopus* oocytes, present in 42S particles, is homologous to elongation factor-1α. *FEBS Lett.* 223:232 (1987).

112. Greenberg JR and Slobin LI: Eukaryotic elongation factor Tu is present in mRNA-protein complexes. *FEBS Lett.* 224:54 (1987).

113. Slobin LI and Greenberg JR: Purification and properties of a protein component of messenger ribonucleoprotein particles that shares a common epitope with eucaryotic elongation factor Tu. *Eur. J. Biochem.* 173:305 (1988).

114. Greenberg JR and Carroll E III: Reconstitution of functional mRNA-protein complexes in a rabbit reticulocyte cell-free translation system. *Mol. Cell. Biol.* 5:342 (1985).

115. Van Ness BG, Howard JB, and Bodley JW: ADP-ribosylation of elongation factor 2 by diphtheria toxin. *J. Biol. Chem.* 255:10710 (1980).

116. Kohno K and Uchida T: Highly frequent single amino acid substitution in mammalian elongation factor 2 (EF-2) results in expression of resistance to EF-2-ADP-ribosylating toxins. *J. Biol. Chem.* 262:12298 (1987).

117. Kohno K, Uchida T, Ohkubo H, Nakanishi S, Nakanishi T, Toshikazu F, Ohtsuka E, Ikehara M, and Ikada Y: Amino acid sequence of mammalian elongation factor 2 deduced from the cDNA sequence: homology with GTP binding proteins. *Proc. Natl. Acad. Sci. U.S.A.* 83:4978 (1986).

118. Toda K, Tasaka M, Mashima K, Kohno K, Uchida T, and Takeuchi I: Structure and expression of elongation factor 2 gene during development of *Dictyostelium discoideum*. *J. Biol. Chem.* 264:15489 (1989).

119. Gill DM, Pappenheimer HMJ, and Baseman JB: Studies on transferase L1 using diphtheria toxin. *Cold Spring Harbor Symp. Quant. Biol.* 34:595 (1969).

120. Omura F, Kohno K, and Uchida T: The histidine residue of codon 715 is essential for function of elongation factor 2. *Eur. J. Biochem.* 180:1 (1989).

121. Nairn AC, Bhagat B, and Palfrey HC: Identification of calmodulin-dependent kinase L11 and its major M_r 100,000 substrate in mammalian tissues. *Proc. Natl. Acad. Sci. U.S.A.* 82:7939 (1985).

122. Nairn AC and Palfrey HC: Identification of the major M_r 100,000 substrate for calmodulin-dependent protein kinase L11 in mammalian cells as elongation factor 2. *J. Biol. Chem.* 262:17299 (1987).

123. Ryazanov AG and Davydova EK: Mechanism of elongation factor 2 (EF-2) inactivation upon phosphorylation. *FEBS Lett.* 251:187 (1989).

124. Oleinikov AV, Jokhadze GG, and Alakhov YB: Primary structure of rat liver elongation factor 2 deduced from the cDNA sequence. *FEBS Lett.* 248:131 (1989).

125. Nilsson L and Nygård O: Localization of the sites of ADP-ribosylation and GTP binding in the eucaryotic elongation factor 2. *Eur. J. Biochem.* 148:299 (1985).

126. Nakanishi T, Kohno K, Ishiura M, Ohashi H, and Uchida T: Complete nucleotide sequence and characterization of the 5'-flanking region of mammalian elongation factor 2 gene. *J. Biol. Chem.* 263:6384 (1988).

127. Lawn RM, Efstratiadis A, O'Connell C, and Maniatis T: The Nucleotide Sequence of the Human β-Globin Gene. *Cell* 21:647 (1980).

128. Gill DM and Dinius LL: The elongation factor 2 content of mammalian cells. *J. Biol. Chem.* 248:654 (1973).

129. Fisher I, Arfin SM, and Moldave K: Regulation of translation. Analysis of intermediary reactions in protein synthesis in exponentially growing and stationary phase Chinese hamster ovary cells in culture. *Biochemistry* 19:1417 (1980).

130. Vargas R and Castañeda M: Role of elongation factor 1 in the translational control of rodent brain protein synthesis. *J. Neurochem.* 37:687 (1981).

131. Cavallius J, Rattan SIS, and Clark BFC: Changes in activity and amount of active elongation factor 1 alpha in aging and immortal human fibroblast cultures. *Exp. Gerontol.* 21:149 (1986).

132. Thomas G, Thomas G, and Luther H: Transcriptional and translational control of cytoplasmic proteins after serum stimulation of quiescent Swiss 3T3 cells. *Proc. Natl. Acad. Sci. U.S.A.* 78:5712 (1981).

133. Thomas G and Thomas G: Translational control of mRNA expression during the early mitogenic response in Swiss mouse 3T3 cells: identification of specific proteins. *J. Cell. Biol.* 103:2137 (1986).

134. Rao TR and Slobin LI: Regulation of the utilization of mRNA for eukaryotic elongation factor Tu in Friend erythroleukemia cells. *Mol. Cell. Biol.* 7:687 (1987).

135. Rao TR and Slobin LI: The stability of mRNA for eucaryotic elongation factor Tu in Friend erythroleukemia cells varies with growth rate. *Mol. Cell. Biol.* 8:1085 (1988).

136. Lu X, Kopun M, and Werner D: Cell cycle phase-specific cDNA libraries reflecting phase-specific gene expression of Ehrlich ascites cells growing *in vivo. Exp. Cell Res.* 174:199 (1988).

137. Krieg PA, Varnym SM, Wormington WM, and Melton DA: The mRNA encoding elongation factor-1α (EF-1α) is a major transcript at the midblastula transition in *Xenopus. Dev. Biol.* 133:93 (1989).

138. Woodland HR: Changes in polysome content of developing *Xenopus laevis* embryos. *Dev. Biol.* 40:90 (1974).

139. Hunt JA: Rate of synthesis and half-life of globin messenger ribonucleic acid. *Biochem. J.* 138:499 (1974).

140. Palmiter RD: Quantitation of the parameters that determine the rate of ovalbumin synthesis. *Cell* 4:189 (1975).

141. Nagata S, Nagashima K, Tsunetsugu-Yokota Y, Fujimura K, Miyazaki M, and Kaziro Y: Polypeptide chain elongation factor 1alpha (EF-1alpha) from yeast: nucleotide sequence of one of the two genes for EF-1alpha from *Saccharomyces cerevisiae. EMBO J.* 3:1825 (1984).

142. Cottrelle P, Thiele D, Price VL, Memet S, Micouin JY, Marck C, Buhler JM, Sentenac A, and Fromageot P: Cloning, nucleotide sequence, and expression of one of two genes coding for yeast elongation factor 1alpha. *J. Biol. Chem.* 260:3090 (1985).

143. Schirmaier F and Philippsen P: Identification of two genes coding for the translation elongation factor EF-1alpha of *S. cerevisiae. EMBO J.* 3:3311 (1984).

144. Linz JE, Katayama C, and Sypherd PS: Three genes for the elongation factor EF-1alpha in *Mucor racemosus. Mol. Cell. Biol.* 6:593 (1986).

145. Sundstrom P, Lira LM, Choi D, Linz JE, and Sypherd PS: Sequence analysis of the EF-1alpha gene family of *Mucor racemosus. Nucleic Acids Res.* 15:9997 (1987).

146. Linz JE, Lira LM, and Sypherd PS: The primary structure and functional domains of an elongation factor EF-1α from *Mucor racemosus. J. Biol. Chem.* 261:15022 (1986b).

147. Pokalsky AR, Hiatt WR, Ridge N, Rasmussen R, Houck CM, and Shewmaker CK: Structure and expression of elongation factor 1 alpha in tomato. *Nucleic Acids Res.* 17:4661 (1989).
148. Lu X and Werner D: The complete cDNA sequence of mouse elongation factor 1 alpha (EF 1 alpha) mRNA. *Nucleic Acids Res.* 17:442 (1989).

Chapter

7

Peptide Chain Termination

Rosaura P. C. Valle and Anne-Lise Haenni
Institut Jacques Monod
Paris, France

7.1. INTRODUCTION

Peptide chain termination is considered as the last step in protein bio-synthesis. It is a complex process during which the various elements of the translation machinery that were coordinately engaged in protein synthesis fall apart. This includes the release of the terminated peptide chain from the peptidyl-tRNA, and release of the tRNA from the ribosome; generally also considered as part of the termination process is release of the mRNA from the ribosome, and dissociation of the ribosomal subunits. This series of events is brought about on the ribosome by the concerted action of specific termination factors, of GTP and of a termination codon on the mRNA.

The era during which the study of termination was in vogue (the 1960s and 1970s) was highlighted by the isolation and characterization of the termination or release factors (RF). Shortly thereafter, interest in this step of protein synthesis dwindled, in particular in connection with eukaryotes. A new surge of interest in this field is now gaining momentum, as certified by this as well as other chapters of this book. Interestingly, it is not the termination step per se that is attracting attention, but rather the strategies that cells have elaborated to circumvent termination as a means of regulating protein synthesis.

Centering on higher eukaryotes (but excluding cell organelles), this review first considers what is known of the basic process of termination. It then focuses on regulation at the level of termination, and on the elements that modulate this process.

Other review articles that deal with termination clearly reflect the oscillatory interest that this step has attracted. Whereas the termination process was reviewed several years ago[1] and only rather briefly touched on in more recent years,[2] there is a cascade of recent review articles that deal with the various mechanisms involved in the regulation of protein synthesis at the level of termination.[3-8]

7.2. REASONS TO TERMINATE: THE BASIC PROCESS

Three codons specify termination: UAG (amber), UGA (opal), and UAA (ochre). Whereas three protein factors have been isolated from *Escherichia coli,* namely, RF1 and RF2 with different codon specificities and RF3 that stimulates the termination process, a single RF has been purified from rabbit reticulocytes (reviewed in Reference 1).

The original assay which led to the isolation of the latter RF is very similar to the one used previously to study peptide chain termination in *E. coli.* It is based on the release of formylmethionine from the fMet-tRNA·ribosome complex *in vitro.* In the reticulocyte system, release is directed by one of the tetranucleotides UAGA, UGAA or UAAA in the presence of GTP and RF. This single factor thus recognizes all three termination codons. Its size is about 105 kDa, and it is composed of two 55 kDa subunits.[9]

After translation of the codon corresponding to the C-terminal amino acid of the polypeptide chain, the peptidyl-tRNA is translocated from the acceptor (A) to the peptidyl (P) site on the ribosome such that the termination codon now occupies the A site. The termination process is mediated by RF and GTP, the latter facilitating binding of the factor to the ribosome. The non-hydrolyzable analog GDPCP, but not GDP, can replace GTP for the binding of RF to the ribosome. Binding of RF with GTP leads to the unmasking of the ribosome-dependent GTPase activity of RF that is enhanced by the presence of a termination codon. As a consequence of GTP hydrolysis to GDP, the RF is dissociated from the ribosome. Ultimately, the ribosome is released from the mRNA and is dissociated into its subunits.

The central event of termination remains the release of the peptidyl moiety from the peptidyl-tRNA. This reaction is catalyzed by the peptidyltransferase contained in the 50S ribosomal subunit. However this event does not appear to be directly linked to GTP hydrolysis. Indeed, antibiotics such as sparsomycin that specifically act on the peptidyltransferase center inhibit release of the peptide chain without interfering with GTP hydrolysis.

7.3. OTHER REASONS TO "TERMINATE"

Although the presence of a termination codon in an mRNA represents the main reason for which a ribosome terminates protein synthesis, other reasons could account for apparent "termination". For example particular secondary and/or tertiary structures in the mRNA, sequences in the mRNA rich in rare codons, or lack of a given aminoacyl-tRNA could lead to stalling of ribosomes. Such ribosomes still carry the nascent peptidyl-tRNA chain, and thus may resume elongation. Alternatively, the peptidyl-tRNA may fall off the ribosome, and it will then be liberated from its tRNA by the peptidyl-tRNA hydrolase. In either case, shortened products are observed upon analysis of the polypeptides synthesized, although no true termination event has occurred. These phenomena are generally designated premature termination.

Although *in vivo* the phenomenon of premature termination has been examined in some detail in prokaryotes (reviewed in Reference 10), it has so far received little attention in eukaryotes.

7.4. REASONS FOR NOT STOPPING

Whereas surprisingly little progress has been made during the past decade in elucidating the termination process itself, the situation is different when it comes to the study of how termination is avoided. This latter phenomenon, known as suppression of termination, has expanded rapidly with the demonstration that several cases of suppression occur among higher eukaryotes, in particular in viral mRNAs, and with the demonstration that this strategy allows two (or more) different proteins with different functions to be produced from a unique initiation event.

Basically, suppression of termination can be achieved either by direct recognition of a termination codon by an aminoacyl-tRNA, or by a shift in reading frame. These two situations are considered below.

7.4.1. Recognition of Termination Codons by tRNAs

This is the phenomenon whereby an aminoacyl-tRNA interacts with a termination codon (also known as nonsense codon), inserting its aminoacyl residue into the growing peptide chain which is now elongated until the translation machinery encounters the next in-frame termination codon. The tRNA endowed with this property is termed nonsense suppressor tRNA, and the protein resulting from suppression is termed "readthrough" protein.

In higher eukaryotes, two types of naturally occurring (*i.e.,* from non-mutagenized cells) suppressor tRNAs have been characterized that can be distinguished from one another depending only on whether they misread the termination codon or whether they exclusively recognize it.

tRNAs that misread termination codons—tRNAs belonging to this category correspond to major cytoplasmic isoaccepting species presumably involved in normal protein synthesis; their anticodon is thus not complementary to the nonsense codon and interaction involves one or two mismatches (reviewed in Reference 6).

All the suppressor tRNAs of this category characterized to date recognize the amber codon. They are the tyrosine, the glutamine, and the leucine tRNAs. In all cases their amber suppression activity has been tested by translating the tobacco mosaic virus (TMV) RNA genome *in vitro*. TMV RNA contains a long open reading frame (ORF) interrupted by a UAG codon. Termination at the amber codon yields a 126 kDa protein, whereas readthrough yields a 183 kDa protein. In some cases, suppression activity has also been shown by coinjecting TMV RNA and the relevant tRNA into *Xenopus laevis* oocytes[11, 12] or by translating the beet necrotic yellow vein virus RNA *in vitro*.[13]

Two closely related tyrosine tRNAs bearing a 5'-GψA-3' anticodon have thus been isolated from uninfected tobacco leaves and shown to suppress the amber codon of TMV RNA *in vitro*. Since both the 126 kDa and the 183 kDa proteins are observed in TMV-infected tobacco protoplasts, the two tRNA^{Tyr} are thought to be responsible for the *in vivo* suppression of the viral UAG codon.[14] Suppressor tRNA$^{Tyr}_{G\psi A}$ have also been isolated from *Drosophila melanogaster*,[11] wheat germ, wheat leaves,[15] and lupin seeds.[12] Interestingly, the tyrosine-accepting species with a 5'-QψA-3' anticodon (Q is a hyper-modified G base) from either wheat germ, lupin seeds, or *D. melanogaster* is incapable of suppressing the amber codon of TMV RNA, suggesting a correlation between the level of base modification in the anticodon and the efficiency of suppression.

Two leucine-accepting tRNA species have been isolated from calf liver that are active in amber suppression *in vitro;* they are the major tRNA$^{Leu}_{CAA}$ and tRNA$^{Leu}_{CAG}$.[16] In addition, the calf liver tRNA^{Tyr} species that possess a 5'-Q*ψA-3' anticodon (Q* is a further modification of Q) are devoid of suppression activity.

Finally, a minor tRNA^{Gln} species bearing a 5'-UmUG-3' anticodon has been isolated from mouse cells; it stimulates readthrough in the TMV RNA *in vitro* suppression assay.[17] This tRNA possibly recognizes the suppressible amber codon of the Moloney murine leukemia virus (Mo-MuLV) RNA genome *in vivo*.

In Mo-MuLV as in a few other retroviruses, the *gag* and the *pol* genes are in the same reading frame, the two ORFs being separated by a UAG codon. Expression of the *pol* ORF occurs via synthesis of a GAG-POL fusion protein that arises from suppression of the UAG codon. Indeed, addition of yeast amber suppressor tRNA to an *in vitro* assay enhances the level of the fusion protein.[18] Direct demonstration for the suppression event *in vivo* has come from the N-terminal sequencing of the viral protease responsible for the maturation of the GAG precursor protein. This protease is encoded by the region of the genome spanning the end of the *gag* gene and the beginning of the *pol* gene. Its fifth amino acid, a glutamine, is inserted in response to the UAG codon.[19] A striking correlation therefore seems to exist between these data and the presence of the tRNA$_{UmUG}^{Gln}$ in mouse cells.

Reports from Kuchino et al.[17] indicate that the level of the minor glutamine tRNA is elevated in Mo-MuLV infected cells, suggesting viral-specific regulation of the suppression event. However, reports from another group[20] indicate no detectable difference either in the distribution of the glutamine isoaccepting species or in the level of aminoacylation activities between uninfected and infected cells. This unsolved discrepancy should stimulate further studies aimed at understanding the parameters that govern efficient suppression.

Apart from the two well-documented examples of natural suppression mentioned so far, namely TMV and Mo-MuLV, numerous eukaryotic (animal and plant) viral mRNAs resort to nonsense suppression as a strategy of gene expression (reviewed in Reference 13). Examination of the nucleotide sequence of viral genomes can often provide a clue as to a possible suppression mechanism: the clearest case is when an ORF is interrupted halfway by a termination codon. Experimentally, however, a suppression event must be demonstrated by translation assays using the appropriate mRNA and suppressor tRNA.

Concerning misreading of the opal codon, the only situation reported concerns rabbit β-globin; this is also the only case of nonviral natural suppression described. A putative opal suppression product of β-globin has been detected *in vivo;* this has led to the partial purification of a tRNA from rabbit reticulocytes. This tRNA is aminoacylatable with tryptophan and suppresses the UGA codon of β-globin mRNA *in vitro;*[21] however, its sequence has not been determined.

Among higher eukaryotes no natural ochre suppressor tRNA species has been isolated. However, evidence exists that mammalian and tobacco cells contain such suppressor activity. Indeed, mature virions are produced when the naturally occurring UAG codon in Mo-MuLV[22] or in TMV,[23] or the UGA codon in Sindbis virus[24] is replaced by a UAA codon.

tRNAs that specifically recognize termination codons—tRNAs belonging to this category correspond to minor species that specifically recognize the UGA codon. Vertebrate cells encode an authentic opal suppressor tRNA gene, whose primary transcript is posttranscriptionally modified to give rise to two minor isoacceptor species harboring a 5'-CmCA-3' and a 5'-NCA-3' anticodon, respectively (N designates an unknown modification of U). Both isoacceptors bear unusual characteristics for a tRNA and also differ in their primary sequences by several pyrimidine transitions. They are aminoacylated with serine and can be further phosphorylated by a specific kinase in the presence of ATP to form phosphoseryl-tRNA^{Ser} (P-Ser-tRNA^{Ser}), and they exclusively recognize the UGA codon (reviewed in References 7 and 8).

Knowledge on these suppressor tRNAs has accumulated for nearly two decades; however, it is only recently that their function has become apparent. The discovery that the mammalian glutathione peroxidase gene contains an in-frame UGA codon encoding the selenocysteine (Se-Cys) found in the active site of the enzyme[25] and the demonstration that the carbon skeleton of such Se-Cys derives from Ser,[26] have provided a hint concerning the function of the opal suppressor tRNA. Lee et al.[27] have shown that the minor tRNA^{Ser} species of eukaryotes which can be converted from Ser-tRNA^{Ser} to P-Ser-tRNA^{Ser}, also exist *in vivo* as Se-Cys-tRNA^{Ser}. It has been suggested that the incorporation of Se-Cys into glutathione peroxidase occurs via conversion of Ser-tRNA^{Ser} to Se-Cys-tRNA^{Ser}.[27] Evidence that P-Ser-tRNA^{Ser} is an obligatory intermediate in this pathway has not yet been fully provided. Studies on an opal suppressor tRNA^{Ser} involved in the insertion of Se-Cys into formate dehydrogenase in *E. coli* confirm this general metabolic pathway for the cotranslational insertion of Se-Cys into proteins.[28]

The opal suppressor tRNAs have been shown to stimulate synthesis of the rabbit β-globin readthrough protein.[29] It would be interesting to examine the nature of the amino acid incorporated in response to the opal codon in β-globin *in vivo:* is it a tryptophan (see above), serine, P-Ser, or Se-Cys residue?

The question remains unanswered as to how the cell discriminates between UGA codons that direct Se-Cys incorporation and those that direct chain termination. As no particular sequence motif can be discerned at the mRNA level, it has been suggested[30] that additional factors and/or features in the tRNA involved in stabilizing codon-anticodon interaction could favor reading of the opal codon by Se-Cys-tRNA. This must be the case since the affinity of this suppressor tRNA for the opal codon is lower than that of RF.[31] Recently, a novel protein factor has been isolated from *E. coli* that forms a specific complex with Se-Cys-tRNA^{Ser}.[32] A eukaryotic homolog of such a factor may exist that could function both as specific elongation factor binding uniquely to Se-Cys-tRNA^{Ser}, and as positive effector for opal codon recognition.

Function and specificity of suppressor tRNAs—As stated above, the nonsence suppressor tRNAs that misread termination codons presumably maintain their ''normal'' function in protein synthesis, which is to insert their amino acid in response to the corresponding sense codon during the elongation process. This has been demonstrated in the case of the yeast tRNA$^{Gln}_{CUG}$ which is encoded by a single gene and can suppress the phenotype of amber mutants when overexpressed.[33] In addition, they are occasionally involved in suppression of termination codons in a large number of eukaryotic viral mRNAs, thereby regulating viral gene expression. On the other hand, authentic nonsense suppressor tRNAs exist as minor species, whose sole function appears to concern suppression of a termination codon. In either case the function of these tRNAs is to regulate the level of synthesis of the ''stopped'' and ''readthrough'' proteins; indeed, recognition of the termination codon which is the outcome of the competition between RF and suppressor tRNA, is not *a priori* a very efficient event. With respect to the insertion of Se-Cys, (where the level of the "stopped" protein has not been investigated) there seems to be no obvious reason why cells should have elaborated such a sophisticated mechanism to introduce Se-Cys, a "21st amino acid", into proteins rather than encode this residue by a specific sense codon.[34]

The most intriguing question is why not all termination codons are recognized by suppressor tRNAs. A like explanation is that the nucleotides surrounding the termination codon affect the level of suppression. This phenomenon known as codon context effect has been studied in considerable detail in prokaryotes (reviewed in References 35 through 37). To date, among eukaryotic systems no thorough comparison has been performed between the contexts of "leaky" vs. "nonleaky" (*i.e.*, "tight") termination codons. Nevertheless, a close examination of the context surrounding the suppressible amber codon in various plant RNA viruses reveals striking conservation patterns (reviewed in Reference 13) that should prompt investigations on the role of these contexts in the efficiency of suppression.

Conversely, one would expect the cell to contain protective codon contexts that would prevent suppression. This supposition is strengthened by the fact that injection of yeast suppressor tRNAs into *X. laevis* oocytes only slightly alters the pattern of proteins synthesized[38] and that mammalian cells constitutively expressing suppressor tRNAs are viable.[39]

Other factors such as the availability of specific tRNAs, base modification, the overall tRNA structure, and/or the ionic environment could influence the level of suppression; these factors have been discussed elsewhere (reviewed in References 6 and 40).

7.4.2. Frameshifting

This is the phenomenon whereby the ribosome shifts from one reading frame to another as it travels along the mRNA, and pursues elongation until it encounters the next termination codon in the new frame. A "stopped" and a "trans-frame" protein are thus synthesized in a regulated fashion. Frameshifting has gained much interest ever since Jacks and Varmus[41] demonstrated that it occurred in Rous sarcoma virus gene expression; it has since been shown to occur in an increasing number of retroviral genomes (reviewed in Reference 13).

In retroviruses the reverse transcriptase as well as a protease are expressed from a GAG-PRO-POL fusion precursor protein that results from in-frame suppression of an amber codon (by misreading, see above), or from (a) -1 frameshift event(s), depending on the genome organization of the virus. In the latter situation, based on whether the protease shares the *gag* frame (Rous sarcoma virus), the *pol* frame (human immunodeficiency virus type 1, HIV-1), or has its own reading frame (mouse mammary tumor virus), retroviruses exist that resort to one or two frameshift events (reviewed in References 6 through 8).

The first indication for a potential frameshift event came from examining the nucleotide sequence of RNA genomes whose long ORFs overlap over short or long distances. Direct evidence was provided by immunological detection of the "stopped" and "trans-frame" proteins synthesized *in vitro*, using *in vitro* transcripts that either corresponded to the *gag-pro* and part of the *pol* gene[41] or that contained the relevant overlap flanked by appropriate reporter genes.[42] A definitive step in demonstrating frameshifting was yielded in some instances by comparing the amino acid sequence of the relevant protein produced *in vivo* to the corresponding nucleotide sequence of the viral genome.[41, 43–45]

A compilation of the demonstrated or suspected frameshift sites in retroviral RNAs has led Jacks et al.[46] to propose the so-called " − 1 simultaneous slippage model". In this model, the frameshift sites all possess heptameric sequences of the type A.AXB.BYZ, in which the 5' base of the codon recognized by the aminoacyl-tRNA in the 0 frame is identical to the 3' base in the preceding codon occupied by the peptidyl-tRNA; likewise the 5' base of the codon occupied by the peptidyl-tRNA is identical to the 3' base of the preceding codon. Simultaneous slippage of the peptidyl-tRNA and the aminoacyl-tRNA by one base in the 5' direction on the mRNA leads to partially correct codon-anticodon pairing in the − 1 frame. Several heptameric frameshift motifs have thus been described. Further demonstration of the involvement of these motifs in frameshift has come from site-directed mutagenesis experiments.[42, 46–47]

Such studies have also shown that not any sequence of the type A.AXB.BYZ can promote efficient frameshift. This has led Jacks et al.[46] to propose that only certain specialized "shifty" tRNAs might participate in a frameshift event. Interestingly, a recent report indicates that some of the tRNA species required for translation within the ribosomal frameshift sequence of HIV-1, bovine leukemia virus and human T-cell leukemia virus type 1 are found in infected cells as hypomodified isoacceptors lacking bulky modified bases in the anticodon loop.[48] Such modified bases are known to increase fidelity of translation and to stabilize codon-anticodon interaction.[35]

Nucleotide deletion experiments have been performed to define the 3' boundary of the mRNA sequences required for efficient frameshift. They have revealed the participation of sequences downstream of the heptameric motif that can fold into a stable stem and loop. Indeed replacement mutagenesis that disrupts part of the stem structure, decreases frameshift efficiency.[42, 46] In the case of infectious bronchitis virus, the observation that even sequences downstream of the stem and loop are also required for efficient frameshift, has led to the finding that these sequences together with the stem and loop can potentially adopt a tertiary configuration in the form of a pseudo-knot.

Here again, replacement mutagenesis that destroys the base-paired region presumably involved in pseudo-knot formation, dramatically reduces frame-shift.[42] A computer-aided search of 22 retroviral and related frameshift regions has revealed that 14 such regions can potentially form a pseudo-knot.[42] Among retroviruses for which frameshifting has been unambiguously demonstrated, HIV-1 stands apart: the region downstream of the frameshift site does not adopt a pseudo-knotted configuration and the complete signal required for frameshift is contained within the first 16 nucleotides of the overlap.[47]

In addition to the retroviruses discussed above and based on nucleotide sequence data, -1 frameshift events have been postulated as a strategy of expression among several plant RNA luteoviruses[49] (reviewed in Reference 13): barley yellow dwarf virus contains a heptameric sequence (reviewed in Reference 13) identical to that of the mouse intracisternal A particle (see Reference 46). In addition, $+1$ frameshift events have been demonstrated in the case of retrotransposons (reviewed in Reference 50); here the mechanism appears to be different and the -1 simultaneous slippage model does not apply.

7.5. WHAT NEXT?

This brief overview reports the recent developments in the study of the regulation of protein synthesis at the level of termination. As already mentioned, if in the last 10 years much work has gone into deciphering these various regulation phenomena, efforts should now be directed towards understanding the molecular mechanisms themselves and the factors involved in the regulation process. In this respect, it is worth mentioning that *E. coli* ribosomes are able to by-pass several nucleotides, "hopping" from one codon to another homologous codon on the mRNA.[51] This phenomenon, as yet undetected among eukaryotes, may shunt a termination codon or may lead to a shift in the reading frame. As already suggested,[52] it would be interesting to determine whether such a phenomenon occurs in the case of the suppressible UAG codon in TMV RNA, since this codon is flanked by two glutamine codons.

It should be stressed that regulated synthesis of two related proteins from a unique mRNA can be achieved via pre-translational mechanisms. For example, in the mRNA encoding human apolipoprotein B-48, an in-frame UAA stop codon is posttranscriptionally converted into a GAA codon by a tissue-specific editing phenomenon, leading to the synthesis of apolipoprotein B-100.[53] In this case, a termination codon is transformed at the mRNA level into a sense codon prior to translation, whereas in suppression a termination codon is recognized as a sense codon by a suppressor tRNA during elongation.

An interesting area of study is the potential relationship between initiation and termination during translation, another way of regulating gene expression.[54] Liu et al.,[55] Peabody and Berg,[56] and Kozak[57] have shown that reading of the second ORF in a dicistronic eukaryotic mRNA is dependent on the presence in the mRNA of an in-frame termination codon upstream of the initiator codon of the second ORF. It thus appears that termination can favor reinitiation, possibly by sliding of the 40S ribosomal subunit along the mRNA until it reaches the AUG of the second ORF where translation resumes.

Given the recent work on eukaryotic translation factors (see other chapters of this book), a major point of interest that needs prompt investigation is the study of the eukaryotic RF at the molecular level. This will definitely provide new tools to dissect the termination step, and should lead to renewed interest in this long neglected area of protein synthesis.

ACKNOWLEDGMENTS

We are indebted to F. Chapeville for his interest and constant encouragements. R.P.C.V. is grateful to the "Société de Secours des Amis des Sciences" and to the "Association pour la Recherche sur le Cancer" for fellowships. This work was supported in part by grants from the "Ligue Nationale Française Contre le Cancer" and from the "Action Concertée MRT: Biologie Moléculaire et Cellulaire Végétale". The Institut Jacques Monod is an "Institut mixte, CNRS—Université Paris VII".

REFERENCES

1. Caskey CT: Peptide chain termination. *Trends Biochem. Sci.* 5:234 (1980).
2. Moldave K: Eukaryotic protein synthesis. *Annu. Rev. Biochem.* 54:1109 (1985).
3. Hatfield D: Suppression of termination codons in higher eukaryotes. *Trends Biochem. Sci.* 10:201 (1985).
4. Morch MD, Valle RPC, and Haenni AL: Regulation of translation of viral mRNAs, in *The Molecular Basis of Viral Replication,* Perez Bercoff R (ed). New York, Plenum Press, 1987, 113.
5. Craigen WJ and Caskey CT: Translational frameshifting: where will it stop? *Cell* 50:1 (1987).
6. Valle RPC and Morch MD: Stop making sense or regulation at the level of termination in eukaryotic protein synthesis. *FEBS Lett.* 235:1 (1988).
7. Hatfield D, Lee BJ, Smith DWE, and Oroszlan S: Role of nonsense, frameshift and missense suppressor tRNAs in mammalian cells. *Prog. Mol. Subcell. Biol.* 11:115 (1990).
8. Hatfield D, Smith DWE, Lee BJ, Worland PJ, and Oroszlan S: Structure and function of suppressor tRNAs in higher eucaryotes. *CRC Crit. Rev. Biochem.* 25:71 (1990).

9. Konecki DS, Aune KC, Tate W, and Caskey CT: Characterization of reticulocyte release factor. *J. Biol. Chem.* 252:4514 (1977).
10. Parker J: Errors and alternatives in reading the universal genetic code. *Microbiol. Rev.* 53:273 (1989).
11. Bienz M and Kübli E: Wild-type tRNA$_G^{Tyr}$ reads the TMV RNA stop codon, but Q base-modified tRNA$_Q^{Tyr}$ does not. *Nature* 294:188 (1981).
12. Barciszewski J, Barciszewska M, Suter B, and Kübli E: Plant tRNA suppressors: *In vivo* readthrough properties and nucleotide sequence of yellow lupin seeds tRNA^{Tyr}. *Plant Sci.* 40:193 (1985).
13. Valle RPC: Etude de la suppression naturelle de codons de terminaison chez les eucaryotes supérieurs: modèles viraux et synthétiques, tRNA suppresseurs et méchanismes de régulation. Thèse de Doctorat, Université Paris 7 (1989).
14. Beier H, Barciszewska M, Krupp G, Mitnacht R, and Gross HJ: UAG readthrough during TMV RNA translation: isolation and sequence of two tRNAs^{Tyr} with suppressor activity from tobacco plants. *EMBO J.* 3:351 (1984).
15. Beier H, Barciszewska M, and Sichinger HD: The molecular basis for the differential translation of TMV RNA in tobacco protoplasts and wheat germ extracts. *EMBO J.* 3:1091 (1984).
16. Valle RPC, Morch MD, and Haenni AL: Novel amber suppressor tRNAs of mammalian origin. *EMBO J.* 6:3049 (1987).
17. Kuchino Y, Beier H, Akita N, and Nishimura S: Natural UAG suppressor glutamine tRNA is elevated in mouse cells infected with Moloney murine leukemia virus. *Proc. Natl. Acad. Sci. U.S.A.* 84:2668 (1987).
18. Philipson L, Andersson P, Olshevsky U, Weinberg R, Baltimore D, and Gesteland R: Translation of MuLV and MSV RNAs in nuclease-treated reticulocyte extracts: enhancement of the gag-pol polypeptide with yeast suppressor tRNA. *Cell* 13:189 (1978).
19. Yoshinaka Y, Katoh I, Copeland TD, and Oroszlan S: Murine leukemia virus protease is encoded by the *gag-pol* gene and is synthesized through suppression of an amber termination codon. *Proc. Natl. Acad. Sci. U.S.A.* 82:1618 (1985).
20. Feng Y-X, Hatfield DL, Rein A, and Levin JG: Translational readthrough of the murine leukemia virus *gag* gene amber codon does not require virus-induced alteration of tRNA. *J. Virol.* 63:2405 (1989).
21. Geller AT and Rich A: A UGA termination suppression tRNA^{Trp} active in rabbit reticulocytes. *Nature* 283:41 (1980).
22. Feng Y-X, Levin J, Hatfield D, Schater T, Gorelick R, and Rein A: Studies of UAA and UGA termination codons in mutant murine leukemia viruses. *J. Virol.* 63:2870 (1989).
23. Ishikawa M, Meshi T, Motoyoshi F, Takamatsu N, and Okada Y: *In vitro* mutagenesis of the putative replicase genes of tobacco mosaic virus. *Nucleic Acids Res.* 14:8291 (1986).
24. Li G and Rice CM: Mutagenesis of the in-frame opal termination codon preceding nsP4 of Sindbis virus: studies of translational readthrough and its effect on virus replication. *J. Virol.* 63:1326 (1989).
25. Chambers I, Frampton J, Goldfarb P, Affara N, McBain W, and Harrison PR: The structure of the mouse glutathione peroxidase gene: the selenocysteine in the active site is encoded by the 'termination' codon, TGA. *EMBO J.* 5:1221 (1986).

26. Sunde RA and Evenson JK: Serine incorporation into the selenocysteine moiety of glutathione peroxidase. *J. Biol. Chem.* 262:933 (1987).

27. Lee BJ, Worland PJ, Davis JN, Stadtman TC, and Hatfield DL: Identification of a selenocysteyl-tRNA^{Ser} in mammalian cells that recognizes the nonsense codon, UGA. *J. Biol. Chem.* 264:9724 (1989).

28. Leinfelder W, Stadtman TC, and Böck A: Occurrence *in vivo* of seleno cysteyl-tRNA$^{Ser}_{UCA}$ in *Escherichia coli*. Effect of *sel* mutations. *J. Biol. Chem.* 264:9720 (1989).

29. Hatfield D, Thorgeirsson SS, Copeland TD, Oroszlan S, and Bustin M: Immunopurification of the suppressor tRNA dependent rabbit β-globin readthrough protein. *Biochemistry* 27:1179 (1988).

30. Schön A, Böck A, Ott G, Sprinzl M, and Söll D: The selenocysteine-inserting opal suppressor serine tRNA from *E. coli* is highly unusual in structure and modification. *Nucleic Acids Res.* 17:7159 (1989).

31. Mitzutani T and Hitaka T: Stronger affinity of reticulocyte release factor than natural suppressor tRNA^{Ser} for the opal termination codon. *FEBS Lett.* 226:227 (1988).

32. Forschhammer K, Leinfelder W, and Böck A: Identification of a novel translation factor necessary for the incorporation of selenocysteine into protein. *Nature* 342:453 (1989).

33. Weiss WA and Friedberg EC: Normal yeast tRNA$_{CAG}$ can suppress amber codons and is encoded by an essential gene. *J. Mol. Biol.* 192:725 (1986).

34. Söll D: Enter a new amino acid. *Nature* 331:662 (1988).

35. Buckingham RH and Grosjean H: The accuracy of mRNA-tRNA recognition, in *Accuracy in Molecular Processes: Its Control and Relevance to Living Systems*. Kirkwood TB, Rosenberg RF, and Galas DJ (Eds). London, Chapman and Hall, 1987, 83.

36. Eggertsson G and Söll D: Transfer ribonucleic acid-mediated suppression of termination codon in *Escherichia coli*. *Microbiol. Rev.* 52:354 (1988).

37. Murgola EJ: tRNA, suppression, and the code. *Annu. Rev. Genet.* 19:57 (1985).

38. Bienz M, Kübli E, Kohli J, De Henau S, Huez G, Marbaix G, and Grosjean H: Usage of three termination codons in a single eukaryotic cell, the *Xenopus laevis* oocyte. *Nucleic Acids Res.* 9:3835 (1981).

39. Sedivy JM, Capone JP, RajBhandary UL, and Sharp PA: An inducible mammalian amber suppressor: propagation of a poliovirus mutant. *Cell* 50:379 (1987).

40. Engelberg-Kulka H and Schoulaker-Schwarz R: Stop is not the end: physiological implications of translational readthrough. *J. Theor. Biol.* 131:477 (1988).

41. Jacks T and Varmus HE: Expression of the Rous sarcoma virus *pol* gene by ribosomal frameshifting. *Science* 230:1237 (1985).

42. Brierley I, Digard P, and Inglis SC: Characterization of an efficient coronavirus ribosomal frameshifting signal: requirement for an RNA pseudoknot. *Cell* 57:537 (1989).

43. Yoshinaka Y, Katoh I, Copeland TD, Smythers GW, and Oroszlan S: Bovine leukemia virus protease: purification, chemical analysis and in vitro processing of *gag* precursor proteins. *J. Virol.* 57:826 (1986).

44. Hizi A, Henderson LE, Copeland TD, Sowder RC, Hixson CV, and Oroszlan S: Characterization of mouse mammary tumor virus *gag-pol* gene products and ribosomal frameshift site by protein sequencing. *Proc. Natl. Acad. Sci. U.S.A.* 84:7041 (1987).

45. Jacks T, Power MD, Masiarz FR, Luciw PA, Barr PJ, and Varmus HE: Characterization of ribosomal frameshifting in HIV-1 *gag-pol* expression. *Nature* 331:280 (1988).

46. Jacks T, Madhani HD, Masiarz FR, and Varmus HE: Signals for ribosomal frameshifting in the Rous sarcoma virus *gag-pol* region. *Cell* 55:447 (1988).

47. Wilson W, Braddock M, Adams SE, Rathjen PD, Kingsman SM, and Kingsman AJ: HIV expression strategies: ribosomal frameshifting is directed by a short sequence in both mammalian and yeast systems. *Cell* 55:1159 (1988).

48. Hatfield D, Feng Y-X, Lee BJ, Rein A, Levin JG, and Oroszlan S: Chromatographic analysis of the aminoacyl-tRNAs which are required for translation of codons at and around the ribosomal frameshift sites of HIV, HTLV-1, and BLV. *Virology* 173:736 (1989).

49. Miller WA, Waterhouse PM, and Gerlach WL: Sequence and organization of barley yellow dwarf virus genomic RNA. *Nucleic Acids Res.* 16:6097 (1988).

50. Kingsman AJ and Kingsman SM: Ty: a retroelement moving forward. *Cell* 53:333 (1988).

51. O'Connor M, Gesteland RF, and Atkins JF: tRNA hopping: enhancement by an expanded anticodon. *EMBO J.* 8:4315 (1989).

52. Weiss RB, Dunn DM, Atkins JF, and Gesteland RF: Slippery runs, shifty stops, backward steps, and forward hops: -2, -1, $+1$, $+2$, $+5$, and $+6$ ribosomal frameshifting. *Cold Spring Harbor Symp. Quant. Biol.* L11:687 (1987).

53. Chen S-H, Habib G, Yang C-Y, Gu Z-W, Lee BR, Weng S-A, Silberman SR, Cai S-J, Deslypere JP, Rosseneu M, Gotto AM, Li W-H, and Chan L: Apolipoprotein B-48 is the product of a messenger RNA with an organ-specific inframe stop codon. *Science* 238:363 (1987).

54. Kozak M: Bifunctional messenger RNAs in eukaryotes. *Cell* 47:481 (1986).

55. Liu C-C, Simonsen CC, and Levinson AD: Initiation of translation at internal AUG codons in mammalian cells. *Nature* 309:82 (1984).

56. Peabody DS and Berg P: Termination-reinitiation occurs in the translation of mammalian cell mRNAs. *Mol. Cell. Biol.* 6:2695 (1986).

57. Kozak M: Effects of intercistronic length on the efficiency of reinitiation by eucaryotic ribosomes. *Mol. Cell. Biol.* 7:3438 (1987).

Regulation of Translation

Chapter

8

Binding of Met-tRNA

Richard J. Jackson
Department of Biochemistry
University of Cambridge, United Kingdom

8.1. INTRODUCTION

To my knowledge, the phosphorylation of eukaryotic initiation factor 2 (eIF-2) was first observed in April 1975 by Paul Farrell, then a graduate student in our group. He had previously established that incubation of crude reticulocyte ribosomes with double-stranded RNA (dsRNA) and ATP generated a potent inhibitor of initiation, which he named DAI (for dsRNA-activated inhibitor, now known to be the dsRNA-activated eIF-2 kinase). This knowledge lead him to examine what proteins were labeled when the reaction was carried out using $[\gamma\text{-}^{32}P]ATP$. He observed labeling of a 35 kDa polypeptide that was dependent on the presence of dsRNA, and was astute enough to guess that this might be the smallest of the three subunits of eIF-2, a supposition which was quickly confirmed by supplementing the reaction with purified eIF-2, that we had acquired from Theo Staehelin and Hans Trachsel. From there it was obvious to test whether HCR (heme-controlled repressor), the inhibitor of initiation generated by incubation of reticulocyte lysates in the absence of hemin, also phosphorylated the α-subunit of eIF-2 (eIFα in the same site(s). This was found to be the case, and indeed the correlation between eIF-2 phosphorylation and inhibition of initiation under different conditions was so close as to lead inevitably to the conclusion that eIF-2 phosphorylation was the fundamental cause of the inhibition.[1] Over subsequent years this view has been sustained, but with some unexpected twists in the story, which is still not completely solved.

In what follows, the effect of phosphorylation of eIF-2α on overall protein synthesis, and on the properties and activity of eIF-2 will be discussed first, mainly in respect of studies using the rabbit reticulocyte lysate system. The relevance of these ideas to the control of initiation in other cells will be debated. The properties and regulation of the kinases known to phosphorylate eIF-2α, and the phosphatase activities carrying out the reverse reaction, will be discussed at the end of the chapter. As this is intended as an overview, I have refrained from citing numerous references that would disrupt the account. For a more detailed bibliography, the reader is recommended to refer to previous reviews that have appeared at regular intervals over the past 10 years.[2-5]

8.2. EFFECT OF eIF-2 PHOSPHORYLATION ON PROTEIN SYNTHESIS IN RETICULOCYTE LYSATES

When rabbit reticulocyte lysates are incubated with dsRNA (leading to activation of DAI), or without hemin or under any of the other conditions listed in Table 8–1 which are believed to cause activation of the heme-controlled kinase, the rate of protein synthesis initiation is the same as in the control assay for the first few minutes, and then drops abruptly to a rate that is at least tenfold lower (Figure 1). This abrupt change in initiation rate is coincident with a precipitous decrease in the steady state level of the 40S/Met-tRNA$_i^{met}$ complexes that are considered to be an obligatory intermediate in the initiation pathway (see Part I, Chapter 4 of this book). As the concentrations of Met-RNA$_i^{met}$ GTP, and 40S subunits in the lysate are unchanged, these observations hint that it is the binding of Met-tRNA$_i^{met}$ to 40S ribosomal subunits that is inhibited.

Pulse-labeling experiments with added [γ-^{32}P]ATP confirm that there is an increase in the phosphorylation of eIF-2α under these conditions of shut-off. As the specific activity of the labeled ATP decreases during the labeling reaction, this type of assay does not give rigorously quantitative data, and although means of maintaining a constant specific activity have been devised,[6] better quantitative methods have now been developed to determine the proportion of eIF-2α phosphorylated, as described below.

The lag in the response (Figure 1) can be explained as a composite of the time required for activation of the kinase, and the time required for phosphorylation of the eIF-2 by the activated kinase, although other factors may contribute to the lag, e.g., the ability of the phosphorylated eIF-2 to catalyze one round of initiation as described below.

The inhibited state can be reversed, or inhibition prevented, by a number of different reagents. As the phosphate group on eIF-2α is being continuously and rapidly turned over through the action of phosphoprotein phosphatases (see Section 8.13), any reagent which blocks kinase action, or which turns off the kinase, will cause reversal of inhibition. Thus the delayed addition of

TABLE 8–1.
Conditions Leading to Shut-Off of Initiation in Reticulocyte Lysates

Conditions Believed to Cause Activation of the Heme-Controlled Kinase (HCR)

Hemin deficiency	Inhibition sets in more slowly and is less severe at lower incubation temperatures.
High temperature	Inhibition manifest when hemin-supplemented lysates are preincubated at 42–44°C, and then assayed under normal conditions (30–34°C).
High hydrostatic pressure	Inhibition manifest when hemin-supplemented lysates are subjected to high hydrostatic pressure at 0°C, then assayed under normal conditions.
Absence of protein thiol reducing system	Caused by the absence of either an NADPH-generating system, or thioredoxin, or thioredoxin reductase. Dithiothreitol can serve as a nonphysiological substitute.
GSSG	GSSG as an inhibitor of initiation per se, apart from its effect in depleting the NADPH and glucose 6-phosphate pools.
Ethanol	$0.4–0.6\ M$
Heavy metal ions	Cd^{2+}, Pb^{2+}, or Hg^{2+} at 10–20 μM; Cu^{2+} at take over 50–100 μM; Fe^{2+} or Zn^{2+} at 300–500 μM.

Conditions Known to Lead to Activation of DAI

dsRNA	Specific for low concentrations (1–100 ng/ml) of dsRNA of minimum size about 50 base-pairs. dsDNA and DNA:RNA hybrids are ineffective.

Conditions Which Inhibit Initiation Without Affecting eIF-2 Phosphorylation

Lack of hexose phosphates	Effect seen in gel-filtered lysates. Glc6P, 2-deoxy-Glc6P, and Fru(1,6)P_2 are effective, 6-phosphogluconate ineffective. The sugar phosphate requirement is additional to the NADPH requirement: if NADPH supplied by another route, sugar phosphates are still needed.

Note: For further details and references see Section 8.11 (HCR activation), Section 8.8 (DAI activation), and Section 8.5 (hexose phosphate effects).

hemin to a hemin-deficient lysate causes protein synthesis to resume after a brief lag which presumably reflects the delay in turning off the kinase (Figure 1). A more rapid reversal is achieved by the addition of fairly high concentrations of a variety of purine compounds, e.g., 2-aminopurine, cAMP, caffeine, etc. These have been shown to inhibit the action of the heme-controlled kinase,[1] most probably by inhibition competitive with respect to ATP. Rescue of protein synthesis activity by 2-aminopurine has come to be considered a diagnostic of inhibition caused by phosphorylation of eIF-2α. While I am not

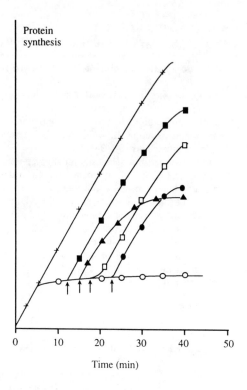

Figure 1. Sketch diagram of the time-course of protein synthesis in rabbit reticulocyte lysates incubated at 30°C, in the absence of hemin, except where stated, and with the following additions; (■———■) eIF-2//GEF complex added at 12.5 min; (▲———▲) eIF-2 added at 15 min in the same molar amount as the added eIF-2/GEF; (□———□) 15μM hemin added at 17.5 min; (●———●) 4mM 2-aminopurine added at 22.5 min; (+——— +) 15μM hemin throughout, i.e. added at zero time. The arrows show the times at which the various additions were made.

aware that this test has lead to any erroneous conclusions in respect of the reticulocyte lysate system, it should be appreciated that these reagents are rather blunt instruments, which may compete with adenine nucleotides in other reactions, and, in the case of intact cell work, might have more subtle effects exerted through transmembrane signaling systems.

Inhibition can also be reversed by the addition of eIF-2 itself (Figure 1), though the efficiency of the rescue is dependent on the purity of the eIF-2 (see below). Surprisingly, no one seems to have achieved reversal by supplementation of the system with additional phosphoprotein phosphatases (see Section 8.13).

While the temporal correlations are indicative that eIF-2 phosphorylation leads to loss of eIF-2 function, which in turn stops 40S/Met-tRNA$_i^{met}$ complex formation and hence blocks initiation, it was soon recognized that there were problems with this simple hypothesis. In the first attempts to quantitate the fraction of eIF-2α phosphorylated, using two-dimensional gel electrophoresis to separate the phosphorylated and nonphosphorylated forms of eIF-2α, it was found that less than 10% of the eIF-2 was phosphorylated in the control conditions of high initiation rates, but no more than 50% seemed to be phosphorylated when protein synthesis was maximally inhibited.[7] This has been confirmed subsequently in many laboratories using the current and less tedious procedure of one-dimensional isoelectric focusing gel electrophoresis under denaturing conditions (urea) followed by immunoblotting, to resolve and quantitate the two forms of eIF-2α: it is generally agreed that protein synthesis initiation can be more than 90% inhibited when only 35 to 40% of the eIF-2α is phosphorylated.[8]

In addition, while it was known that addition of purified eIF-2 could rescue protein synthesis activity in lysates incubated without hemin, it was puzzling that the rescue was less effective the purer the preparation of eIF-2. The final conundrum was that when the activity of purified eIF-2 phosphorylated by purified HCR was tested in a fully fractionated initiation assay system, phosphorylation was found to have no effect on its ability to promote initiation, a result which did not appear to be explicable by dephosphorylation of eIF-2 of the factor during the incubation.[1, 9]

The solution to this conundrum was to emerge in the early 1980s, when it was recognized that the recycling of eIF-2 between successive initiation events required interaction with another moiety, GEF (guanine nucleotide exchange factor, eIF-2B). It is this interaction which is actually blocked by phosphorylation of eIF-2α as described in Section 8.4. In contrast, a single cycle of eIF-2 action in initiation seems to be unaffected by phosphorylation: ternary complex formation and the binding of the ternary complex to 40S subunits in a way that allows subsequent mRNA binding and completion of the initiation process (see Part I, Chapter 4 of this book), all proceed normally. In the fully fractionated system for studying initiation complex formation, it seems likely that each molecule of input eIF-2 functions not more than once, with no recycling,[9] in which case the lack of any effect of eIF-2α phosphorylation can be readily explained. As for the rescue of protein synthesis activity brought about by addition of eIF-2 to inhibited crude lysates, the explanation is that the less highly purified preparations of eIF-2, which showed the greater rescue activity, contained some GEF (as an eIF-2/GEF complex). It is generally considered that when highly purified eIF-2 is added to a lysate in which protein synthesis is inhibited through hemin deficiency, it promotes a stoichiometric stimulation of initiation (i.e., each molecule added functions only once), whereas the addition of eIF-2/GEF complex effects a catalytic rescue.[10]

8.3. PHOSPHORYLATION SITES IN eIF-2

Direct protein sequencing has shown that both the dsRNA-activated kinase and the heme-controlled kinase phosphorylate a single site on the α-subunit of eIF-2, identified as Ser-51 rather than Ser-48.[11] This has been confirmed by expressing mutations of the human eIF-2α cDNA either in transfected COS cells, or by *in vitro* translation: in both cases the Ala-48/Ser-51 mutant was phosphorylated but the Ser-48/Ala-51 mutant was not.[12, 13] These observations on systems expressing only one of the three subunits of eIF-2 raise the question of whether there was phosphorylation of the separate subunit, or whether it was only the eIF-2 assimilated into the complete three subunit structure that was phosphorylated. It is surprising that although the Ala-48/Ser-51 mutant was phosphorylatable, it showed some properties closer to those of the Ser-48/Ala-51 mutant than to wild type: *both* mutants showed the capacity to reverse, *in vivo,* an inhibition of protein synthesis believed to be caused by dsRNA.[13, 14]

The β-subunit of eIF-2 can be rapidly phosphorylated *in vitro* by several different kinases, with casein kinase 2 and protein kinase C phosphorylating distinct sites,[15] but as yet no functional significance has been ascribed to these phosphorylations, nor is it clear how the *in vitro* phosphorylation relates to *in vivo* events. In growing HeLa cells the β-subunit is not rapidly labeled on incubation of the cells with radioactive phosphate, and it migrates as a single spot on isoelectric focusing gels.[16] This may be a phosphorylated species, since under some conditions traces of a less acidic form are seen, as well as small amounts of a more acidic form.[16, 17] It seems not impossible that the rapid phosphorylation seen *in vitro* has no *in vivo* relevance.

8.4. HOW DOES eIF-2 PHOSPHORYLATION INTERFERE WITH GEF ACTIVITY?

A valid model for the action of GEF is an essential prerequisite for understanding how the phosphorylation of eIF-2 might affect GEF activity. There is ample evidence that under physiological conditions eIF-2 binds GDP with much higher affinity than GTP. The relative affinities are often said to differ by a factor of 100, but a recent and more systematic analysis has suggested a value nearer 400.[18] As eIF-2 dissociates from the initiating ribosome at the step when GTP hydrolysis is known to take place (see Part I, Chapters 4 and 5 of this book), the assumption, given the binding data, is that it dissociates as an eIF-2/GDP complex. This proposition does not appear to have been tested directly, but it seems far from implausible. If it is true, then the affinity measurements tell us that dissociation of the eIF-2/GDP complex would be the rate-limiting step in the recycling of eIF-2 to make it available for ternary complex formation (with GTP and Met-tRNA$_i^{met}$) and a further round of initiation.

The role of GEF is to promote this recycling by catalyzing the dissociation of the GDP from eIF-2/GDP. Indeed the standard assay for GEF activity exploits the fact that complexes of eIF-2 with guanine nucleotides bind to nitrocellulose filters, and determines the rate of displacement of labeled [³H or ³²P]GDP from preformed eIF-2/GDP complex on incubation with GEF and unlabeled GDP or GTP.[18, 19] The presence of GEF greatly stimulates this displacement process, provided the eIF-2 is not phosphorylated.

How does GEF catalyze the dissociation of the bound GDP? A number of different models were proposed in the first part of this decade and are summarized by Pain.[4] Subsequently there have been two more detailed studies, but regrettably the conclusions are in conflict. Panniers and colleagues studied the kinetics of labeled GDP displacement from eIF-2 using the nitrocellulose filter binding assay and GEF/eIF-2 complex as the source of GEF (because this is much more stable than free GEF). They deduced an enzyme displacement mechanism[19] as shown in Figure 2A. In contrast, Wahba's group studied the displacement of [³H]GDP or fluorescent GDP derivates from eIF-2 catalyzed by free GEF, assayed by nitrocellulose filter binding, sucrose gradients, and fluorescence measurements. They found that there was no displacement

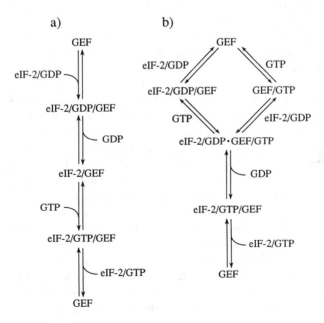

Figure 2. Models for the mechanism of action of GEF in the dissociation of eIF-2/GDP complexes: (a) according to Reference 19, (b) according to Reference 20.

of labeled GDP unless GTP was present, and proposed a sequential mechanism[20] as depicted in Figure 2B. Consistent with this mechanism, they found that GEF could bind GTP (with moderate affinity), this binding being to the 40 kDa subunit.[21]

Notwithstanding this divergence of opinion, there is general agreement that phosphorylation of eIF-2 impairs the GEF catalyzed exchange reaction. What is the mechanism of this inhibition? Early models (reviewed in Reference 4) suggested the sequestration of GEF as a very tight, an essentially undissociatable eIF-2(P)/GEF complex. The analysis made by Rowlands et al.[19] implies that GEF binding to eIF-2(P)/GDP is not irreversible, but is very much tighter than to eIF-2/GDP. Thus eIF-2(P)/GDP can be regarded as a competitive inhibitor of GEF interaction with eIF-2/GDP, but as they point out, the magnitude of the difference in affinities is so great that under physiological conditions this type of competitive binding would effectively mimic sequestration.[22]

According to these ideas, phosphorylation of eIF-2 inhibits initiation by 'sequestering' endogenous GEF into a tight complex with eIF-2(P), leaving no free GEF to promote the recycling of eIF-2 by catalyzing the dissociation of eIF-2/GDP. Thus a near complete shut-off of initiation would be achieved once the concentration of phosphorylated eIF-2 exceeded the concentration of GEF. It should be noted that a mechanism of this type would lead to a very nonlinear relationship between the fraction of eIF-2α phosphorylated and the severity of the inhibition of initiation: a small increase in eIF-2 phosphorylation that took it over the critical threshold would result in a drastic reduction in initiation rates. Therefore we need to know the relative concentrations of GEF and eIF-2 in any system in order to predict what proportion of the eIF-2 must be phosphorylated to effect inhibition. While good antibodies against eIF-2 are available for quantitation by ELISA or immunoblotting, there are as yet no generally available good antibodies against any of the five subunits of GEF. Thus we must fall back on the rather less reliable data of purification studies. These have suggested that in reticulocyte lysates, the concentration of GEF is between 15% and 25% of that of eIF-2, on a molar basis.[3, 22] A somewhat different approach relies on the observation that eIF-2/GEF complex does not bind to CM-Sephadex at about 100 mM salt, whereas free eIF-2 binds tightly. Thus by quantitating by immunoblotting the relative proportion of eIF-2 in the bound and unbound fractions, an estimate of the ratio of GEF to eIF-2 can be deduced.[22] This method assumes that all the cellular content of GEF is found as eIF-2/GEF complex, an assumption apparently supported by the fact that free GEF has never been detected in cell-free extracts, but as free GEF (obtained by dissociation of eIF-2/GEF complex) is known to be highly unstable, one may question the significance of the failure to detect free GEF. The method has given an estimate of the GEF pool as 15% (on a molar basis) of the eIF-2 pool in the case of reticulocyte lysates, and about 50% in Ehrlich ascites cells.[22]

The studies discussed above concerning GEF interaction with eIF-2/GDP and the influence of eIF-2 phosphorylation on this process have been carried out with purified components (eIF-2/GEF complex or separated eIF-2 and GEF), and it is pertinent to ask whether the model is consistent with such evidence as can be gathered by studying the crude systems in which the control of initiation is actually operating. One approach being used with increasing frequency is to assay the endogenous GEF activity of a crude lysate by adding preformed eIF-2/[^{3}H or ^{32}P]GDP to the crude lysate together with unlabeled GDP, and determining the rate of dissociation of the labeled GDP using the nitrocellulose filter binding method.[23] Thus GEF activity is assayed under conditions similar to those used for cell-free protein synthesis, except that in most cases,[22, 23] but not all,[24] the lysate is diluted for the GEF activity assay. These assays invariably show a much slower rate of labeled GDP displacement in lysates in which protein synthesis has been inhibited by heme-deficiency or by dsRNA, than in control uninhibited lysates. It is presumed that the reason for this slower dissociation is that most of the endogenous GEF in the inhibited lysates is tightly bound, one might even say sequestered, to the phosphorylated eIF-2. However, it should be appreciated that the results merely show reduced GEF activity, and cannot distinguish between a decrease in the amount of available GEF (caused by sequestration with phosphorylated eIF-2) and a decrease in the intrinsic activity of the GEF for some other reason. Panniers and Henshaw view eIF-2(P)/GDP as a competitive inhibitor of GEF interaction with the eIF-2/[^{3}H]GDP substrate, and therefore argue that high concentrations of this labeled substrate provide a measure of the total GEF in the reaction mix, whether previously 'sequestered' or not, while low concentrations provide a measure of just the 'unsequestered' GEF.[22]

8.5. OTHER WAYS OF MODULATING GEF ACTIVITY

It is clear from the above discussion that any modulation of GEF activity by processes other than tight binding to phosphorylated eIF-2 could result in an inhibition of initiation with characteristics similar to that seen when eIF-2 kinases are activated: 40S/Met-tRNA$_i^{met}$ complexes would decrease because the recycling of eIF-2 would become rate limiting, and addition of either eIF-2 or, better, eIF-2/GEF complex would rescue protein synthesis.

GEF is a complex protein, composed of five polypeptide chains (approximately 82, 65, 55, 40, and 34 kDa). It has been reported that the factor contains up to 2 mol/mol bound NADPH and that NADP inhibits its activity in the GDP exchange assay,[25] but as this effect was only seen with extremely high concentrations of NADP (about 0.4 mM for 50% inhibition), its physiological significance is very doubtful. Moreover, removal of the bound NADPH reduced the stability rather than the activity of the factor. One would like to know whether, and under what conditions, the bound NADPH can be oxidized *in situ,* and whether this has any effect on the GEF activity.

It has been reported that the 65 kDa subunit of GEF can be phosphorylated *in vitro* by what appeared to be an autophosphorylation mechanism,[10] but this observation does not seem to have been followed up. The largest of the five subunits of reticulocyte GEF (82 kDa) can be phosphorylated *in vitro* by casein kinase 2, which results in a fivefold increase in the activity of the factor in the GDP exchange assay.[26] The low abundance of GEF and the absence of antibodies suitable for Western blotting have so far precluded any attempt to determine whether the phosphorylation of this subunit is regulated *in vivo*. Besides possessing a GTP binding site (on the 40 kDa subunit), there appears to be at least one ATP binding site, but its functional significance is unclear.[21]

Another possible regulator of GEF activity is hexose phosphate, specifically glucose-6-phosphate (Glc6P). Experiments with gel-filtered reticulocyte lysates reveal that Glc6P, or other sugar phosphates, are required for the maintenance of high initiation rates, even though hemin is also present. Glc6P fulfills two roles: it acts as an NADPH generating system, and it is required per se to fill some type of cofactor role.[27, 28] Even if the reducing power is supplied by another route, for example by dithiothreitol or by isocitrate acting via NADPH-linked isocitrate dehydrogenase, there is still a requirement for sugar phosphates, which can be satisfied by several hexose phosphates: Glc6*P* is the most effective at low concentrations, followed by 2-deoxyGlc6*P,* and then fructose-1,6-bisphosphate. Whereas the absence of reducing power leads to activation of an eIF-2 kinase activity, probably a reversible type of HCR activity, the absence of sugar phosphates has no effect on eIF-2 phosphorylation, yet leads to inhibition of initiation with the same characteristics, loss of 40S/Met-tRNA$_i^{met}$ complexes.[28] Sugar phosphates are therefore the only regulator affecting this step in initiation which do not influence the phosphorylation state of eIF-2.

It has recently been shown that the presence or absence of sugar phosphates affects the endogenous GEF (eIF-2B) activity in reticulocyte lysates, suggesting that the inhibition of initiation caused by the absence of hexose phosphates is due to low GEF activity.[24] This is consistent with an earlier report that the sugar phosphate requirement could be substituted by supplementary GEF.[29] However, there is a problem with this hypothesis in that the guanine nucleotide exchange activity of purified eIF-2/eIF-2B complex is uninfluenced by the presence or absence of sugar phosphates. This hints at a significant difference between the GEF catalyzed reaction in crude systems and in purified reactions. There may still be an unknown dimension to the guanine nucleotide exchange reaction.

Although earlier reports from one group showed a strong stimulation of protein synthesis initiation in cell-free systems from HeLa, ascites, and L-cells by sugar phosphates,[30] this does not seem to be the general experience in other laboratories (see Section 8.12), nor has the original report been followed up sufficiently to ascertain whether sugar phosphates might be acting in the same way in these other systems as in reticulocyte lysates.

8.6. ALTERNATIVE INTERPRETATIONS

While the model developed so far represents what I believe to be common ground, it would be misleading to imply that there are no discrepancies or alternative opinions. One controversy concerns whether the binding of Met-tRNA$_i^{met}$ to 40S subunits is the only step in the initiation pathway that is defective under conditions in which eIF-2 is phosphorylated. The other main problem relates to whether the distribution of eIF-2 and GEF in the inhibited lysate is as predicted: all the GEF should accumulate in the cytoplasm as a tight GEF/eIF-2(P) complex containing some 15 to 20% of the total eIF-2 pool, the rest of which should exist in both the phosphorylated and non-phosphorylated forms complexed with GDP, and also most likely to be found in the cytoplasm.

Apart from a small amount associated with ribosomes, all the GEF in an inhibited lysate was found as a 15S GEF/eIF-2(P) complex, as expected, which contained bound GDP in a form readily exchangable with free GDP.[31] This complex was said to contain all the phosphorylated eIF-2 that was not associated with ribosomes. Of the ribosome-associated eIF-2, only a small proportion was bound to native 40S subunits. In two reports, the amount of eIF-2 bound to 40S subunits was very similar in inhibited lysates as in the controls supplemented with hemin,[32, 33] but a more recent study by one of these groups showed a decrease in eIF-2 bound to native 40S subunits in the inhibited lysate, as expected.[34] However, the amount of eIF-2 bound to 40S subunits only decreased by 50%, compared with a threefold decrease in the amount of bound Met-tRNA$_i^{met}$.

The most striking change occurring at or just before the onset of inhibition was a substantial increase in the amount of eIF-2 bound to native 60S subunits and 80S ribosomes, accompanied by a stoichiometric increase in bound GDP, which exchanged very slowly with free GTP/GDP.[32, 33] In contrast, the exchange of guanine nucleotides (and presumably also the eIF-2) associated with native 60S subunits in hemin-supplemented lysates, or heme-deficient lysates supplemented with GEF, was rapid.[33] Both groups also find eIF-2 bound to the polyribosomes of a hemin-supplemented lysate, and to the small polysomes present in inhibited lysates,[32, 34] the eIF-2 actually being associated with the 60S subunits of these ribosomes.[32] Thomas et al.[32] explain these results by suggesting that the eIF-2 bound to native 40S subunits at the first step of the initiation pathway (see Part I, Chapter 4 of this book), may be transferred at the subunit joining step to the 60S subunit as an eIF-2/GDP complex, where it remains until its release is catalyzed by GEF. On the other hand, Gross et al.[34] propose that when subunit joining occurs with a 60S subunit that has tightly bound eIF-2(P)/GDP, it generates an aberrant 80S/Met-tRNA$_i^{met}$/mRNA complex which has a low probability of continuing on to successful polypeptide synthesis, and a high probability of dissociating to give free 60S subunits, and 40S/Met-tRNA$_i^{met}$/mRNA complexes (48S complexes) in which the Met-tRNA$_i^{met}$ is susceptible to deacylation as explained below.

This binding of eIF-2 to 60S subunits is most unexpected, and it is hard to know what to make of these results. They could conceivably be explained as a nonspecific and irrelevant binding, perhaps promoted by the particular ionic conditions used for the sucrose gradient centrifugation. This possibility is suggested by the fact that the binding sites do not seem to be saturable: the binding of added eIF-2 to the 60S subunits and 80S ribosomes in an inhibited lysate increases linearly with the amount added, even over a wide concentration range.[33] However, this type of explanation can only account for the results if the nonspecific binding of eIF-2/GDP, and especially eIF-2(P)/GDP, to 60S subunits and 80S ribosomes is considerably tighter than the binding of other forms of eIF-2.

Gross has long argued that in the absence of hemin or the presence of dsRNA, when the rate of initiation is 10% or less than that of the control (plus hemin), the rate of Met-tRNA$_i^{met}$ binding to 40S subunits is only reduced to about 30% of the control.[35] Most of the 40S/Met-tRNA$_i^{met}$ complexes so formed are thought to give an aberrant 80S complex that has a high probability of dissociating and reverting to a 40S/Met-tRNA$_i^{met}$/mRNA complex, in which the bound Met-tRNA$_i^{met}$ is susceptible to hydrolysis by a deacylase.[34, 35] Apart from the evidence discussed above, he bases this on the following observations: the rate of Met-tRNA$_i^{met}$ consumption or turnover in the inhibited state is as high as 30 to 40% of the control rate;[35] incubation of inhibited lysates with [^{35}S]Met-tRNA$_i^{met}$ leads to labeling of a 48S intermediate identified as a 40S/Met-tRNA$_i^{met}$/mRNA complex, with a yield some 15% of that of the 40S/Met-tRNA$_i^{met}$ complexes formed in parallel controls containing hemin;[35] supplementation of the lysate with purified deacylase decreases the yield of labeled 48S complexes (but doesn't affect the 40S/Met-tRNA$_i^{met}$ complexes in the controls), without affecting protein synthesis rates;[36] supplementation with antibodies against the deacylase increases the yield of labeled 48S complexes without affecting the overall rate of translation.[37] A further point, not inconsistent with this model, is that the translation activity of his heme-deficient lysates can be rescued not only by addition of eIF-2 or GEF, but by what appears to be eIF-5 (the initiation factor promoting subunit joining), although his preparation is admittedly not homogeneous.[34] As formyl-methionyl-tRNA$_i^{met}$ is not hydrolyzed by the deacylase, Gross[35] has argued that his model could provide an explanation for the long-standing but per-plexing observation that the synthesis of chains initiated by the formylated initiator tRNA is much less susceptible to inhibition under conditions of eIF-2 phosphorylation than is normal protein synthesis. However, this explanation is not consistent with his more recent ideas which view deacylation of Met-tRNA$_i^{met}$ as a consequence of other lesions in the initiation pathway, rather than the cause of these lesions.[34] While the data from this group have been largely self-consistent over many years, they do not seem to be in accord with general experience: the 48S complex (40S/Met-tRNA$_i^{met}$ mRNA complex) does not seem to be a prominent feature in other studies, while other assays of Met-tRNA$_i^{met}$ turnover rates are not consistent with a significant rate of deacylation.[38]

Complexes between 40S subunits, Met-tRNA$_i^{met}$ and mRNA stable to sucrose density gradient centrifugation under high salt conditions have been seen in another study where labeled VSV mRNA was added to lysates that had been preincubated with dsRNA.[39] These complexes contained bound GDP, and also eIF-2α(P), though in respect of the latter it is not clear whether the fraction studied might have also included some 60S subunits. On addition of 2-aminopurine (to block further eIF-2 kinase action), dephosphorylation of eIF-2 occurred, and the 40S/Met-tRNA$_i^{met}$/mRNA complexes were con-verted to 80S complexes without dissociation of the mRNA or Met-tRNA$_i^{met}$. This conversion was rather slow, much slower than the assimilation of free mRNA into 80S complexes, reflecting the fact that the eIF-2(P) as-sociated with the 40S complexes was dephosphorylated more slowly than the bulk of the eIF-2(P) in the lysate.[40]

The existence of such 'stuck' but potentially initiation competent 40S/Met-tRNA$_i^{met}$/mRNA is clearly not consistent with the simple model elaborated previously, which would predict that any such complexes formed by 'leaky' inhibition should be competent to proceed to the 80S complex stage.

8.7. IS PHOSPHORYLATION OF eIF-2 A UBIQUITOUS CONTROL MECHANISM?

There are numerous instances of physiologically relevant conditions in which there is a shut-down of protein synthesis that is either entirely general (for all types of mRNA), or relevant to the majority of mRNA species, and is accompanied by a disappearance of polysomes that is resistant to cyclo-heximide pretreatment, a diagnostic test of inhibited initiation. These cases are comprehensively reviewed in Part II, Chapter 12 of this book, and the present discussion is confined to a critical review of those instances where an increase in the phosphorylation of eIF-2α has been demonstrated, not just by pulse labeling of cell-extracts using [γ-^{32}P]ATP, but by the use of iso-electric focusing coupled with immunoblotting (see Section 8.2) to obtain quantitative data on the increase in the proportion of eIF-2α that is phospho-rylated (Table 8–2).

The critical question is whether the increases in eIF-2α phosphorylation recorded in Table 8–2, some of which are quite modest, are sufficient to account for the inhibition of initiation. It is at this stage that fairly solid data vanish into a veritable quagmire of difficulties of interpretation. There can be no universally valid answer to the question, for, as explained in Section 8.4, the degree of phosphorylation of eIF-2α necessary to effect a near-complete shut-down of protein synthesis is dependent on the intracellular molar ratio of eIF-2 to GEF, for which reliable data are hard to come by. Apart from the case of reticulocyte lysates, the only published estimates concern Ehrlich ascites cells, where, according to the CM-Sephadex frac-tionation method (see Section 8.4), the GEF pool is as high as 50% of the eIF-2 pool on a molar basis.[22] If this value also applies to other tissue culture cells such as HeLa or CHO cells, then in many of the conditions listed in Table 8–2, the increase in eIF-2 phosphorylation would be quite insufficient to explain the inhibition of initiation. However, there are reasons to suspect that Ehrlich ascites cells may be atypical, for they show a far higher, and quite variable, level of eIF-2 phosphorylation under control, unstressed con-ditions, than other cells (Table 8–2). The GEF:eIF-2 ratio in HeLa, CHO, or L-cells therefore remains an open question, and so we can only hazard a guess as to the extent of eIF-2 phosphorylation needed to cause severe re-duction in initiation rates in these cells.

It is common practice to attempt to substantiate the importance of eIF-2 phosphorylation in these conditions by examining the GEF activity in extracts from stressed and control cells. This is done by adding preformed eIF-2/ [^{3}H]GDP complex together with unlabeled GDP to the cell extracts and as-saying the rate of displacement of labeled GDP, using the nitrocellulose filter binding method (see Section 8.4). However, as discussed in Section 8.4, there are serious doubts as to whether this assay can distinguish between a low GEF activity caused by its 'sequestration' as a tight eIF-2(P)/GEF com-plex, and a low activity caused for other reasons.

Some recent studies of the response of Ehrlich ascites cells to heat-shock and glutamine starvation serve to illustrate the problems in the interpretation of this type of data.[8, 22] On severe heat-shock (43°C), the proportion of eIF-2α in the phosphorylated state rose to over 60% from a level of about 25%. A comparison of the time course of this increase in eIF-2α phosphorylation and the kinetics of protein synthesis suggests that a sharp shut-down of ini-tiation occurs once the phosphorylation level exceeds 45 to 50%, consistent with the estimated 1:2 molar ratio of GEF:eIF-2 in these cells. Moreover, when the heat-shock was reversed, the proportion of eIF-2α in the phospho-

TABLE 8–2.
Phosphorylation of eIF-2α In Tissue Culture Cells

Cell Type and Experimental Conditions	% eIF-2α Phosphorylated In		
	Control Cells	Stressed Cells	Ref.
Uninfected Cells			
HeLa cells (monolayers)			
Severe heat-shock (45°C)	<5	25–50	17
Serum depleted	<5	15	16
Serum depleted, refed 2 h	<5	9	16
Hypertonic shock	<5	<5	103
12% dimethyl sulphoxide	<5	<5	103
10 mM NaF	<5	30	103
75 μM sodium arsenite	<5	20–30	104
500 μM diamide	<5	20–30	104
10 mM azetidine carboxylic acid	<5	<5–15 (variable)	104
HeLa cells (suspension and monolayers)			
Moderate heat-shock (41–42°C)	<5	<5	41
Severe heat-shock (45°C)	<5	25–50	41
CHO cells (suspension)			
Aminoacyl-tRNA depletion (ts activating enzyme mutant incubated at 39.5°C)	3–8	13–21	99
Ehrlich ascites cells (suspension)			
Severe heat-shock (43°C)	20–25	60–65	8
Glutamine deprivation	20–25	35–40	8
Glucose deprivation	38	48	8
Emetine	23	16	8
Moderate heat-shock (41°C)	25	30	22
Severe heat-shock (44°C)	25	45	22
Glutamine deprivation	31	40	22
Serum depleted	19	45	22
Serum depleted, refed 3 h	19	25	22
Virus Infected Cells			
	Uninfected Cells	Infected Cells	
Reovirus infection of L-cell monolayers			
Untreated cells	5–10	15–20	105
Interferon-treated cells	18–20	26–30	105
Encephalomyocarditis virus infection of HeLa cells (suspension)			
Untreated cells	Approx. 5	Approx. 5	56
Interferon-treated cell	5	40	56
Adenovirus infection of HeLa cells (suspension)			
Wild type virus	<10	25–30	72
dl331 mutant virus	<10	90	72

rylated form fell gradually and when it decreased below 45 to 50% there was a sharp increase in protein synthesis rates.[8] Assays of GEF activity in extracts from heat-shocked cells showed a greatly reduced activity when low concentrations of eIF-2/[³H]GDP were used as substrate but not at high input, which the authors interpret as showing that most of the GEF must have been sequestered by tight, but not completely irreversible, binding to some component, presumably eIF-2(P).[22]

In contrast, glutamine starvation of the same cells resulted in a sharp drop in initiation rates even though the degree of phosphorylation of eIF-2α only reached 40 to 45%, which is slightly below the critical level for shut-off deduced from the heat-shock experiments.[8] This suggests that sequestration of endogenous GEF by phosphorylated eIF-2 may not be the whole explanation. In support of this argument, assays of the endogenous GDP exchange activity indicated that the catalytic activity of GEF might be impaired in a way quite distinct from binding to eIF-2(P), since the full (control) rate of exchange was not observed when saturating levels of eIF-2/[³H]GDP were added.[22]

An increase in the degree of phosphorylation (25 to 50% phosphorylated varying between individual experiments) of eIF-2α in HeLa cells was noted on severe heat-shock (45°C for 30 min), and correlated with functional assays suggesting that eIF-2 from such stressed cells was defective in activity.[17] In addition the timing of the increase in eIF-2α phosphorylation which occurred sharply between 5 and 10 min post heat-shock, coincided with the abrupt 90% inhibition in translation.[41] Although these results are at first sight entirely consistent with those obtained using Ehrlich ascites cells, a recent study from this group sounds a note of caution. While confirming that a severe heat-shock resulted in more than 25% phosphorylation of eIF-2α, it was noted that on temperature shift-down the dephosphorylation of eIF-2α was a fairly rapid process complete in 30 min, whereas the recovery of protein synthesis was slower, requiring 90 to 180 min for completion.[41] In addition, a less severe heat-shock (e.g., 41 to 42°C) caused no phosphorylation of eIF-2, even though the rate of initiation was inhibited by more than 70%.[41] The inevitable conclusion is that heat shock can cause inhibition of initiation without increasing eIF-2 phosphorylation.

It is regrettable that the analyses of other conditions of reduced initiation rates correlated with a modest increase in eIF-2α phosphorylation (Table 8–2) have not been as detailed as these studies of stressed Ehrlich ascites cells and heat-shocked HeLa cells. Given that initiation in intact cells can be inhibited by hypertonic shock, mild heat-shock, and several other conditions (Table 8–2), without any detectable increase in the degree of eIF-2α phosphorylation, and that the inhibition in Ehrlich ascites cells caused by glutamine starvation doesn't seem to be completely explicable in terms of the conventional eIF-2 phosphorylation model, one may question whether the rather modest increase in eIF-2α phosphorylation seen in other conditions or other cells has a causal significance. Inhibition of protein synthesis for reasons quite unrelated to eIF-2 function could cause small changes in the proportion of eIF-2 phosphorylated as a result of the redistribution of eIF-2 between different states (see Section 8.14). This type of explanation may apply to the small change, admittedly a decrease in phosphorylation, caused by incubating Ehrlich ascites cells with emetine (Table 8–2), an elongation inhibitor not known to interfere with eIF-2 function or initiation.

Whatever the outcome of this debate, there remains the question of what kinase is responsible for the increase in eIF-2α phosphorylation in these cells. Of the two eIF-2 kinases found in reticulocytes, one controlled by dsRNA and the other by heme deficiency as well as some other conditions, the former is definitely present in many cell types, and is likely to be the enzyme responsible for increased eIF-2α phosphorylation observed on infection by some viruses (see Section 8.10). It is hard to see how this kinase, which shows a fairly strict requirement for dsRNA for activation, could also be involved in the response to heat-shock, amino acid starvation, etc. On the other hand the status of the heme-controlled kinase as a universal entity common to most mammalian cell types is highly controversial, and no other well-characterized eIF-2 kinase has been described so far (see Section 8.12).

8.8. THE dsRNA-ACTIVATED PROTEIN KINASE

The eIF-2 kinase activated by dsRNA was first identified during investigations into why low concentrations of dsRNA inhibit initiation in reticulocyte lysates with the same characteristics as are seen in heme deficiency. Inhibition is absolutely specific for dsRNA; dsDNA, ssRNA, or DNA:RNA hybrids are all ineffective.[42] In the original work all types of dsRNA appeared equally effective as inhibitors,[42] but more systematic studies have shown that natural dsRNAs are more potent than those made from homopolymers, e.g., poly I:poly C.[43] However, this should not be taken to imply that there is a specific nucleotide sequence required for dsRNA-induced inhibition. However, there is a requirement for a minimum length of about 50 base pairs of perfect, or near-perfect duplex.[42, 44]

An important and paradoxical feature of this inhibition is that whereas low concentrations of dsRNA (10 to 100 ng/ml) are inhibitory to protein synthesis and promote activation of DAI, high concentrations (10 to 100 μg/ml) neither activate DAI nor inhibit initiation.[1, 42] At high concentrations, dsRNA is not simply a nonactivator of the kinase, but has a specific anti-activation effect. This anti-activation property can be demonstrated thanks to the fact that short lengths of dsRNA, or imperfect duplex structures such as the adenovirus VA RNAs, can act as anti-activators but not as activators.[45]

Thus high concentrations of a short dsRNA or of VA RNA prevent the activation of DAI, or the inhibition of initiation, promoted by low concentrations (10 to 100 ng/ml) of a long dsRNA species, but cannot block the eIF-2 kinase activity of DAI which has already been activated. It is interesting that although adenovirus VA$_I$ RNA (approximately 160 nucleotides long) has base-paired regions, a systematic mutagenesis suggested that some of the unpaired loops were at least as important as the duplex regions for the anti-activation properties.[46] (Most strains of adenovirus code for an additional RNA of similar size, VA$_{II}$ RNA, which is thought to exhibit the same properties, but has been studied less than VA$_I$ RNA, because it is much less abundant in infected cells.)

Unactivated DAI (pro-kinase) is found almost entirely associated with crude ribosomes when these are isolated under low salt conditions.[1,47] Preincubation of crude reticulocyte ribosomes with both dsRNA and ATP was found to generate (1) an eIF-2 kinase activity, and (2) an activity which blocked initiation in reticulocyte lysates, both activities being unaffected by the presence of high concentrations of dsRNA in the actual assay step.[1] The generation of these activities had an absolute requirement for ATP, which could not be fulfilled by GTP or nonhydrolyzable analogs of ATP, and when [γ-^{32}P]ATP was used in the preincubation step, labeling of a 67 to 70 kDa polypeptide (commonly known as p68) was observed.[1] The characteristics of this labeling suggested that p68 was the kinase itself, or a component of the kinase, and that its phosphorylation was an integral part of the activation mechanism.

Although DAI was first recognized in reticulocyte lysates, most of the further studies have focused on the enzyme from tissue culture cells, in which it is induced by interferon (see Section 8.9). There appear to be no fundamental differences between the tissue culture cell enzyme and its reticulocyte counterpart. DAI has been purified from several sources, and consists of a single polypeptide chain of about 68 kDa. Binding studies with purified pro-kinase and labeled dsRNA, detecting the complexes by their retention on nitrocellulose filters, gave a linear Scatchard plot indicative of a single type of high affinity binding site.[45] When the rate of labeling of the enzyme in the presence of dsRNA and [γ-^{32}P]ATP was studied over a range of pro-kinase concentrations, it was found to be related to the square of this concentration,[45] suggesting a model in which phosphorylation (and hence activation) involves two molecules of DAI binding at neighboring sites on the dsRNA, with one phosphorylating the other. Since it seems that the binding of individual DAI molecules to dsRNA is not cooperative, this model proposes that high concentrations of dsRNA fail to activate the enzyme because the DAI molecules would be bound too far apart. It also offers an explanation of why short dsRNA, probably insufficient in length to accommodate two DAI molecules simultaneously, fail to cause activation but can act as anti-activators.

As the binding of DAI to dsRNA seems uninfluenced by ATP, although ATP binding to the enzyme is dependent on dsRNA,[47] the strict ATP requirement for activation is thought to reflect the fact that phosphorylation of p68 is necessary for activation of DAI. This is supported by the finding that a type 1 protein phosphatase purified from reticulocytes will dephosphorylate p68, which results in a loss of eIF-2 kinase activity that can be recovered on reincubation with dsRNA and ATP.[48] In these studies the reticulocyte type 2A phosphatase was much less active towards DAI than the type 1, but there is another report indicating the contrary.[49] The phosphorylation site on DAI has not yet been identified, but the possibility of more than one site is raised by the resolution of different forms of phosphorylated p68 on denaturing gel electrophoresis.[50, 51]

Once activated by dsRNA and ATP, the enzyme seems to remain tightly bound to the dsRNA, at least under physiological salt concentrations, and it is presumably in this form that the enzyme phosphorylates eIF-2. Although DAI and HCR phosphorylate eIF-2α on the same site (Ser-51), DAI seems to have a more extended substrate range. It will phosphorylate certain histones, as well as an unidentified 90 kDa protein found in reticulocyte lysates,[52] which has not been recognized as a substrate for phosphorylation by HCR.

Surprisingly, it has been reported that sulfated polysaccharides, particularly heparin, can cause activation of DAI as witnessed by increased p68 phosphorylation.[53] However, almost all this work was done with antibody-absorbed enzyme, and one may speculate whether the binding of the prokinase to the antibody may not have altered the activation properties of the enzyme, particularly as the same studies showed activation by dsDNA at concentrations which definitely don't inhibit protein synthesis or activate eIF-2 kinase in reticulocyte lysates.

8.9. INTERFERON TREATMENT INDUCES (LATENT) DAI

Interferon treatment of responsive cells induces the synthesis but not the activation of DAI (reviewed in Reference 54). This induction is seen not only as an increase in potential DAI activity, but also by immunoblotting assays.[47] The increase in DAI is typically quoted as between five- and tenfold, but differs according to cell type: the basal uninduced levels of DAI differ between different cells types, as well as the fully induced levels. The earlier supposition that uninduced cells invariably have no potential DAI activity is certainly not universally correct, and in our hands L-cells not knowingly treated with interferon yield extracts which have high DAI activity.

This induction would increase the sensitivity of the cell's response to any dsRNA produced during viral infection, and is consistent with the known anti-viral action of interferon. Given the extraordinary sensitivity of DAI activation to very low concentrations of dsRNA, it is not hard to imagine sufficient dsRNA being produced by dsRNA viruses, or by ssRNA viruses as a result of complementary strand synthesis. Even DNA viruses such as adenoviruses seem to produce some dsRNA probably as a result of the overlap between transcription units located on opposite strands of the DNA.[55] This raises a question of whether the resulting inhibition of initiation shows any discrimination between host cell and viral mRNAs, and, if so, what is the explanation for this discrimination. In many virus-host systems there seems no reason to postulate discrimination: virus infection of the interferon-treated cells results in a total block on all protein synthesis as exemplified by the studies of Rice et al.[56] One could also imagine that if DAI activation were to be less extreme or less persistent, the host-cell might survive a brief period of indiscriminately reduced translation, while the block on viral RNA translation at a critical stage could abort the infection.

If there is any discrimination between host and viral mRNA translation, the argument usually advanced is that eIF-2 phosphorylation might be localized to regions in the cell in the vicinity of the dsRNA, i.e., in the vicinity of the viral mRNA. If activated DAI remains tightly bound to the dsRNA, eIF-2 phosphorylation would occur mainly in close proximity to the viral RNA, and if the eIF-2 phosphatases are as active as in reticulocyte lysates, it is far from implausible that phosphorylated eIF-2 might not be evenly distributed but be localized to the proximity of the viral RNA.

The first evidence for this proximity effect was obtained in the reticulocyte-lysate system, which is perhaps surprising as cell-free systems are usually considered to be uncompartmentalized. When lysates were incubated with two differentially labeled mRNA preparations, one of which had poly(U) hybridized to its poly(A) tail, the mRNA lacking the dsRNA tail was efficiently assimilated into 80S initiation complexes, whereas that with the dsRNA tail was not.[57] This discrimination was ascribed to localized activation of DAI, on the grounds that inhibitors of DAI activation allowed both types of mRNA to enter 80S complexes. It should be noted that the mRNA preparation with the dsRNA tail, though excluded from 80S complexes, did enter the 48S complexes (40S/mRNA/Met-tRNA$_i^{met}$) that are a distinctive feature of the inhibited state in this group's work (see Section 8.6).

Further evidence comes from COS cell transfection experiments of genes carried in certain vectors that are thought to give rise to extensive dsRNA synthesis, since the transfected gene is poorly expressed unless the cell is co-transfected with adenovirus VA$_I$RNA[58] or with a mutant eIF-2α cDNA lacking the Ser-51 phosphorylation site.[13] The curious feature is that it is only the expression of the transfected gene, not the translation of the endogenous cell mRNA that is affected.[58] On the assumption, as yet not rigorously proven, that the poor translation of the mRNA coded by the transfected gene is due to some dsRNA transcribed from the plasmid DNA, probably from the vector sequences, this argues that the inhibitory effects of the dsRNA were partitioned within the cell, so that the translation of other mRNAs was not significantly reduced.

8.10. DAI AND VIRUS INFECTION

Although it is an open question whether DAI might be activated by the production of dsRNA of cellular origin, possibly at certain stages of the cell cycle or during differentiation, we would expect this kinase to be most relevant to virus infection. Table 8–2 gives the results of experiments in which the percentage phosphorylation of eIF-2α in virus-infected cells has been rigorously determined by the immunoblotting methods described in Section 8.2. A nonquantitated increase in phosphorylation shown by [γ-^{32}P]ATP labeling of cell extracts has been noted in other cases, or inferred by the fact that addition of eIF-2 and/or GEF stimulates translation in extracts of VSV-infected cells but not control extracts.[59]

Except in some virus-host systems, the increase in eIF-2α phosphorylation caused by virus infection is quite modest (Table 8–2). This fact, and the very success of viral infections, implies that viruses may have evolved means of avoiding DAI activation or of escaping its consequences, which are largely effective except in interferon-treated cells, where the higher levels of DAI increase the sensitivity of the system to dsRNA. These aspects of virus-host interactions have been recently reviewed by Schneider and Shenk.[60]

Infection by poliovirus results in a selective inhibition of host cell mRNA translation through inactivation of the cap-binding factor eIF-4F (see Part II, Chapter 11 of this book) mediated indirectly by the poliovirus 2A gene product which is known to be a protease. Viruses with a mutation in the 2A gene grow inefficiently because there is no degradation of eIF-4F and therefore no selective shut-off of host-cell translation, but rather there is a gradual and general inhibition of all protein synthesis, both host and viral.[61] It was suggested that this general shut-off may be caused through phosphorylation of eIF-2 by activated DAI, and it was argued that wild type virus must have a means of opposing or avoiding this general inhibition. However, more recent work using another mutant in the 2A gene suggests a more complicated explanation, as there was equal eIF-2α phosphorylation on infection with the wild type and mutant viruses,[62] even though the protein synthesis data resembled those of Bernstein et al.[61]

When antibodies against p68 were used to quantitate the level of DAI and examine its state of phosphorylation in cells infected with (wild type) poliovirus, a progressive disappearance/destruction of p68 was observed which eventually reduced the amount to only about 10% of the uninfected cell level.

However, the surviving p68/DAI was extensively phosphorylated and activated, and eIF-2 in the infected cells was more highly phosphorylated than in the controls.[63] These observations suggest that destruction of DAI may allow wild type poliovirus to partially avoid the restrictions that would be caused by DAI activation, but viral protein synthesis may eventually become limited by the accumulation of sufficient phosphorylated eIF-2.[62]

A decrease in the amount of DAI/p68 accompanied by an increase in the state of phosphorylation (and presumed activation) of the remaining DAI has also been noted in other viral infections, particularly with encephalomyocarditis virus (EMCV), or vaccinia, and to a lesser extent VSV.[64] This decrease is not simply the result of cessation of further p68 synthesis, for the 6 to 7 h half-life of p68 in uninfected HeLa cells was reduced to 2 to 3 h in EMCV infected cells, an accelerated rate of turnover dependent on viral gene expression.[64] Other studies have shown that whereas a successful EMCV infection of control HeLa cells involves little or no increase in eIF-2α phosphorylation, the abortive infection of interferon-treated cells is accompanied by a progressive increase in the proportion of eIF-2α phosphorylated, eventually reaching 40%.[56]

The observation, described above, of p68 destruction coupled with an increase in the phosphorylation state and activity of the remaining DAI in interferon-treated HeLa cells infected with vaccinia is somewhat at variance with other studies. In most cell types, the growth of vaccinia is remarkably resistant to the anti-viral effects of interferon; little or no activation of DAI, or phosphorylation of either p68 or eIF-2α can be detected.[65] However, destruction of p68/DAI is unlikely to be the complete explanation for this since active DAI could be recovered from extracts of infected interferon-treated cells by absorbing it on to poly I:poly C-cellulose.[65] The explanation appears to be that an early vaccinia gene product acts as an inhibitor of DAI activation, probably by binding to any dsRNA and 'hiding' it from the DAI.[66] The efficacy of this system can be seen by the fact that in some cell types, notably L-cells, EMCV and VSV will grow successfully in interferon-treated cells, provided the cells are co-infected with vaccinia.[67, 68] Reovirus also seems to code for dsRNA binding protein, $\sigma3$,[69] although a slight increase in eIF-2 phosphorylation has been observed in reovirus infected cells (Table 8–2). Another virus which seems to negate the potentially inhibitory effects of DAI is influenza: in cells infected with this virus, an as yet uncharacterized inhibitor of DAI activation appears after about 2 h, the appearance being dependent on viral gene expression.[70]

The adenovirus VA RNAs (approximately 160 nucleotides in length) represent perhaps the most striking mechanism whereby a virus combats the inhibitory potential of DAI. Of the two VA RNAs encoded in most strains of adenovirus, VA RNA$_I$ is synthesized in much greater yield in infected cells, and has been most studied, but all the indications are that VA RNA$_{II}$

has the same function. The significance of these virus-coded RNAs first came to light in studies of mutants such as dl331, which carries a deletion of the VA$_I$ RNA gene and grows very poorly, mainly because late viral mRNA translation is inefficient. At 6 to 24 h post-infection, cells infected with dl331 were found to have a high activity of DAI,[71] and an exceptionally large fraction of the eIF-2α was phosphorylated,[72] as compared with cells infected by wild type virus (see Table 8–2). The poor translation activity shown by extracts of dl331 infected cells could be rescued by the addition of GEF or eIF-2/GEF complex.[73, 74] These observations inevitably lead to the suggestion that VA$_I$ RNA must prevent DAI activation in cells infected by wild type virus. It is found that VA$_I$ RNA at relatively high concentrations prevents the activation of DAI by low concentrations of *bona fide* dsRNA, but does not itself activate DAI even though it binds to the enzyme.[45, 46]

Even in infections of HeLa cells with wild type adenovirus, the VA RNA system is not completely successful in supressing DAI activation, since the proportion of phosphorylated eIF-2α rises to 30% (Table 8–2), approaching the level thought sufficient for total inhibition of protein synthesis. That this might be at least partly responsible for the shut-off of host cell mRNA translation is shown by the fact that adenovirus infection of cell lines which have a very low endogenous DAI content does not inhibit host cell mRNA translation.[72] The same cell lines support efficient replication of the dl331 mutant virus. It is regrettable that the degree of phosphorylation of eIF-2α in these cell lines was not determined.

Finally there is the case of the TAR sequences, an imperfectly base-paired stem-loop structure found in all HIV-1 mRNAs, and rendering these genes poorly expressed in the absence of HIV-1 TAT (*trans*-activating) protein. It has recently been shown that mRNAs with the TAR element at the 5'-end are translated inefficiently in reticulocyte lysates, because the TAR sequence activates DAI even though its double-stranded segments are shorter than what had previously been considered the lower limit for activation.[75] These experiments used mRNAs generated by *in vitro* transcription of cloned DNA, a procedure which is notorious for generating traces of dsRNA contaminants through a low level of nonspecific transcription from the complementary DNA strand. However, the controls to ensure that DAI activation was caused by the TAR sequence rather than random dsRNA contaminants seem adequate, and therefore the implication is that the TAT protein may play a role in counteracting the activation of DAI by the TAR segment. This, however, is unlikely to be the whole story with respect to TAT/TAR interaction, for it is known that the activating effect of the TAT protein requires at least some nuclear events.[76]

8.11. THE HEME-CONTROLLED KINASE, HCR

Remarkably little work has been done in this area since I reviewed it almost 10 years ago.[2] I will therefore summarize the conclusions drawn then, as modified by such progress as has been made in the meantime, and refer the reader to that review for more details of the evidence and the bibliography.

HCR exists in the postribosomal supernatant of fresh lysates as an inactive eIF-2 kinase, or a latent inhibitor of initiation, often called the pro-inhibitor. Activation can occur under a wide variety of conditions, with the complication that several different forms of the activated kinase/inhibitor have been identified. The 'reversible' inhibitor is generally prepared by incubation of crude lysates or partially purified pro-inhibitor in the absence of hemin for 60 min or longer, and is characterized by the fact that it is readily converted back to pro-inhibitor on incubation with hemin. Irreversible inhibitor is prepared by much more prolonged incubation in the absence of hemin, or by brief incubation with *N*-ethylmaleimide, and is distinguished by the fact that neither incubation with hemin nor any other conditions convert it back to pro-inhibitor. A third form, 'intermediate' inhibitor has been identified by the fact that incubation of partially purified intermediate inhibitor with hemin does not convert it to pro-inhibitor, yet incubation in the lysate under protein synthesis conditions does. As its name implies, this is considered an intermediate between the reversible and irreversible forms, but what appears to be intermediate inhibitor can be formed (in the presence of hemin) under other conditions such as preincubation of lysates at temperatures above 42°C, or by subjecting the lysate to high hydrostatic pressure (see Table 8–1). In practice, the distinctions between these forms of the inhibitor often seem somewhat blurred, but this may be because any preparation of predominantly one form always contains some of the other two forms.

All these different types of inhibitor are considered to be slightly different conformations or different states of the same entity, since they exhibit very similar properties in terms of size, kinase activity, kinase specificity, and inhibition of initiation. Moreover, antibodies raised against one form cross-react with the other forms of activated kinase, but, surprisingly, don't seem to cross-react with the pro-inhibitor.[77]

Apart from inhibition of initiation in lysates, and phosphorylation of eIF-2, one property shown by all types of activated inhibitor, but not by the pro-inhibitor, is autophosphorylation of a component of approximately 85 to 90 kDa. This is a true self-phosphorylation reaction, and almost certainly involves phosphorylation of multiple sites. The assignment of the 85 kDa component to the HCR itself is supported by the fact that incubation of partially purified incompletely phosphorylated HCR with ATP causes the inhibitory activity to bind more tightly to anion exchange columns.[78] All forms of the inhibitor appear to be the same size by gel-filtration and sucrose density gradient centrifugation. The data suggest a highly elongated molecule of about 170 kDa in the native state, and this is supported by the fact that chemical cross-linking of the phosphorylated inhibitor gives a product of 170 kDa.[79, 80] However, it is still not clear whether the kinase is a homodimer of the self-phosphorylatable 85 kDa subunit, or a heterodimer.

When fresh lysate or postribosomal supernatant is fractionated by gel-filtration and the position of the pro-inhibitor located by activating it in each fraction, pro-inhibitor is found to be the same size as the activated inhibitor. This suggests that activation must involve a conformational change rather than the dissociation of some regulatory subunit. The fact that the rate of activation on incubation in the absence of haemin is independent of dilution or the presence of low molecular weight metabolites is consistent with this view. The failure of antibodies raised against the activated inhibitor to cross-react with the pro-inhibitor[77] suggests that this conformational change must be quite radical. Although the activated forms of the inhibitor can be self-phosphorylated, this does not seem an integral feature of the activation process: activation is neither dependent on ATP nor accelerated by its presence, in contrast to the case of DAI activation. Even in crude lysates, which have many phosphatase activities, the phosphate groups on the 85 kDa component turn over very slowly indeed. As a result, a preparation of HCR that is already phosphorylated (with unlabeled phosphate) shows very little labeling on subsequent incubation with [γ-^{32}P]ATP. Thus labeling of the 85 kDa component is not an infallible test of whether a given eIF-2 kinase is HCR or not.

As mentioned above, the irreversible state of the inhibitor can be rapidly formed by incubation with *N*-ethylmaleimide, and several other means of activation also implicate protein $-SH$ groups as critical in the mechanism of activation. For example, incubation of lysates with oxidized glutathione (GSSG) or with heavy metal ions leads to activation of an inhibitor/eIF-2-kinase which appears identical to HCR in all its characteristics including cross-reaction with anti-HCR antibodies.[77, 81] In addition, it is known that the maintenance of high initiation rates requires a functional NADPH generating system, thioredoxin reductase and thioredoxin. The absence of any one of these components leads to some oxidation of $-SH$ groups in lysate proteins, an inhibition of initiation, and an increase in the phosphorylation of eIF-2.[27, 28] As antibodies against HCR prevent this inhibition,[24] it appears that oxidation of some as yet unidentified $-SH$ groups has activated HCR, albeit in a highly reversible form.[27] Other oxidizing, or NADPH-depleting, conditions seem to cause HCR activation, presumably by the same mechanism. Thus the inhibition of initiation and phosphorylation of eIF-2α in reticulocyte lysates caused by either arachidonic acid, or by unsaturated phospholipids together with Ca^{2+}, seems related to the formation of liperoxides, and can be prevented by the addition of NADPH or Glc6*P*.[82, 83] It is not clear whether inhibition promoted by ethanol (Table 8–1) has a similar explanation, but it seems to operate at least partly through activation of HCR, since antibodies against HCR give partial relief of inhibition.[77]

It has recently been suggested that hemin might maintain the inhibitor in its latent inactive state by promoting disulfide bond formation between the two ca. 85 kDa subunits. This was based on the finding that incubation of activated phosphorylated (labeled) kinase with hemin caused formation of the ca. 180 kDa cross-linked species.[80] While these experiments could be criticized on the grounds that the effects of 10 μM hemin on a solution of 5 μg/ml of purified HCR cannot be equated with the effects of 10 to 20 μM hemin in crude lysates (protein concentration more than 50 mg/ml), the idea is worth pursuing further.

While activating conditions such as high temperature, high pressure, heavy metal ions, GSSG, or lack of reducing systems can readily be imagined as inducers of conformational changes and/or changes in the status of the sulfhydryl groups of HCR, one parameter controlling activation, namely, the GTP/ATP ratio, doesn't obviously fit into this category. High concentrations of GTP (1 to 2 mM) prevent activation, and ATP, while having no effect on its own, antagonizes this GTP effect.[2] GTP is particularly effective in preventing the activation, or rapidly reversing activation, caused by heme-deficiency, preincubation at high temperature, or high hydrostatic pressure, conditions which all lead to the formation of intermediate inhibitor. Gross has partially purified a cytoplasmic protein from reticulocyte lysates which carries out a GTP-dependent inactivation of intermediate inhibitor, presumably converting it back to pro-inhibitor.[84] Thus the GTP effect is likely to be more complicated than a GTP-HCR binary interaction, while the basis of the antagonistic effect of ATP remains unknown. As pointed out previously,[2] the action of GTP in completely overcoming these particular inhibitory conditions should not be confused with the rather weak effect of GTP in partially antagonizing inhibition caused by other conditions such as inhibition by dsRNA or irreversible HCR, which probably has a different explanation.

While the above account would be accepted by most of the groups that have worked on this topic, there are some who propose a much more complicated scenario. Hardesty's group find three types of polypeptide associated with their HCR preparations: a self-phosphorylatable component (100 kDa in their hands), a component of 95 kDa detected only by photoaffinity labeling using 8-azido-ATP, and an approximately 90 kDa protein originally identified as a degradation product of β-spectrin by several criteria, including cross reaction with monoclonal antibodies against spectrin, but now claimed to be hsp 90, the 90 kDa heat-shock protein.[85-87] However, they agree that the translation inhibitory activity of HCR behaves as an elongated protein of native molecular weight about 180,000,[86] so it clearly cannot contain all three types of polypeptide. Most other groups regard the 90 kDa component (which was the only stainable protein detectable after gel electrophoresis) as an irrelevant contaminant because it doesn't exactly co-purify with kinase or inhibitory activity, but Hardesty's group clearly consider it a potential regulator (see below). They also view the self-phosphorylated component as either an irrelevance or as a regulatory subunit of HCR, but not the actual catalytic subunit. One argument used was that autophosphorylation labeled a 100 kDa polypeptide and 8-azido-ATP a 95 kDa, but this overlooks the possibility that it could be the same protein whose mobility on gel electrophoresis was affected by the different labeling conditions. The other argument was that the purified HCR preparation could be resolved, by binding to a monoclonal anti-spectrin-Sepharose column, into a minor fraction which showed eIF-2 kinase activity but no 100 kDa autophosphorylated component, and a major fraction with both activities.[86] However, as explained above, the inability to label an autophosphorylated component is not an infallible indication that an eIF-2 kinase actually lacks this component. Thus the case for the self-phosphorylated component not being the essential catalytic part of HCR seems less than watertight.

It is harder to explain their results on the regulation of the kinase activity by the 90 kDa polypeptides except that the results aren't very self-consistent. Some monoclonal antibodies against β-spectrin were reported to stimulate the eIF-2 kinase activity of the purified HCR,[85] but in a subsequent report α- or β-spectrin itself also activated the kinase activity.[86] On the other hand, the inhibition of translation seen when β-spectrin was added to heme-supplemented lysates did not show characteristics typical of a block caused by eIF-2 phosphorylation.[88] More recently, it has been claimed that hsp 90 stimulates the eIF-2 kinase activity of purified HCR, and inhibits reticulocyte lysate protein synthesis with characteristics consistent with an inhibition caused by eIF-2 phosphorylation.[87]

8.12. HCR-LIKE KINASES IN OTHER CELL TYPES

Apart from DAI, the status of eIF-2 kinases in cells other than reticulocytes remains controversial. A common approach is to test extracts from other cells for eIF-2 kinase activity as exemplified by work on skeletal muscle,[89] or for their ability to cause inhibition of initiation and eIF-2 phosphorylation when added to hemin-supplemented reticulocyte lysates, as shown by extracts of Ehrlich ascites cells and heat-shocked HeLa cells.[17, 90] Attention needs to be paid to the possibility that a kinase active against purified eIF-2 may not necessarily be effective under the conditions of protein synthesis in crude extracts, or may not necessarily phosphorylate eIF-2α on Ser-51, while an entity that promotes eIF-2 phosphorylation in reticulocyte lysates could operate indirectly via activation of endogenous HCR. There have been several claims for HCR-like activities in other cells (see references in References 4,

5, and 91) but none have been pursued to the point of meeting even these preliminary criteria.

A recent study of mouse erythroleukemic cells (MEL cells) is particularly instructive.[91] A kinase activity capable of phosphorylating purified eIF-2 was found in both uninduced and induced cells. The activities from the two sources behaved similarly through the first steps of purification, but clearly differed in the subsequent steps. The kinase from induced (differentiated) MEL cells showed properties very similar to reticulocyte HCR, and inhibited initiation in a rabbit reticulocyte cell-free system. The kinase from uninduced cells was clearly quite distinct from HCR, and did not inhibit initiation *in vitro*. The question of whether this kinase from uninduced cells actually phosphorylated the eIF-2 in the protein synthesis assay system was not addressed, nor was the site of phosphorylation on eIF-2α examined. These results suggest that HCR may be unique to terminally differentiated erythroid cells. They also serve as a warning that not every entity capable of phosphorylating the α-subunit of purified eIF-2 will necessarily act as an inhibitor of initiation in cell-free systems.

The one strong claim of an activity in nonerythroid cells virtually indistinguishable from HCR concerns HeLa cells subjected to a moderate heat-shock (42.5°C), conditions under which initiation was severely inhibited after 10 min and then subsequently recovered to about 50% of control rates. Phosphorylation of eIF-2α occurred under these conditions, with maximum labeling observed at 10 min after heat-shock, but the fraction of eIF-2α phosphorylated was not determined.[92] Incubation of extracts of these heat-shocked cells with [γ^{32}P]ATP lead to the labeling of both eIF-2α and a 90 kDa polypeptide, which could be inhibited by hemin or by antibodies raised against reticulocyte HCR, though it should be noted that the antigen was HCR prepared by Hardesty's group, which seems somewhat atypical (see Section 8.11). Moreover, heat-shock of the HeLa cells in the presence of extracellular hemin reduced the severity of the inhibition of translation.[92] This represents the best claim for HCR in other cells, but it worth noting that this group has consistently made HeLa cell-free extracts closer in properties to reticulocyte lysates than other laboratories, in that hemin and sugar phosphates cause a greater stimulation of initiation.[30] This might be a peculiarity of the particular HeLa cell line, but as they find similar effects with extracts from other cells (myeloma, L-cells) the difference may be one of methodology.

A final comment is that the existence of irreversible forms of reticulocyte HCR may have misleadingly conditioned us to expect that all such entities will be evident as dominant inhibitors of translation when added to other cell-free systems. It should be recalled that under some conditions, such as the absence of reducing systems, initiation in reticulocyte lysates is severely inhibited and eIF-2α phosphorylated[28] through the action of a form of HCR,[24] yet it is very difficult to demonstrate the presence of a dominant inhibitor of initiation in such lysates.[27] Are there similarly elusive kinases present in tissue culture cells?

8.13. DEPHOSPHORYLATION OF eIF-2α(P)

As already discussed in Section 8.2, the dephosphorylation of eIF-2 and the rescue of protein synthesis in heme-deficient reticulocyte lysates when 2-aminopurine is added to inhibit HCR bears witness to the presence of high phosphoprotein phosphatase activity in the lysate. Turnover rates are usually measured by adding ^{32}P-labeled eIF-2, or ^{32}P-labeled crude ribosomes (containing eIF-2) to the lysate and following the rate of loss of the labeled phosphate group. Half-lives of 20 sec[93] to over 2 min[2, 28] have been reported, and the rate is often nonuniform. The discrepancies and the nonuniformity may reflect the fact that the dephosphorylation rate can be influenced by the binding of eIF-2 to other ligands (see Section 8.14), and hence the rate may depend critically on the amount of eIF-2 added or the particular conditions used.

The presumed phosphatase has been purified from reticulocyte lysates using phosphorylated eIF-2 as the substrate,[94] and has been classified as a type 2A$_2$ phosphatase on the basis of its subunit composition (60 kDa and 35 kDa), substrate specificity and insensitivity to phosphatase inhibitors 1 and 2.[95] However, lysates have at least five distinct phosphatase activities, two type 1 enzymes and three type 2,[96] and type 1 phosphatases seem as active as the type 2A$_5$ against eIF-2(P),[97] though there is a more recent report to the contrary.[48] There are two reasons for questioning whether the 2A$_2$ enzyme is the physiologically relevant species. Supplementation of crude lysates with quite high concentrations of this purified phosphatase caused only a slight increase in the rate of dephosphorylation of eIF-2α(P).[94] Second, the 99% loss in activity experienced in an early step of the purification raises the question of whether it is only a minor, physiologically unimportant, activity that has been purified. It has also been noted that addition of the type 1 phosphatase inhibitor-2 to hemin supplemented lysates causes inhibition of initiation and increased phosphorylation of eIF-2α, which was tentatively explained as an inhibition of the physiologically relevant phosphatase (i.e., a type 1 enzyme) coupled with a low level of basal eIF-2 kinase activity.[98] It is also remarkable that no one has yet achieved a rescue of protein synthesis activity on supplementing a heme-deficient lysate with extra phosphatase, a result which would be expected if the physiologically relevant phosphatase was used.

To date there has been no evidence that the dephosphorylation step is regulated *in vivo* or *in vitro*. The small effects on dephosphorylation rate that have been observed[93] are most likely explained by the fact that the binding of various ligands to eIF-2 can have a profound effect on the rate of dephosphorylation (see Section 8.14).

Phosphatases active against eIF-2 must be presumed to be present in other cells, and indeed both the type 1 and type 2A$_2$ enzymes seem widespread.

However, it is a curious observation that cell-free extracts from both CHO cells,[99] and L-cells (unpublished observations) have very little eIF-2 phosphatase activity in comparison with reticulocyte lysates.

8.14. ACCESSIBILITY OF eIF-2 TO KINASES AND PHOSPHATASES

It is evident that eIF-2 can interact with an enormously wide range of ligands, both individually and in combinations: GTP, GDP, Met-tRNA$_i^{met}$, GEF, 40S ribosomal subunits, and probably also 60S subunits, 80S ribosomes, as well as mRNA and ATP.[100] Thus eIF-2 may exist in the cell in a plethora of different states, and it is pertinent to ask whether all these forms are equivalent as substrates for the phosphorylation and dephosphorylation reactions, and if not, which is the physiologically relevant form of eIF-2 for these reactions.

It is already clear that the different forms are not equivalent as substrates. In the phosphorylation reaction (by HCR) eIF-2 in complex with GEF (and GDP) was said to be poorly phosphorylated unless excess GDP is present,[31] and eIF-2 bound to the 60S subunits of polyribosomes may be phosphorylated more rapidly than free eIF-2.[32] In assays with purified components, the rate of phosphorylation of the α-subunit by HCR was slower with eIF-2/GEF complexes than free eIF-2, and phosphorylation of the β-subunit by casein kinase 2 was completely blocked.[101] Interaction of eIF-2 with Met-tRNA$_i^{met}$ and GTP (in ternary complexes), or with GEF, reduced the rate of dephosphorylation of the α-subunit by about fourfold, and blocked β-subunit dephosphorylation.[101] In the reticulocyte lysate, eIF-2 (probably complexed with GDP) bound to 40S subunits was dephosphorylated more slowly than free eIF-2(P).[40] In addition, a 67 kDa protein has been described which appears to bind to eIF-2 since it copurifies through several steps, and apparently inhibits the phosphorylation of eIF-2α by HCR.[102]

We therefore need to bear in mind that there is a problem over what are the true physiologically relevant substrates for the kinase and phosphatase. It seems more than likely that the use of purified eIF-2 as the substrate for purifying and studying kinases and phosphatases is not entirely appropriate. It is also worth bearing in mind that any inhibition of translation for whatever reason might lead to a redistribution of the eIF-2 between the different states, and if there is a low basal kinase activity present in the system, this redistribution could lead to a change in the degree of phosphorylation of eIF-2α which would be the consequence of the inhibition rather than its cause. This re-emphasizes the need for further information on the 'life cycles' of eIF-2 and GEF during on-going protein synthesis, and on the GEF:eIF-2 ratio in tissue culture cells.

ACKNOWLEDGMENT

Work carried out in the author's laboratory was supported by grants from the Medical Research Council.

REFERENCES

1. Farrell PJ, Balkow K, Hunt T, Jackson RJ, and Trachsel H: Phosphorylation of initiation factor eIF-2 and the control of reticulocyte protein synthesis. *Cell* 11:187 (1977).
2. Jackson RJ: The cytoplasmic control of protein synthesis, in Perez-Bercoff R (Ed): *Protein Biosynthesis in Eukaryotes*, New York, Plenum Press, 1982, 363.
3. Ochoa S: Regulation of protein synthesis initiation in eukaryotes. *Arch Biochem. Biophys.* 223:325 (1983).
4. Pain VM: Initiation of protein synthesis in mammalian cells. *Biochem. J.* 235:625 (1986).
5. Sarre TF: The phosphorylation of eukaryotic initiation factor 2: a principle of translational control in mammalian cells. *BioSystems* 22:311 (1989).
6. Jackson RJ and Hunt T: The use of hexose phosphates to support protein synthesis and generate [γ-^{32}P]ATP in reticulocyte lysates. *FEBS Lett.* 93:235 (1978).
7. Farrell PJ, Hunt T, and Jackson RJ: Analysis of phosphorylation of protein synthesis initiation factor eIF-2 by two-dimensional gel electrophoresis. *Eur. J. Biochem.* 89:517 (1978).
8. Scorsone KA, Panniers R, Rowlands AG, and Henshaw EC: Phosphorylation of eukaryotic initiation factor 2 during physiological stresses which affect protein synthesis. *J. Biol. Chem.* 262:14538 (1987).
9. Trachsel H and Staehelin T: Binding and release of eukaryotic initiation factor eIF-2 and GTP during protein synthesis initiation. *Proc. Natl. Acad. Sci. U.S.A.* 75:204 (1978).
10. Konieczny A and Safer B: Purification of the eukaryotic initiation factor-2 eukaryotic initiation factor 2B complex and characterization of its guanine nucleotide exchange activity during protein synthesis initiation. *J. Biol. Chem.* 258:3402 (1983).
11. Colthurst DR, Campbell DG, and Proud CG: Structure and regulation of eukaryotic initiation factor eIF-2: sequence of the site in the α subunit phosphorylated by the haem-controlled repressor and by the double-stranded RNA-activated inhibitor. *Eur. J. Biochem.* 166:357 (1987).
12. Pathak VK, Schindler D, and Hershey JWB: Generation of a mutant form of protein synthesis initiation factor eIF-2 lacking the site of phosphorylation by eIF-2 kinases. *Mol. Cell. Biol.* 8:993 (1988).
13. Kaufman RJ, Davies MV, Pathak VK, and Hershey JWB: The phosphorylation of eukaryotic initiation factor 2 alters translational efficiency of specific mRNAs. *Mol. Cell. Biol.* 9:946 (1989).

14. Davies MV, Furado M, Hershey JWB, Thimmappaya B, and Kaufman RJ: Complementation of adenovirus virus-associated RNA_I gene deletion by expression of a mutant translation initiation factor. *Proc. Natl. Acad. Sci. U.S.A.* 86:9162 (1989).

15. Clark SJ, Colthurst DR, and Proud CG: Structure and phosphorylation of eukaryotic initiation factor 2. Casein kinase 2 and protein kinase C phosphorylate distinct but adjacent sites in the β subunit. *Biochim. Biophys. Acta* 968:211 (1988).

16. Duncan R and Hershey JWB: Regulation of initiation factors during translational repression caused by serum deprivation: covalent modifications. *J. Biol. Chem.* 260:5493 (1985).

17. Duncan R and Hershey JWB: Heat shock-induced translational alterations in HeLa cells. *J. Biol. Chem.* 259:11882 (1984).

18. Panniers R, Rowlands AG, and Henshaw EC: The effect of Mg^{2+} and guanine nucleotide exchange factor on the binding of guanine nucleotides to eukaryotic initiation factor 2. *J. Biol. Chem.* 263:5519 (1988).

19. Rowlands AG, Panniers R, and Henshaw EC: The catalytic mechanism of guanine nucleotide exchange factor action and competitive inhibition by phosphorylated eukaryotic initiation factor 2. *J. Biol. Chem.* 263:5526 (1988).

20. Dohlakia JN and Wahba AJ: Mechanism of the nucleotide exchange reaction in eukaryotic polypeptide chain initiation. *J. Biol. Chem.* 264:546 (1989).

21. Dohlakia JN, Francis BR, Haley BE, and Wahba AJ: Photoaffinity labelling of the rabbit reticulocyte guanine nucleotide exchange factor and eukaryotic initiation factor 2 with 8-azidopurine nucleotides. *J. Biol. Chem.* 264:20638 (1989).

22. Rowlands AG, Montine KS, Henshaw EC, and Panniers R: Physiological stresses inhibit guanine nucleotide exchange factor in Ehrlich cells. *Eur. J. Biochem.* 175:93 (1988).

23. Matts RL and London IM: The regulation of initiation of protein synthesis by phosphorylation of eIF-2(α) and the role of reversing factor in the recycling of eIF-2. *J. Biol. Chem.* 259:6708 (1984).

24. Gross M, Rubino MS, and Skarn TK: Regulation of protein synthesis in rabbit reticulocyte lysate: glucose 6-phosphate is required to maintain the activity of initiation factor eIF-2B by a mechanism that is independent of the phosphorylation of eIF-2α. *J. Biol. Chem.* 263:12486 (1988).

25. Dohlakia JN, Mueser TC, Woodley CL, Parkhurst LJ, and Wahba AJ: The association of NADPH with the guanine nucleotide exchange factor from rabbit reticulocytes: a role of pyridine dinucleotides in eukaryotic polypeptide chain initiation. *Proc. Natl. Acad. Sci. U.S.A.* 83:6746 (1986).

26. Dohlakia JN and Wahba AJ: Phosphorylation of the guanine nucleotide exchange factor from rabbit reticulocytes regulates its activity in polypeptide chain initiation. *Proc. Natl. Acad. Sci. U.S.A.* 85:51 (1988).

27. Jackson RJ, Campbell EA, Herbert P, and Jackson RJ: The preparation and properties of gel-filtered rabbit-reticulocyte lysate protein-synthesis systems. *Eur. J. Biochem.* 131:289 (1983).

28. Jackson RJ, Herbert P, Campbell EA, and Jackson RJ: The roles of sugar phosphates and thiol-reducing systems in the control of reticulocyte protein synthesis. *Eur. J. Biochem.* 131:313 (1983).

29. Jackson RJ and Hardon EM: Control of polypeptide chain initiation by sugar phosphates. *Biochem. Soc. Trans.* 13:673 (1985).

30. Lenz JR, Chatterjee GE, Maroney PA, and Baglioni C: Phosphorylated sugars stimulate protein synthesis and Met-tRNA$_f$ binding activity in extracts of mammalian cells. *Biochemistry* 17:80 (1978).

31. Thomas NSB, Matts RL, Petryshyn R, and London IM: Distribution of reversing factor in reticulocyte lysates during active protein synthesis and on inhibition by heme deprivation or double-stranded RNA. *Proc. Natl. Acad. Sci. U.S.A.* 81:6998 (1984).

32. Thomas NSB, Matts RL, Levin DH, and London IM: The 60S ribosomal subunit as a carrier of eukaryotic initiation factor 2 and the site of reversing factor activity during protein synthesis. *J. Biol. Chem.* 260:9860 (1985).

33. Gross M, Redman R, and Kaplansky DA: Evidence that the primary effect of phosphorylation of initiation factor 2 (α) in rabbit reticulocyte lysate is inhibition of the release of eukaryotic initiation factor 2/GDP from 60S ribosomal subunits. *J. Biol. Chem.* 260:9491 (1985).

34. Gross M, Wing, M, Rundquist C, and Rubino MS: Evidence that phosphorylation of eIF-2(α) prevents the eIF-2B-mediated dissociation of eIF-2/GDP complexes from the 60S subunit of complete initiation complexes. *J. Biol. Chem.* 262:6899 (1987).

35. Gross M: Control of protein synthesis by hemin: evidence that the hemin-controlled translational repressor inhibits formation of 80S initiation complexes from 48S intermediate initiation complexes. *J. Biol. Chem.* 254:2370 (1979).

36. Kaplansky DA, Kwan A, and Gross M: Effect of Met-tRNA$_f$ deacylase on polypeptide chain initiation in rabbit reticulocyte lysates. *J. Biol. Chem.* 257:5722 (1982).

37. Gross M, Nguyen T, Redman R, and Rosen J: Use of an antibody to characterise and determine the role of the major Met-tRNA$_f$ deacylase from rabbit reticulocyte ribosomes. *Biochim. Biophys. Acta* 867:220 (1986).

38. Jackson RJ and Hunt T: The turnover of methionine in the Met-tRNA pool and the control of protein synthesis in reticulocyte lysates: no evidence that haem-deficiency promotes deacylation of Met-tRNA$_f$. *FEBS Lett.* 143:301 (1982).

39. De Benedetti A and Baglioni C: Phosphorylation of initiation factor eIF-2α, binding of mRNA to 48S complexes, and its reutilisation in initiation of protein synthesis. *J. Biol. Chem.* 258:14556 (1983).

40. De Benedetti A and Baglioni C: Kinetics of dephosphorylation of eIF-2(αP) and reutilisation of mRNA. *J. Biol. Chem.* 260:3135 (1985).

41. Duncan R and Hershey JWB: Protein synthesis and protein phosphorylation during heat stress, recovery, and adaptation. *J. Cell Biol.* 109:1467 (1989).

42. Hunter T, Hunt T, Jackson RJ, and Robertson HD: The characteristics of inhibition of protein synthesis by double-stranded ribonucleic acid in reticulocyte lysates. *J. Biol. Chem.* 250:409 (1975).

43. Content J, Lebleu B, and De Clercq E: Differential effects of various double-stranded RNAs on protein synthesis in rabbit reticulocyte lysates. *Biochemistry* 17:88 (1978).

44. Minks MA, West DA, Benvin S, and Baglioni C: Structural requirements of double-stranded RNA for the activation of $2',5'$-oligo(A) polymerase and protein kinase of interferon-treated HeLa cells. *J. Biol. Chem.* 254:10180 (1979).

45. Kostura M and Mathews MB: Purification and activation of the double-stranded RNA-dependent eIF-2 kinase DAI. *Mol. Cell. Biol.* 9:1576 (1989).

46. Mellits KH and Mathews MB: Effect of mutations in stem and loop regions on the structure and function of adenovirus VA RNA$_I$. *EMBO J.* 7:2849 (1988).

47. Galabru J and Hovanessian A: Autophosphorylation of the protein kinase dependent on double-stranded RNA. *J. Biol. Chem.* 262:15538 (1987).

48. Szyszka R, Kudlicki W, Kramer G, Hardesty B, Galabru J, and Hovanessian A: A type 1 phosphoprotein phosphatase active with phosphorylated M_r = 68,000 initiation factor 2 kinase. *J. Biol. Chem.* 264:3827 (1989).

49. Petryshyn R, Levin DH, and London IM: Regulation of double-stranded RNA-activated eukaryotic initiation factor 2α kinase by type 2 protein phosphatase in reticulocyte lysates. *Proc. Natl. Acad. Sci. U.S.A.* 79:6512 (1982).

50. Levin DH, Petryshyn R, and London IM: Characterisation of purified double-stranded RNA-activated eIF-2α kinase from rabbit reticulocytes. *J. Biol. Chem.* 256:7638 (1981).

51. Krust B, Galabru J, and Hovanessian AG: Further characterisation of the protein kinase activity mediated by interferon in mouse and human cells. *J. Biol. Chem.* 259:8494 (1984).

52. Rice AP, Kostura M, and Mathews MB: Identification of a 90 kDa polypeptide which associates with adenovirus VA RNA$_I$ and is phosphorylated by the double-stranded RNA-dependent protein kinase. *J. Biol. Chem.* 264:20632 (1989).

53. Hovanessian AG and Galabru J: The double-stranded RNA-dependent protein kinase is also activated by heparin. *Eur. J. Biochem.* 167:467 (1987).

54. Pestka S, Langer JA, Zoon KC, and Samuel CE: Interferons and their actions. *Annu. Rev. Biochem* 56:727 (1987).

55. Maran A and Mathews MB: Characterisation of the double-stranded RNA implicated in the inhibition of protein synthesis in cells infected with a mutant adenovirus defective for VA RNA$_I$. *Virology* 164:106 (1988).

56. Rice AP, Duncan R, Hershey JWB, and Kerr IM: Double-stranded RNA-dependent protein kinase and 2-5A system are both activated in interferon-treated encephalomyocarditis virus-infected HeLa cells. *J. Virol.* 54:894 (1985).

57. De Benedetti A and Baglioni C: Inhibition of mRNA binding to ribosomes by localised activation of dsRNA-dependent protein kinase. *Nature* 311:79 (1984).

58. Kaufman RJ and Murtha P: Translational control mediated by eukaryotic initiation factor-2 is restricted to specific mRNAs in transfected cells. *Mol. Cell. Biol.* 7:1568 (1987).

59. Dratewka-Kos E, Kiss I, Lucas-Lenard J, Mehta HB, Woodley CL, and Wahba AJ: Catalytic utilisation of eIF-2 and mRNA binding proteins are limiting in lysates from vesicular stomatitis virus infected L cells. *Biochemistry* 23:6184 (1984).

60. Schneider RJ and Shenk T: Impact of virus infection on host cell protein synthesis. *Annu. Rev. Biochem.* 56:317 (1987).

61. Bernstein HD, Sonenberg N, and Baltimore D: Poliovirus mutant that does not selectively inhibit host cell protein synthesis. *Mol. Cell. Biol.* 5:2913 (1985).

62. O'Neill RE and Racaniello V: Inhibition of translation in cells infected with a poliovirus 2A^{pro} mutant correlates with phosphorylation of the alpha subunit of eukaryotic initiation factor 2. *J. Virol.* 63:5069 (1989).

63. Black TL, Safer B, Hovanessian A, and Katze M: The cellular 68,000-M_r protein kinase is highly autophosphorylated and activated yet significantly degraded during poliovirus infection: implications for translational regulation. *J. Virol.* 63:2244 (1989).

64. Hovanessian AG, Galabru J, Meurs E, Buffet-Janvresse C, Svab J, and Robert N: Rapid decrease in the levels of the double-stranded RNA-dependent protein kinase during virus infections. *Virology* 159:126 (1987).

65. Rice AP and Kerr IM: The interferon-mediated double-stranded RNA-dependent protein kinase is inhibited in extracts from vaccinia virus-infected cells. *J. Virol.* 50:229 (1984).

66. Whitaker-Dowling P and Youngner J: Characterization of a specific kinase inhibitory factor produced by vaccinia virus which inhibits the interferon-induced protein kinase. *Virology* 137:171 (1984).

67. Whitaker-Dowling P and Youngner J: Vaccinia rescue of VSV from interferon-induced resistance: reversal of translation block and inhibition of protein kinase activity. *Virology* 131:128 (1983).

68. Whitaker-Dowling P and Youngner J: Vaccinia-mediated rescue of encephalomyocarditis virus from the inhibitory effects of interferon. *Virology* 152:50 (1986).

69. Imami F and Jacobs BL: Inhibitory activity for the interferon-induced protein kinase is associated with the reovirus serotype 1 σ3 protein. *Proc. Natl. Acad. Sci. U.S.A.* 85:7887 (1988).

70. Katze MG, Tomita J, Black T, Krug RM, Safer B, and Hovanessian A: Influenza virus regulates protein synthesis during infection by repressing autophosphorylation and activity of the cellular 68,000-M_r protein kinase. *J. Virol.* 62:3710 (1988).

71. O'Malley RP, Mariano TM, Siekierka J, and Mathews MB: A mechanism for the control of protein synthesis by adenovirus VA RNA$_I$. *Cell* 44:391 (1986).

72. O'Malley RP, Duncan RF, Hershey JWB, and Mathews MB: Modification of protein synthesis initiation factors and the shut-off of host protein synthesis in adenovirus-infected cells. *Virology* 168:112 (1989).

73. Reichel PA, Merrick WC, Siekierka J, and Mathews MB: Regulation of a protein synthesis initiation factor by adenovirus virus-associated RNA. *Nature* 313:196 (1985).

74. Katze MG, DeCorato D, Safer B, Galabru J, and Hovanessian AG: Adenovirus VA$_I$ RNA complexes with the 68,000 M_r protein kinase to regulate its autophosphorylation and activity. *EMBO J.* 6:689 (1987).

75. Edery I, Petryshyn R, and Sonenberg N: Activation of double-stranded RNA-dependent kinase (dsI) by the TAR region of HIV-1 mRNA: a novel translational control mechanism. *Cell* 56:303 (1989).

76. Braddock M, Chambers A, Wilson W, Esnouf MP, Adams SE, Kingsman AJ, and Kingsman SM: HIV-1 TAT "activates" presynthesized RNA in the nucleus. *Cell* 58:269 (1989).

77. Gross M and Redman R: Effect of antibody to the hemin-controlled translational repressor in rabbit reticulocyte lysate. *Biochim. Biophys. Acta* 908:123 (1987).

78. Jackson RJ and Hunt T: A novel approach to the isolation of rabbit reticulocyte haem-controlled eIF-2α protein kinase. *Biochim. Biophys. Acta* 826:224 (1985).

79. Hunt T: The control of protein synthesis in rabbit reticulocyte lysates. *Miami Winter Symp.* 16:321 (1979).

80. Chen, J-J, Yang JM, Petryshyn R, Kosower N, and London IM: Disulfide bond formation in the regulation of eIF-2α kinase by haem. *J. Biol. Chem.* 264:9559 (1989).

81. Hurst R, Schatz JR, and Matts RL: Inhibition of rabbit reticulocyte lysate protein synthesis by heavy metal ions involves the phosphorylation of the α subunit of the eukaryotic initiation factor 2. *J. Biol. Chem.* 262:15939 (1987).

82. De Haro C, De Herreros AG, and Ochoa S: Activation of the heme-stabilised translational inhibitor of reticulocyte lysates by calcium ions and phospholipid. *Proc. Natl. Acad. Sci. U.S.A.* 80:6843 (1983).

83. De Herreros AG, De Haro C, and Ochoa S: Mechanism of activation of the heme-stabilised translational inhibitor of reticulocyte lysates by calcium ions and phospholipid. *Proc. Natl. Acad. Sci. U.S.A.* 82:3119 (1985).

84. Gross M, Watt-Morse P, and Kaplansky DA: Characterization of a rabbit reticulocyte supernatant factor that reverses the translational inhibition of hemin deficiency. *Biochim. Biophys. Acta* 654:219 (1981).

85. Kudlicki W, Fullilove S, Kramer G, and Hardesty B: The 90-kDa component of reticulocyte heme-regulated eIF-2α (initiation factor 2 α-subunit) kinase is derived from the β subunit of spectrin. *Proc. Natl. Acad. Sci. U.S.A.* 82:5332 (1985).

86. Kudlicki W, Fullilove S, Read R, Kramer G, and Hardesty B: Identification of spectrin-related peptides associated with the reticulocyte haem-controlled α subunit of eukaryotic translational initiation factor 2 kinase and of a M_r 95,000 peptide that appears to be the catalytic subunit. *J. Biol. Chem.* 262:9695 (1987).

87. Rose DW, Welch WJ, Kramer G, and Hardesty B: Possible involvement of the 90 kDa heat shock protein in the regulation of protein synthesis. *J. Biol. Chem.* 264:6239 (1989).

88. Kudlicki W, Kramer G, and Hardesty B: Inhibition of protein synthesis by the β-subunit of spectrin. *FEBS Lett.* 200:271 (1986).

89. Proud CG and Pain VM: Purification and phosphorylation of initiation factor eIF-2 from rabbit skeletal muscle. *FEBS Lett.* 143:55 (1982).

90. Austin SA and Clemens MJ: The effects of haem on translational control of protein synthesis in cell-free extracts from fed and lysine-deprived Ehrlich ascites tumour cells. *Eur. J. Biochem.* 117:601 (1981).

91. Sarre TF, Hermann M, and Bader M: Differential effect of hemin-controlled eIF-2α kinases from mouse erythroleukemia cells on protein synthesis. *Eur. J. Biochem.* 183:137 (1989).

92. De Benedetti A and Baglioni C: Activation of hemin regulated initiation factor-2 kinase in heat-shocked HeLa cells. *J. Biol. Chem.* 261:338 (1986).

93. Safer B and Jagus R: Control of eIF-2 phosphatase activity in rabbit reticulocyte lysate. *Proc. Natl. Acad. Sci. U.S.A.* 76:1094 (1979).

94. Crouch D and Safer B: Purification and properties of eIF-2 phosphatase. *J. Biol. Chem.* 255:7918 (1980).

95. Pato MD, Adelstein RS, Crouch D, Safer B, Ingebritsen TS, and Cohen P: The protein phosphatases involved in cellular regulation. *Eur. J. Biochem.* 132:283 (1983).

96. Foulkes JG, Ernst V, and Levin DH: Separation and identification of type 1 and type 2 protein phosphatases from rabbit reticulocytes. *J. Biol. Chem.* 258:1439 (1983).

97. Stewart AA, Crouch D, Cohen P, and Safer B: Classification of an eIF-2 phosphatase as a type-2 protein phosphatase. *FEBS Lett.* 119:16 (1980).

98. Ernst V, Levin DH, Foulkes JG, and London IM: Effects of skeletal muscle protein phosphatase inhibitor-2 on protein synthesis and protein phosphorylation in rabbit reticulocyte lysates. *Proc. Natl. Acad. Sci. U.S.A.* 79:7092 (1982).

99. Clemens MJ, Galpine A, Austin SA, Panniers R, Henshaw EC, Duncan R, Hershey JWB, and Pollard JW: Regulation of polypeptide chain initiation in chinese hamster ovary cells with temperature-sensitive leucyl-tRNA synthetase: changes in phosphorylation of initiation factor eIF-2 and in the activity of guanine nucleotide exchange factor GEF. *J. Biol. Chem.* 262:767 (1987).

100. Kaempfer R: Regulation of eukaryotic translation, in Frankel-Conrat H and Wagner RR (Eds): *Comprehensive Virology* Vol. 19, New York, Plenum Press, pp 99–175 (1984).

101. Crouch D and Safer B: The association of eIF-2 with Met-tRNA$_i$ or eIF-2B alters the specificity of eIF-2 phosphatase. *J. Biol. Chem.* 259:10363 (1984).

102. Datta B, Chakrabarti D, Roy AL, and Gupta NK: Roles of a 67-kDa polypeptide in reversal of protein synthesis inhibition in heme-deficient reticulocyte lysate. *Proc. Natl. Acad. Sci. U.S.A.* 85:3324 (1988).

103. Duncan R and Hershey JWB: Initiation factor protein modifications and inhibition of protein synthesis. *Mol. Cell. Biol.* 7:1293 (1987).

104. Duncan R and Hershey JWB: Translational repression caused by chemical inducers of the stress response occurs by different pathways. *Arch. Biochem. Biophys.* 256:651 (1987).

105. Samuel CE, Duncan R, Knutson GS, and Hershey JWB: Mechanism of interferon action: increased phosphorylation of protein synthesis initiation factor eIF-2α in interferon-treated, reovirus-infected mouse L929 fibroblasts in vitro and in vivo. *J. Biol. Chem.* 259:13451 (1984).

Covalent Modification of Translational Components

Markus Hümbelin
F. Hoffmann-La Roche, Ltd
Basel, Switzerland

and

George Thomas
Friedrich Miescher Institute
Basel, Switzerland

9.1. INTRODUCTION

As reviewed in earlier chapters of this book, a large number of molecular components are involved in controlling gene expression at the translational level. Furthermore, in many well characterized systems, such as mitosis, meiosis, wound healing, starvation and the activation of cell growth by oncogenes and growth factors, control has been shown to be exerted at the level of initiation of protein synthesis. In the case of initiation, control could be regulated by modulating the level of certain translational components or by a reversible posttranslational modification of existing molecules. Results obtained in studies investigating the levels of specific initiation factors during different states of growth[1] and on eIF-2 phosphorylation (see Part II, Chapter 8 of this book) support the latter model. In addition, it is becoming evident from a number of other reports that phosphorylation/dephosphorylation is the major reversible intracellular posttranslational regulatory mechanism employed by eukaryotic organisms.[2] These two findings have recently focused attention on the possible role of phosphorylation in regulating protein synthesis through translation components other than eIF-2. In this chapter we shall concentrate on the reversible phosphorylation of specific proteins of the translational apparatus, and other modifications which may be involved in regulating translation.

9.2. PHOSPHORYLATION OF eIF-4F

The initiation factor eIF-4F is made up of three distinct subunits: eIF-4E, eIF-4A, and p220, which have molecular weights of 25, 45, and 220 kDa, respectively. It should be noted, however, that there is still some discussion regarding whether eIF-4E is a subunit of eIF-4F (see Part I, Chapter 5 of this book). Of the three proteins, eIF-4E and p220 have been shown to be phosphorylated *in vivo*. eIF-4E is thought to be the first protein of the translational apparatus with which mRNA interacts. In erythrocytes, five distinct electrophoretic species have been observed.[3] However, this may be due to deamination of the protein in aged erythrocytes, since in HeLa cells and young reticulocytes only two forms are predominantly observed[3-5] with the more acidic form containing phosphate. The major site of phosphorylation in both HeLa cells and reticulocytes is Ser_{53};[6] however, recent results suggest the existence of additional sites (see below). The phosphorylated and nonphosphorylated forms of eIF-4E occur in about equal amounts in normally growing cells, and both forms bind equally well to m^7GTP-Sepharose.[3-5] Recently, Ser_{53} was genetically altered to Ala and the cDNA transcribed and translated *in vitro*. The variant form of eIF-4E was retained on a m^7GTP-Sepharose and could be specifically eluted with m^7GTP.[7] Examination of the variant $eIF-4E^{Ala}$ by isoelectric polyacrylamide gel electrophoresis revealed two species which had the same pI as phosphorylated and nonphosphorylated parent $eIF-4E^{Ser}$. These preliminary results suggest second site phosphorylation and that phosphorylation at site Ser_{53} appears to play no role in capbinding, in agreement with earlier findings.[3-5] However, in the assay established by Hiremath et al.,[8] the variant $eIF-4E^{Ala}$, unlike the parent $eIF-4E^{Ser}$, was unable to form 48S ribosome initiation complexes in the presence of GMP-PNP. Furthermore, in the 48S ribosomal complex formed with $eIF-4E^{Ser}$ the predominant form was phosphorylated, suggesting a role for eIF-4E phosphorylation in enhancing the transfer of mRNA to the 48S ribosome initiation complex.

In vitro, a number of kinases have been tested for their ability to phosphorylate eIF-4E in the eIF-4F complex, including the cyclic AMP-dependent protein kinase, protease activated kinase I and II (the latter is also referred to as S6 kinase) and protein kinase C.[9] Only protein kinase C effectively phosphorylated eIF-4E *in vitro*.[9] The amount of phosphate incorporated is approximately 1 mol/mol of eIF-4E. More recent results show that phosphopeptide maps derived from eIF-4E phosphorylated either *in vitro* with protein kinase C, or *in vivo* by treating reticulocytes[10] or 3T3-LI cells (Morley and Traugh, personal communication) with phorbol-12-myristate-13-acetate (TPA), are equivalent. Furthermore, down-regulation of protein kinase C by long-term TPA treatment blocks subsequent insulin-induced phosphorylation of eIF-4E in 3T3-LI cells (Morley and Traugh, personal communication). Together the results strongly indicate that protein kinase C may be regulating the level of eIF-4E phosphorylation. It should be noted, however, that a second kinase has recently been identified from reticulocytes which also phosphorylates eIF-4E (Haas and Hagedorn, personal communication). The kinase has a M_r of 35 kDa and is distinct from protein kinase C. Future studies should resolve whether this enzyme plays a role in regulating eIF-4E phosphorylation *in vivo*.

Consistent with *in vitro* findings, studies *in vivo* show a rough correlation between the level of eIF-4E phosphorylation and the rate of protein synthesis. Earlier it was shown that the rate of protein synthesis is reduced during mitosis.[11] More recently, it has been shown that eIF-4E is less phosphorylated during this time, and that addition of eIF-4E to extracts prepared from mitotic cells stimulates protein synthesis *in vitro*.[12] In contrast, addition of eIF-4E to interphase cell extracts has no corresponding effect.[12] In agreement with these findings, heat shock, which leads to a shut-off of protein synthesis, also induces eIF-4E dephosphorylation;[5] and stimulation of either quiescent 3T3 LI cells with insulin, or Swiss 3T3 cells with serum or TPA leads to an increase in protein synthesis which is paralleled by increased eIF-4E phosphorylation (Morley and Traugh, personal communication; and Reference 13). It should be noted, however, that down-regulation of protein kinase C, which abrogates subsequent eIF-4E phosphorylation by TPA or insulin, does not block the mitogenic response or the activation of protein synthesis in quiescent Swiss 3T3 cells.[14, 15] This would suggest that if eIF-4E phosphorylation is mediated by protein kinase C it is not an obligatory step in the activation of protein synthesis when quiescent cells are stimulated to proliferate.

Much less is known about the phosphorylation of the p220 subunit of eIF-4F. Phosphorylation of this protein increases fivefold in reticulocytes treated with TPA.[10] and to the same extent when quiescent 3T3-LI cells are stimulated with insulin (Morley and Traugh, personal communication). The phosphopeptide map generated from the p220 subunit of reticulocytes treated with TPA appears similar to that induced by protein kinase C *in vitro* (Morley and Traugh, personal communication). In contrast, the phosphopeptide map derived from p220 of 3T3-LI cells stimulated with insulin contain a number of phosphopeptides distinct from those regulated by protein kinase C (Morley and Traugh, personal communication). Indeed, unlike eIF-4E phosphorylation, in protein kinase C down-regulated 3T3-LI cells, subsequent insulin stimulation still induces a fivefold increase in p220 phosphorylation, suggesting that this event is mediated by a protein kinase C-independent pathway.

9.3. PHOSPHORYLATION OF eIF-4B

As many as 12 to 20 distinct isoelectric forms of eIF-4B have been described.[16] Recently, sufficient amounts of the protein were obtained to generate amino acid sequence information and to clone the gene for this protein (Milburn et al., personal communication). Partial amino acid sequence data reveals that there may be two forms of eIF-4B. This would agree with recent Northern analyses which revealed the existence of two distinct size classes of mRNA. It has yet to be proven whether they represent transcripts from distinct genes or whether the two forms have been generated by alternative splicing events or different polyadenylation site usage. The amino acid sequence deduced from the longer cDNA clone reveals a protein with a mass of 70 kDa, which matches the peptide sequence obtained above, confirming the cDNA clone as eIF-4B. Three potential regulatory domains have been identified in the protein: an RNA binding domain, a very hydrophilic region, and a stretch of amino acids rich in serine residues. The RNA binding domain contains the highly conserved octapeptides and pentapeptides that have been recently described in RNA binding proteins.[17, 18] The middle domain contains 51% of alternating basic and acidic amino acids. The many isoforms of eIF-4B have been shown to be due to differential phosphorylation,[16] and the phosphorylation appears to be random, not ordered (Duncan and Hershey, personal communication). It might be speculated that the phosphorylation sites in eIF-4B reside in the area rich in serine residues at the carboxy-tail of the protein, as with 40S ribosomal protein S6 (see below).

A number of kinases have been shown to phosphorylate eIF-4B, including protein kinase C, an S6 kinase from rat liver, protease activated kinase II, cAMP-dependent protein kinase and casein kinase I.[10] TPA induces eIF-4B phosphorylation in both reticulocytes[9] and 3T3-LI cells (Morley and Traugh, personal communication), pointing to protein kinase C as a possible mediator of eIF-4B phosphorylation. Down-regulation of protein kinase C by long-term TPA treatment, however, only leads to a partial attenuation of the insulin-induced eIF-4B response in 3T3-LI cells. This would suggest that kinases other than protein kinase C also mediate eIF-4B phosphorylation. As suggested by the results with insulin in 3T3-LI cells, activation of cell growth has been shown to lead to an increase in eIF-4B phosphorylation in HeLa cells.[16] This increase roughly parallels the increase in protein synthesis.[16] Furthermore, stimulation of cell growth in quiescent 3T3 cells is accompanied by alterations in mRNA usage which are controlled at the level of translation as well as transcription.[19, 20] Thus, phosphorylation of eIF-4B could alter its activity towards mRNA in general or towards a specific subclass of mRNAs.

9.4. S6 PHOSPHORYLATION

S6 is an integral protein of the small 40S subunit, has a molecular weight of 28,661 Da, and, like the other 40S ribosomal proteins, is present in one copy per ribosome (for a review see Reference 21). The protein has been localized to the mRNA binding site of the 40S subunit by a number of methods, including crosslinking to synthetic mRNA,[22] protection experiments,[23] and immuno-electron microscopy.[24] During maturation of *Xenopus* oocytes,[25] partial hepotectomy,[26] refeeding of starved animals[27] or activation of quiescent cells to proliferate by mitogens or oncogenes,[28] up to 5 mol of phosphate are incorporated per mole of 56 protein. By phosphopeptide analysis the incorporation of phosphate into seryl residues has been shown to follow a specific order.[29] This suggests that the phosphorylation of one site could regulate the phosphorylation of the next, as described for glycogen synthetase.[30] All the sites of phosphorylation have been shown to reside in a 32-amino acid cyanogen bromide peptide located at the carboxyl end of the molecule.[31] Furthermore, phosphorylation appears to proceed in the order $Ser_{236} \rightarrow Ser_{235} \rightarrow Ser_{240} \rightarrow Ser_{244} \rightarrow Ser_{247}$. The functional impact of each of these sites on protein synthesis remains to be established, although from *in vivo* and *in vitro* studies (see below), it would appear to be the late sites which alter rates of protein synthesis.

As the structure of the ribosome begins to unravel[32] it is clear that S6 may occupy a functional site involved in the binding of tRNA and mRNA as well as interact with the larger 60S subunit.[33] This may explain earlier results indicating that phosphorylation of S6 causes conformational changes in proteins[34] as well as in the rRNAs[35] of the small and large subunits.[34, 35] Initial *in vitro* studies failed to show any functional role of S6 phosphorylation in protein synthesis,[36–38] even though *in vivo* studies indicated that ribosomes containing highly phosphorylated S6 had a selective advantage in entering polysomes.[28, 39] Since this selection only appears to reside in the tri- to pentaphosphorylated derivatives of S6, it may be that the 40S ribosomes used in earlier studies were not as extensively phosphorylated, and consequently may have lacked phosphate in these positions. This conclusion is consistent with recent findings from Palen and Traugh[40] demonstrating that in a reconstituted protein synthesizing system from reticulocytes, incorporation of 1.5 mol of phosphate per mole of S6 by the cAMP-dependent kinase has no effect on the ability of the ribosomes to bind or translate mRNAs for α- and β-globin. From two-dimensional polyacrylamide gels[40] and earlier studies of Wettenhall and Morgan,[41] it is most likely that in these experiments S6 was phosphorylated in only the first two sites, Ser_{235} and Ser_{236}. Employing a protease-activated kinase from liver, Palen and Traugh incorporated an average of 2.5 mol of phosphate per mole of S6. Under these conditions, they claimed an overall fourfold stimulation of protein synthesis and selective effects on α-

and β-globin synthesis in the reconstituted system. Based on two-dimensional polyacrylamide gel analysis, the sites of phosphorylation presumably include the later sites, Ser_{240}, Ser_{244} and Ser_{247}, as well as the early sites, Ser_{235} and Ser_{236}. This is an important finding which has yet to be substantiated by other laboratories. Consistent with these results, it has recently been shown that extracts can be prepared from *Xenopus* oocytes which faithfully mimic the progesterone-induced twofold increase observed in protein synthesis *in vivo*.[42] The increase cannot be mimicked by the addition of initiation factors eIF-4F, 2, 2B, or 4A but in preliminary studies can be by addition of salt-washed ribosomes (Patrick and Pain, personal communication). The increase in progesterone-induced protein synthesis in oocytes is closely associated with increased S6 phosphorylation.[25] It will be of interest to test the ability of the S6 kinase (see below) to activate protein synthesis in such extracts.

Many kinases have been reported to phosphorylate S6 *in vitro*, including the cyclic AMP- and GMP-dependent protein kinases,[43] protein kinase C,[44, 45] calmodulin-dependent protein kinase,[46] casein kinase I,[47] and the protease-activated kinases PAK II[48] and H4P kinase.[49] The extent to which most of these enzymes phosphorylate S6 is limited and their modes of activation would not be consistent with a mitogen-induced phosphorylation cascade. A few years ago this led to the search for such an enzyme, and the subsequent finding that extracts from serum-stimulated 3T3 cells were up to 25-fold more potent in phosphorylating S6 than equivalent extracts from quiescent cells.[50] To recover full kinase activity from cell extracts required the presence of phosphatase inhibitors, which argued that phosphorylation controlled the activity of the kinase or some regulatory component of the kinase. The kinase responsible for phosphorylating S6 *in vitro* was recently purified to homogeneity from Swiss 3T3 cells and identified as a single polypeptide of M_r 70 kDa.[51] The kinase itself is activated by serine/threonine phosphorylation and selectively inactivated by a type 2A phosphatase.[52] *In vitro* it autophosphorylates, is highly specific for S6 and phosphorylates all the same S6 phosphopeptides as observed *in vivo*. The kinase is a rare cellular protein, and the amount of material recovered from 3T3 cells was insufficient to begin antibody production or sequence studies. However, knowing the identity of the mitogen-activated S6 kinase allowed the screening of whole cell animal systems in which S6 phosphorylation can be induced to high levels. This led to the identification of an activated S6 kinase from livers of cycloheximide-treated rats.[27] The chromatographic behavior, molecular weight, and peptide pattern of the autophosphorylated enzyme are equivalent to those obtained from the S6 kinase of 3T3 cells. Employing extraction procedures similar to those described above, a number of laboratories have detected the existence of an equivalent activity in other biological systems. Only one of these has been purified to homogeneity: a 92 kDa kinase from *Xenopus* eggs.[53] The 3T3 cell enzyme has a number of features in common with the Xenopus egg enzyme, including its Km for ATP, requirement for Mg^{2+}, and restricted substrate specificity.[54, 55] However, there is a large difference in M_r, abundance,[54, 55] and sensitivity to Mn^{2+}. Furthermore, preliminary sequence data shows that the 70 kDa protein (Ferrari and Totty, personal communication) is distinct from the 92 kDa protein. It will be of interest to determine whether the 70 kDa protein is also present in *Xenopus* oocytes and whether it becomes activated during maturation.

Since the kinase is activated by serine/threonine phosphorylation it argues that there must be at least one kinase between the receptor tyrosine kinase and the S6 kinase, a putative S6 kinase kinase. Recent results show that the S6 kinase is biphasically activated, with the second phase under the control of protein kinase C. Loss of the second phase by down-regulation of protein kinase C leads to a decrease in the ability of epidermal growth factor (EGF) to induce S6 phosphorylation, the shift of inactive 80S ribosomes into polysomes, and cell growth.[15] The biphasic nature of kinase activation is shared with a number of mitogens, which suggests that the activation of the S6 kinase may be more complicated than the simple protein phosphorylation cascade model presented above.

9.5. AMINOACYL-tRNA SYNTHETASES

The aminoacyl-tRNA synthetases are a family of enzymes that play a central role in protein biosynthesis: aminoacylation of tRNA. Several synthetases also catalyze the synthesis of P^1, P^4-bis (5′-adenosyl)-tetraphosphate $(A_{p4}A)$, which may play a central role in the G_1/S transition. Most of the synthetases examined to date are phosphorylated *in vivo*.[56, 57] The phosphorylation of glutaminyl- and aspartyl-tRNA synthetases, two of eight synthetases in the high M_r synthetase complex, as well as threonyl-tRNA synthetase, is stimulated *in vivo* in response to treatment of reticulocytes with 8-bromo-cAMP.[58, 59] The glutamyl-tRNA synthetase component of the complex is selectively phosphorylated in response to phorbol esters (Venema and Traugh, personal communication).

Casein kinase I copurifies with the high M_r synthetase complex containing several of the synthetases, and can phosphorylate up to four of them *in vitro*. Phosphorylation of these synthetases by casein kinase I can result in up to a 40% decrease in their ability to aminoacylate certain tRNAs.[60] In addition, cyclic AMP-dependent protein kinase, protein kinase C and the protease activated kinase I all phosphorylate different sets of synthetases *in vitro*.[56] However, more striking is the recent report that phosphorylation of threonyl- and seryl-tRNA synthetases by the cyclic AMP-dependent protein kinase can induce up to a sixfold increase in their ability to synthesize $A_{p4}A$.[59] With the recent assertion that cyclic AMP may play a positive role in the activation of cell growth,[61] it would be of interest to learn whether the phosphorylation state of these translational components change in response to specific growth factors and oncogenes.

9.6. HYPUSINATION OF eIF-4D

Hypusination is a two-step posttranslational modification of lysine residues. First, the 4-amino butyl moiety of spermidine is transferred to the ε-amino group of lysine and the resulting deoxyhypusine is then hydroxylated to give hypusine.[62-64] Initiation factor eIF-4D is the only known protein to be hypusinated.[65] The factor is highly conserved among eukaryotes,[66] showing 61% identity and 84% homology between mammals and yeast (Schnier, personal communication). Recently, Šmit-McBride et al.[67] showed that nonhypusinated eIF-4D fails to stimulate methionyl-puromycin synthesis *in vitro*. It also does not inhibit the native factor, suggesting that hypusination is required for the interaction of the eIF-4D with the 80S initiation complex. Consistent with these findings, when Lys_{50} was changed to arginine and transfected into cells it had no effect on protein synthesis. Over-expression of the wild type form of eIF-4D also had no effect, suggesting that normal cellular levels of eIF-4D are saturating. In a second approach, Park[68] purified two isoelectrically distinct non-hypusinated precursors of eIF-4D from spermidine-depleted cells. As in the results above, neither form of eIF-4D was able to stimulate methionyl-puromycin synthesis *in vitro*. Thus, in normally growing cells hypusination of eIF-4D does not appear limiting. However, it is possible that under as yet undescribed conditions of growth hypusination could be limiting, and thus regulate translation.

REFERENCES

1. Duncan R and Hershey JWB: Regulation of initiation factors during translational repression caused by serum depletion, abundance, synthesis and turnover rates. *J. Biol. Chem.* 260:5486 (1985).
2. Edelman AM, Blumenthal DK, and Krebs EG: Protein serine/threonine kinases. *Annu. Rev. Biochem.* 56:567 (1987).
3. Rychlik W, Gardner PR, Vanamom TC, and Rhoads RE: Structural analysis of the messenger RNA cap-binding protein. *J. Biol. Chem.* 261:71 (1986).
4. Buckley R and Ehrenfeld E: Two-dimensional gel analysis of the 24-kDa cap-binding protein from polio virus-infected and uninfected Hela cells. *Virology* 152:497 (1986).
5. Duncan R and Hershey JWB: Regulated phosphorylation and low abundance of Hela cell initiation factor eIF-4F suggest a role in translational control. *J. Biol. Chem.* 262:380 (1987).

6. Rychlik W, Russ MA, and Rhoads RE: Phosphorylation site of eukaryotic initiation factor 4E. *J. Biol. Chem.* 262:10434 (1987).

7. Joshi-Barve S, Rychlik W, and Rhoads RE: Alteration of the major phosphorylation site of eukaryotic protein synthesis initiation factor 4E prevents its association with the 48S initiation complex. *J. Biol. Chem.* 265:2979–2983 (1990).

8. Hiremath LS, Hiremath ST, Rychlik W, Joshi S, Domier LL, and Rhoads RE: *In vitro* synthesis, phosphorylation and localization of 48S initiation complexes of human protein synthesis initiation factor 4E. *J. Biol. Chem.* 264:1132 (1989).

9. Tuazon PT, Merrick WC, and Traugh JA: Comparative analysis of phosphorylation of translational initiation and elongation factors by seven protein kinases. *J. Biol. Chem.* 264:2773 (1989).

10. Morley SJ and Traugh JA: Phorbol esters stimulate phosphorylation of eukaryotic initiation factors 3, 4B and 4F. *J. Biol. Chem.* 264:2401 (1989).

11. Fan H and Penmann S: Regulation of protein synthesis in mammalian cells. *J. Mol. Biol.* 50:655 (1970).

12. Bonneau A-M and Sonenberg N: Involvement of the 24-kDa cap-binding protein in regulation of protein synthesis in mitosis. *J. Biol. Chem.* 262:11134 (1987).

13. Kaspar RL, White MW, Rychlik W, Rhoads RE, and Morris DL: Simultaneous cytoplasmic redistribution of ribosomal protein L32 mRNA and phosphorylation of eIF-4E after mitogenic stimulation of Swiss 3T3 cells. *J. Biol. Chem.* (in press).

14. Coughlin SR, Lee WMF, Williams PW, Giels GN, and Williams LT: c-MYC gene expression is stimulated by agents that activate protein kinase C and does not account for the mitogenic effect of PDGF. *Cell* 43:243 (1985).

15. Šuša M, Olivier AR, Fabbro D, and Thomas G: EGF induces biphasic S6 kinase activation: late phase is protein kinase C-dependent and contributes to mitogenicity. *Cell* 57:817 (1989).

16. Duncan R and Hershey JWB: Regulation of initiation factors during translational repression caused by serum deprivation. Covalent modification. *J. Biol. Chem.* 260:5493 (1985).

17. Dreyfuss G, Swanson MS, and Piñol-Rona S: Heterogeneous nuclear ribonucleoprotein peptides and the pathway of mRNA formation. *Trends Biochem. Sci.* 13:86 (1988).

18. Bandziulis RJ, Swanson MS, and Dreyfuss G: RNA-binding proteins as developmental regulators. *Genes Dev.* 3:431 (1989).

19. Thomas G, Thomas G, and Luther H: Transcriptional and translational control of cytoplasmic proteins after serum stimulation of quiescent Swiss 3T3 cells. *Proc. Natl. Acad. Sci. U.S.A.* 78:5712 (1981).

20. Thomas G and Thomas G: Translational control of mRNA expression during the early mitogenic response in Swiss mouse 3T3 cells: identification of specific proteins. *J. Cell. Biol.* 103:2137 (1986).

21. Kozma SC, Ferrari S, and Thomas G: Unmasking a growth factor/oncogene-activated S6 phosphorylation cascade. *Cell. Signalling* 1:219 (1989).

22. Terao K and Ogata K: Crosslinker between poly U and ribosomal proteins in 40S subunits induced by UV irradiation. *J. Biochem.* 86:605 (1979).

23. Terao K and Ogata K: Effects of preincubation of poly U with 40S subunits on the interactions of 40S subunit proteins with aurintricarboxylic acid and with N^1,N^1-P-phenylenedimaleinide. *J. Biochem.* 86:597 (1979).
24. Bommer U-A, Noll F, Lutsch G, and Bielka H: Immunochemical detection of proteins of the small subunit of rat liver ribosomes involved in binding of the ternary initiation complex. *FEBS Lett.* 111:171 (1980).
25. Nielsen PJ, Thomas G, and Maller J: Increased phosphorylation of ribosomal protein S6 during meiotic maturation of *Xenopus* oocytes. *Proc. Natl. Acad. Sci. U.S.A.* 79:2937 (1982).
26. Gressner AM and Wool IG: The phosphorylation of liver ribosomal proteins *in vivo*. *J. Biol. Chem.* 249:6917 (1974).
27. Kozma SC, Lane HA, Ferrari S, Luther H, Siegmann M, and Thomas G: Purification of an activated 40S ribosomal S6 kinase from rat liver: identity with the mitogen-activated S6 kinase. *EMBO J.* 13:4125 (1989).
28. Thomas G, Martin-Pérez, J, Siegmann M, and Otto AM: The effect of serum EGF, PGF_{2a} and insulin on S6 phosphorylation and the initiation of protein and DNA synthesis. *Cell* 30:235 (1982).
29. Martin-Pérez J and Thomas G: Ordered phosphorylation of 40S ribosomal protein S6 after serum stimulation of quiescent 3T3 cells. *Proc. Natl. Acad. Sci. U.S.A.* 80:926 (1983).
30. Fiol CJ, Mahrenholz AM, Wang Y, Roeske R, and Roach P: Formation of protein kinase recognition sites by covalent modification of the substrate. *J. Biol. Chem.* 262:14042 (1987).
31. Krieg J, Hofsteenge J, and Thomas G: Identification of the 40S ribosomal protein S6 phosphorylation sites induced by cycloheximide. *J. Biol. Chem.* 263:11473 (1988).
32. Stern S, Powers T, Changchien LM, and Noller HF: RNA-protein interactions in 30S ribosomal subunits: folding and function of 16S rRNA. *Science* 244:783 (1989).
33. Nygard O and Nika H: Identification by RNA-protein crosslinking of ribosomal proteins located at the interface between the small and large subunits of mammalian ribosomes. *EMBO J.* 1:357 (1982).
34. Kisilevsky R, Treloar MA, and Weiler L: Ribosomal conformational changes associated with protein S6 phosphorylation. *J. Biol. Chem.* 259:1351 (1984).
35. Hallberg RL, Wilson PG, and Sutton C: Regulation of ribosome phosphorylation and antibiotic sensitivity in *Tetrahymena thernophila:* a correlation. *Cell* 26:47 (1981).
36. Leader DP, Thomas A, and Voorma HO: The protein synthetic activity *in vitro* of ribosomes. Differing in the extent of phosphorylation of their ribosomal proteins. *Biochim. Biophys. Acta* 656:69 (1981).
37. Mastropaolo W and Henshaw EG: Phosphorylation of ribosomal protein S6 in the Ehrlich ascites tumor cell. Lack of effect of phosphorylation upon ribosome function in vitro. *Biochim. Biophys. Acta* 656:246 (1981).
38. Tas PWL and Martini OHW: Are highly phosphorylated 40S subunits preferentially utilized during protein synthesis in a cell-free system from Hela cells? *Eur. J. Biochem.* 163:553 (1987).

39. Duncan R and McConkey E: Preferential utilization of phosphorylated 40S ribosomal subunits during initiation complex formation. *Eur. J. Biochem.* 123:535 (1982).

40. Palen E and Traugh JA: Phosphorylation of ribosomal protein S6 by cAMP-dependent protein kinase and mitogen-activated S6 kinase differentially alters translation of globin mRNA. *J. Biol. Chem.* 262:3518 (1987).

41. Wettenhall REH and Morgan FJ: Phosphorylation of hepatic ribosomal protein S6 on 80 and 40S ribosomes. *J. Biol. Chem.* 259:2084 (1984).

42. Patrick T, Lewer CE, and Pain J: Preparation and characterization of cell-free protein synthesis systems from oocytes and eggs of *Xenopus laevis. Development* 106:1 (1989).

43. DelGrande R and Traugh J: Phosphorylation of 40S ribosomal subunits by cAMP-dependent, cGMP-dependent and protease-activated protein kinases. *Eur. J. Biochem.* 123:421 (1982).

44. LePuch CJ, Ballester R, and Rosen OM: Purified rat brain calcium- and phospholipid-dependent protein kinase phosphorylates 40S ribosomal protein S6. *Proc. Natl. Acad. Sci. U.S.A.* 80:6858 (1983).

45. Parker PK, Katan M, Waterfield MA, and Leader DP: The phosphorylation of eukaryotic ribosomal protein S6 by protein kinase C. *Eur. J. Biochem.* 148:579 (1985).

46. Gorlick F, Cohn J, Freedman S, Delahunt N, Gershoni J, and Jamieson J: Calmodulin-stimulated protein kinase activity from rat pancreas. *J. Cell Biol.* 97:1294 (1983).

47. Cobb NH and Rosen OM: Description of a protein kinase derived from insulin-treated 3T3-LI cells that catalyzes the phosphorylation of ribosomal protein S6 and casein. *J. Biol. Chem.* 258:12472 (1983).

48. Lubben TH and Traugh JA: Cyclic nucleotide-independent protein kinase from rabbit reticulocytes. *J. Biol. Chem.* 258:13992 (1983).

49. Donahue M and Masaracchia RA: Phosphorylation of ribosomal protein S6 at multiple sites by a cyclic AMP-independent protein kinase from lymphoid cells. *J. Biol. Chem.* 259:435 (1984).

50. Novak-Hofer I and Thomas G: An activated S6 kinase in extracts from serum and epidermal growth factor-stimulated Swiss 3T3 cells. *J. Biol. Chem.* 259:5995 (1984).

51. Jenö P, Ballou LM, Novak-Hofer I, and Thomas G: Identification and characterization of a mitogen-activated S6 kinase. *Proc. Natl. Acad. Sci. U.S.A.* 85:406 (1988).

52. Ballou LM, Siegmann M, and Thomas G: S6 kinase in quiescent mouse 3T3 cells is activated by phosphorylation in response to serum treatment. *Proc. Natl. Acad. Sci. U.S.A.* 85:7154 (1988).

53. Erikson E and Maller JL: A protein kinase from *Xenopus* eggs specific for ribosomal protein S6. *Proc. Natl. Acad. Sci. U.S.A.* 82:742 (1985).

54. Erikson E and Maller JL: Purification and characterization of a protein kinase from *Xenopus* eggs highly specific for ribosomal protein S6. *J. Biol. Chem.* 261:350 (1986).

55. Jenö P, Jäggi N, Luther H, Siegmann M, and Thomas G: Purification and characterization of a 40S ribosomal protein S6 kinase from vanadate-stimulated Swiss 3T3 cells. *J. Biol. Chem.* 264:1293 (1989).

56. Traugh JA and Pendergast AM: Regulation of protein synthesis by phosphorylation of ribosomal protein S6 and aminoacyl-tRNA synthetases, in *Progress in Nucleic Acid Research and Molecular Biology*, Vol. 33, New York, Academic Press, 1986, 195.

57. Traugh JA: Approaches to examine the role of multiple serine protein kinases in the coordinate regulation of cell growth, in Mond JJ, Cambier JC, and Weiss A (Eds): *Regulation of Cell Growth and Activation* Vol. I, New York, Raven Press, 1989, 173.

58. Pendergast AM, Venema RC, and Traugh JA: Regulation of phosphorylation of aminoacyl-tRNA synthetases in the high molecular weight core complex in reticulocytes. *J. Biol. Chem.* 262:5939 (1987).

59. Dang CV and Traugh JA: Phosphorylation of threonyl- and seryl-tRNA synthetase by cAMP-dependent protein kinase. *J. Biol. Chem.* 264:5861 (1989).

60. Pendergast AM and Traugh JA: Alteration of aminoacyl-tRNA synthetase activities by phosphorylation with casein kinase I. *J. Biol. Chem.* 260:11769 (1985).

61. Rozengurt E: Early signals of the mitogenic response. *Science* 234:161 (1986).

62. Park MH, Cooper HL, and Folk JE: Identification of hypusine, an unusual amino acid, in a protein from human lymphocytes and of spermidine as its biosynthetic precursor. *Proc. Natl. Acad. Sci. U.S.A.* 78:2869 (1981).

63. Park MH, Cooper HL, and Folk JE: The biosynthesis of protein-bound hypusine (N^E-(4-amino-2-hydroxybutyl)lysine); lysine as the amino acid precursor and the intermediate role of deoxyhypusine (N^E-(4-aminobutyl)lysine). *J. Biol. Chem.* 257:7217 (1982).

64. Park MH, Liberato DJ, Yergey AL, and Folk JE: The biosynthesis of a protein-bound hypusine (N^E-(4-amino-2-hydroxybutyl)lysine); alignment of the butylamine segment and source of the secondary amino nitrogen. *J. Biol. Chem.* 259:12123 (1984).

65. Cooper HL, Park MH, Folk JE, Safer B, and Braverman R: Identification of the hypusine-containing protein Hy$^+$ as translation initiation factor eIF-4D. *Proc. Natl. Acad. Sci. U.S.A.* 80:1854 (1983).

66. Gordon ED, Mora R, Meredith S, Lee C, and Lindquist SL: Eukaryotic initiation factor 4D, the hypusine-containing protein, is conserved among eukaryotes. *J. Biol. Chem.* 262:16585 (1987).

67. Šmit-McBride Ž, Schnier J, Kaufman RJ, and Hershey JWB: Protein synthesis initiation factor eIF-4D; functional comparison of native and unhypusinated forms of the protein. *J. Biol. Chem.* 264:18527 (1989).

68. Park MH: The essential role of hypusine in eukaryotic translation initiation factor 4D (eIF-4D); purification of eIF-4D and its precursors and comparison of their activities. *J. Biol. Chem.* 264:18531 (1989).

Examples of Eukaryotic Translational Control: GCN4 and Ferritin

Alan G. Hinnebusch and Richard D. Klausner
Laboratory of Molecular Genetics
and
Cell Biology and Metabolism Branch
National Institute of Child Health and Human Development
National Institutes of Health
Bethesda, Maryland

10.1. INTRODUCTION

The controlled regulation of the expression of genetic information is central to all forms of life. Over the past decade, mechanisms underlying gene regulation have begun to yield to experimental inquiry, and an increasing number of loci of regulation have become apparent. Currently, most available information relates to the regulation at early (initiation of DNA transcription) and late (posttranslational) stages in gene expression. Comparatively little is known about the regulation of the complex life of the RNA molecules that mediate the conversion of the genetic code into protein products. These events can be divided into the posttranscriptional production of mature cytoplasmic mRNA and the fate of that mature mRNA. The former includes the splicing of transcripts, posttranscriptional covalent modifications (capping, polyadenylation, sequence editing) and nuclear-cytoplasmic transport. While it is clear that some, if not all, of these processes can be regulated, little is known about the molecular mechanisms underlying such regulation. The fate of mature mRNA can be regulated by controlling either the stability of a message or its translation. Studies in a variety of systems have pointed to two general classes of specifically translationally regulated mRNAs. These can be distinguished by the sequence/structural elements contained within the 5' untranslated region of the messages which determine the ability of these messages to be translationally regulated. One group of messages contain multiple upstream open reading frames. The best studied example of this comes from the yeast *Saccharomyces cerevisiae* and concerns the translational regulation of GCN4. The second class of mRNAs contain particular sequence/structural motifs that appear to be the binding site for specific regulatory proteins. The best studied example of the latter comes from work on the regulated expression of proteins involved in eukaryotic iron metabolism. In particular, the translational control of ferritin has provided the clearest model for this second, *cis-trans,* model of translational control. In this chapter we will review translational control by examining these two different eukaryotic systems.

10.2. GENE-SPECIFIC TRANSLATIONAL CONTROL IN *S. CEREVISIAE* BY UPSTREAM OPEN-READING FRAMES

In most eukaryotic transcripts, including those of the yeast *S. cerevisiae,* the initiation codon for synthesis of the encoded protein is the 5' proximal AUG triplet.[1, 2] Insertion of an additional AUG codon into the mRNA leader generally leads to reduced translation of the downstream coding sequences, even when a stop codon located in-frame with the first AUG codon precedes the second initiation site. The strong preference for 5' proximal AUG codons as translational start sites has been explained by proposing that the initiation complex assembles at the 5' end of the mRNA and scans in the 3' direction until an AUG codon is encountered, whereupon translation begins. The fact that an in-frame stop codon generally does not abolish the inhibitory effect of an upstream AUG codon suggests that reinitiation at internal start sites is usually inefficient.[3-6]

Translational inhibition by an upstream open-reading-frame (uORF) can be reduced by changing the sequence context of its AUG codon in a way expected to lower initiation at that site. Presumably, these alterations allow the initiation complex to bypass the first AUG codon and initiate translation at a downstream start site.[7] The inhibitory effect of an uORF has also been reduced by moving it further upstream from the start site for the protein-coding sequences. To account for the latter effect, it was suggested that the efficiency of reinitiation is elevated in response to a greater separation between the two coding sequences, perhaps because of increased time available to reassemble an initiation complex before the second start site is encountered.[8] Recent results show that both the length and the sequences that surround the stop codon of an uORF also affect its inhibitory effect on translation downstream, possibly by influencing the probability of resumed scanning and reinitiation following termination at the uORF.[9]

Described below are two transcripts in *S. cerevisiae* that represent the first known cases in which the inhibitory effects of naturally occurring uORFs are modulated in response to the physiological state of the cell. The uORFs in these mRNAs therefore function in translational control of gene expression. Analysis of the sequence requirements for the regulatory roles of these elements suggests that the control mechanisms involved are different in the two cases. A single uORF regulates translation of *CPA1* mRNA, and it appears that the small uORF-encoded peptide plays an important role in modulating the flow of ribosomes to the *CPA1* start codon according to the availability of arginine. By contrast, multiple uORFs are required for translational control of *GCN4* expression. The available evidence suggests that prior translation of the 5'-proximal uORF, coupled with modifications of the translational apparatus under conditions of amino acid starvation, allows ribosomes to pass through the translational barrier at the 3'-proximal uORFs and initiate GCN4 protein synthesis.

10.2.1 *CPA1* Expression is Regulated by a Single uORF

CPA1 encodes an enzyme in the arginine biosynthetic pathway that is repressed by arginine about fivefold. Little or no change in the steady-state level of *CPA1* mRNA was observed in response to an arginine supplement by RNA blot-hybridization analysis, suggesting that enzyme repression occurs at the translational level.[10] The *CPA1* transcript contains a single uORF of 25 codons.[11, 12] Removal of the uORF start codon leads to derepression of *CPA1* enzyme expression in the presence of excess arginine, while apparently having little effect on the *CPA1* mRNA level. Interestingly, a variety of missense and nonsense mutations mapping throughout the upstream ORF also lead to derepressed *CPA1* expression in a *cis*-dominant fashion (*CPA1-O* mutations)[13] (Figure 1A).

Given their various positions within the uORF, it seems unlikely that the *CPA1-O* mutations impair regulation by interfering with recognition of the uORF AUG codon as a translational start site. An alternative explanation is that the peptide product of the 25-codon uORF itself functions in translational repression. Because the *CPA1-O* mutations are only expressed in *cis,* the putative leader peptide would necessarily be confined to regulating the translation of its own mRNA.

Another important aspect of the *CPA1* upstream ORF is that removal of its AUG codon does not lead to increased *CPA1* expression under derepressing conditions, implying that during growth in media lacking arginine, the *CPA1* upstream AUG codon has little effect on initiation at the *CPA1* start site. This

A.

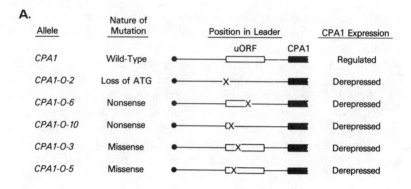

B.

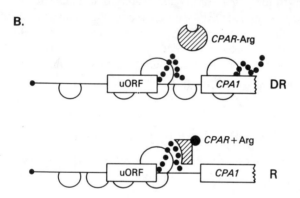

Figure 1. Translational control of *CPA1* expression. (A) The effects of point mutations in the uORF on expression of *CPA1* enzyme activity. The positions of point mutations in the uORF are indicated by "X"s. (B) Hypothetical model for uORF-mediated translational repression. Most scanning 40S subunits bypass the uORF start codon but initiate efficiently at the *CPA1* coding sequences. Translation of the uORF is without consequence in the absence of arginine (− Arg). In the presence of arginine, a *trans*-acting factor designated CPAR is altered and acts to arrest the progression of ribosomes translating the uORF, which in turn, blocks the advance of scanning 40S subunits to the *CPA1* start codon. The structure of the peptide plays a critical role in the interaction of CPAR + Arg with ribosomes.

finding was interpreted to indicate that most ribosomes bypass the uORF AUG codon and initiate at the *CPA1* start site downstream. To account for the regulatory function of the leader peptide it was proposed that, in the presence of arginine, the small amount of leader peptide produced blocks initiation at the *CPA1* coding sequences by the majority of 40S subunits that bypass the uORF AUG codon[13] (Figure 1B). The nature of the interaction between the uORF peptide and the ribosome was left unspecified. It remains to be determined whether missense mutations in the uORF will be tolerated by the regulatory mechanism, as might be expected if the only function of the uORF is to encode the leader peptide.

10.2.2. Translational Control of *GCN4* Expression by Multiple uORFs

Regulation of *GCN4* expression—The GCN4 protein of *S. cerevisiae* is a positive regulator of 30 to 40 unlinked genes encoding enzymes in eleven different amino acid biosynthetic pathways. GCN4 stimulates transcription of these genes in response to starvation for any one of at least ten amino acids by binding to a short nucleotide sequence present upstream from each gene under its control (reviewed in Reference 14).

Transcriptional activation by GCN4 increases in response to amino acid starvation because synthesis of GCN4 protein is elevated under these conditions. Analysis of *GCN4-lacZ* fusions suggests that GCN4 expression increases 10- to 50-fold upon starvation for a single amino acid.[15, 16] This derepression requires the products of the positive regulatory genes *GCN2* and *GCN3* that are needed for derepression of the structural genes under GCN4 control.[17] Negative regulatory factors in this system encoded by *GCD1, GCD2, GCD10, GCD11,* and *GCD13* function as repressors of *GCN4* expression under nonstarvation conditions.[17, 18] GCN2 and GCN3 are thought to function by inactivating one or more of these GCD factors because mutational inactivation of a *GCD* gene makes GCN2 and GCN3 completely dispensable for high-level *GCN4* expression.

Regulation of *GCN4* expression by these *trans*-acting factors occurs largely at the translational level. One piece of evidence leading to this conclusion is that *gcn2* and *gcn3* mutations impair the derepression of *GCN4-lacZ* enzyme activity under starvation conditions without reducing the steady-state amount of the fusion transcript.[16, 17] A second indication is that regulation of *GCN4-lacZ* enzyme expression by *GCN* and *GCD* factors remains intact following replacement of the *GCN4* promoter with that of the *GAL1* gene.[15, 17, 18] (*GAL1* transcription is not subject to general amino acid control.)

A third piece of evidence for translational control is that the *cis*-acting sequences required to regulate *GCN4* expression are present in the leader of *GCN4* mRNA. These sequences include four short open-reading-frames (uORFs) of only two or three codons in length (Figure 2A).[15, 16] Removal of the four uORFs from the *GCN4-lacZ* transcript, either by deletion or point mutations in the four ATG start codons, results in constitutively derepressed fusion enzyme expression, independent of amino acid availability (Figure 2B). As expected, amino acid biosynthetic enzymes under GCN4 control are derepressed when the same mutations are introduced upstream from the authentic *GCN4* protein-coding sequences. These mutations have little or no effect on the steady-state levels of *GCN4* or *GCN4-lacZ* mRNAs, showing that the uORFs affect *GCN4* expression at the translational level.[15–17]

Mutations in *GCD* genes cause little additional increase in *GCN4-lacZ* enzyme expression once the uORFs are removed, suggesting that *GCD* factors

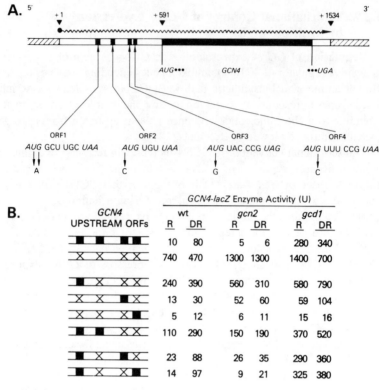

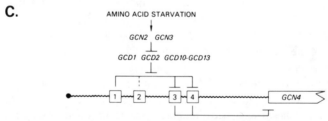

Figure 2. Regulation of *GCN4* expression by upstream open reading frames in *GCN4* mRNA. (A) Schematic of the *GCN4* transcription unit (wavy arrow) shown above the protein-coding sequences (large solid rectangle) and the four short open reading frames in the leader region (small solid boxes, uORFs 1-4). The sequences of the uORFs are given below along with the point mutations constructed in their ATG codons. (B) The effects of point mutations in the upstream ATG codons (''X''s) on expression of *GCN4-lacZ* fusion enzyme activity in wild type (wt), *gcn2* and *gcd1* mutant cells under nonstarvation conditions (repressing, R) or under conditions of histidine starvation (derepressing, DR). (C) A model summarizing the functional interactions between the uORFs in the *GCN4* mRNA leader. uORFs 3 and 4 are needed for efficient repression of *GCN4* expression under nonstarvation conditions. In response to starvation, recognition of uORF1, and to a lesser extent uORF2, overcomes the barrier to initiation downstream exerted by uORFs 3 and 4, leading to increased *GCN4* expression. The *GCD* factors prevent this interaction between the uORFs in nonstarvation conditions; the *GCD* factors are antagonized or repressed in starvation conditions (amino acid starvation) by the products of *GCN2* and *GCN3*.

are required primarily to promote the inhibitory effects of the uORFs under nonstarvation conditions. In *gcn2* and *gcn3* mutants, removing the uORFs produces the same high levels of *GCN4* expression that it does in wild type cells, as would be expected if GCN2 and GCN3 function to overcome the inhibitory effects of uORFs under starvation conditions by antagonism of GCD factors[17, 18] (Figure 2B). The target of the *GCN* and *GCD* factors was narrowed down to the sequence interval in *GCN4* mRNA containing the four uORFs by showing that regulation typical of *GCN4* could be conferred upon a heterologous yeast mRNA by insertion of only this segment of the mRNA leader into the 5' end of the heterologous transcript.[18] Thus, the *GCN4* uORFs constitute a translational control element that allows ribosomes to reach an AUG start codon downstream in the mRNA only when cells are starved for an amino acid.

mRNA sequence requirements for *GCN4* translational control—By constructing alleles containing different combinations of upstream ATG codons, it became clear that the various uORFs play different roles in the translational control mechanism. The third or fourth uORF (counting from the 5' end) is necessary for efficient repression of *GCN4* expression under nonstarvation conditions. By contrast, the first and second uORFs are relatively weak negative elements when present alone in the mRNA leader. In fact, when uORF 3 or 4 is present, uORF1 acts as a positive control element, being required for efficient *GCN4* expression under starvation conditions. (uORF2 also acts as a positive element in this situation, but to a much lesser degree). Only when uORF1 is present upstream from uORF 3 or 4, is there a strong requirement for *GCD* gene products to efficiently repress *GCN4* expression (Figure 2B). These findings led to the suggestion, summarized in Figure 2C, that (1) recognition of uORFs 1 and 2 overcomes the inhibitory effects of initiation at uORFs 3 and 4; (2) this functional interaction between the uORFs is prevented from occurring under nonstarvation conditions by the *GCD* regulatory factors.[18, 19]

In large part, deletions that remove a subset of the four uORFs have the same phenotype as point mutations in the corresponding ATG codons: (1) deletions of uORFs 3–4 increase *GCN4-lacZ* expression under repressing conditions but have much less effect under derepressing conditions, as expected if they remove *cis*-acting negative elements;[20] (2) deletions of uORF1, or a combination of uORFs 1–2, reduce *GCN4-lacZ* expression only under derepressing conditions, as expected if they remove a positive control site. Deletions of uORFs 2 and 3 have greater quantitative effects on *GCN4* expression than removal of ATG codons 2–3 by point mutations. However, even when all but 25 bp that normally separate uORFs 1 and 4 are deleted, which reduces the derepression ratio to a value of about 5, additionally removing the uORF1 ATG codon completely abolishes the ability to derepress *GCN4* expression under starvation conditions.[20] Thus, uORF1 can function as a positive control site in the absence of most of the nucleotides normally present between uORFs 1 and 4. While the latter sequences make an important quantitative contribution to the derepression ratio, they are dispensable for regulation per se. The fact that large deletions of leader sequences can be made without destroying *GCN4* translational control suggests that long-range mRNA structure is not an important aspect of this regulatory mechanism.

Although much of the leader between uORF4 and the *GCN4* start site can be deleted without any effect on the derepression ratio, the inhibitory effect of the uORFs is increased by this deletion, resulting in lower absolute levels of *GCN4* expression under both repressing and derepressing conditions. Similarly, uORF1 is slightly more inhibitory as a solitary uORF when moved closer to the *GCN4* start site,[20, 21] and uORF4 inhibits *GCN4* expression somewhat less when introduced farther upstream in the position normally occupied by uORF1.[9] These observations are reminiscent of results obtained on preproinsulin mRNA in which inhibition of protein synthesis by an uORF decreased as the uORF was moved progressively farther upstream from the preproinsulin start codon. This trend could indicate that reinitiation following translation of the uORF is more likely with greater separations between the uORF stop codon and the preproinsulin start site.[8] Alternatively, ribosomes translating the uORF may sterically hinder initiation at the next start site when the two AUG codons are very close together. The latter mechanism is improbable for *GCN4* mRNA because the effects of uORF-proximity on *GCN4* expression have been observed for separations between the two initiation sites of 100 nucleotides or more. Thus, it seems likely that reinitiation at downstream start sites is possible on *GCN4* mRNA following translation of uORF. As discussed further below, the efficiency of reinitiation seems to differ greatly for the different uORFs, occurring at high levels only following translation of uORF1.

10.2.3 Translation Control with Heterologous uORFs

The 3′-proximal uORFs were substituted with heterologous coding sequences without destroying the important qualitative features of *GCN4* translational control. For example, a completely heterologous 43-codon uORF (uORF5) inserted in place of uORFs 3 and 4 reduced *GCN4-lacZ* expression to a low constitutive level. Introduction of authentic uORF1 upstream from uORF5 reduced the inhibitory effect of the latter, increasing *GCN4-lacZ* expression ca. fivefold under derepressing conditions (Figure 3A). In the same circumstances, uORF1 overrides the inhibitory effect of authentic uORF3 to about the same extent.[22] Similar results were obtained with a 9-codon heterologous uORF (uORF6 in Figure 3A) inserted downstream from uORF1.[21]

The positive regulatory function of uORFs 1–2 was also reconstituted with a heterologous uORF.[20] A segment containing uORFs 1–2 was deleted

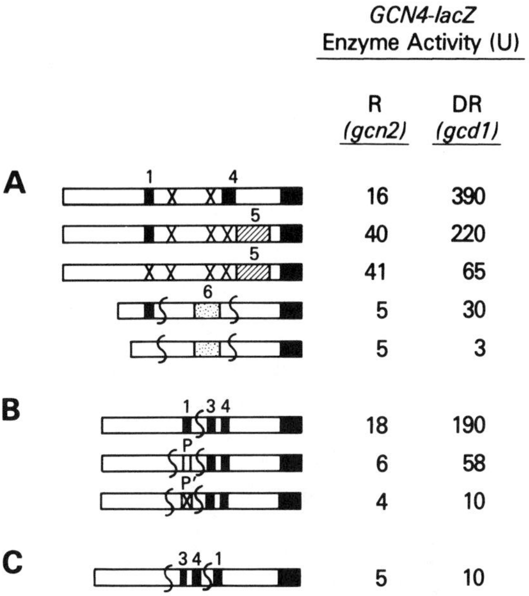

Figure 3. Sequence and positional requirements of the uORFs for *GCN4* translational control. The uORFs (and point mutations in their ATG codons, "X"s) are shown in the leader as in Figure 2B. *GCN4-lacZ* enzyme activity was measured in *gcn2* transformants (repressed conditions, R) or in *gcd1* transformants (derepressed conditions, DR). (A) uORF5 (hatched) is a 43-codon uORF constructed by insertion of a linker containing an ATG codon downstream from the normal location of uORF4. uORF6 (stippled) is an eight-codon uORF contained in a ~100 bp fragment from a sea urchin tubulin cDNA inserted in place of uORFs 1-4 or 2-4, the latter being removed by deletions. (B) P designates a three-codon uORF containing the initiation region of the *Saccharomyces PGK1* gene inserted on a 20 bp fragment in place of uORF1, the latter being removed by a deletion. P′ is identical to the P sequence except that it contains a point mutation in the ATG codon of the P uORF. (C) A rearrangement that places a ~60 bp segment containing uORF1 downstream from uORFs 3-4.

and replaced by a synthetic oligonucleotide containing an uORF with the first three codons and −1 to −7 nucleotides found at the *PGK1* gene. These sequences correspond to the consensus sequence for initiation regions of highly expressed yeast genes.[23] When present singly in the leader, this heterologous uORF substantially reduced *GCN4* expression under repressing conditions; however, it mimicked uORF1 and stimulated *GCN4* expression under derepressing conditions when inserted upstream from uORFs 3–4 (Figure 3B).

The fact that heterologous uORFs can reconstitute the regulatory functions of the authentic *GCN4* uORFs supports the idea that these elements participate in the regulatory mechanism as translated coding sequences. It also implies that no strict requirement exists for the nucleotide sequences or secondary structures of the uORFs; rather, regulation appears to be a more general consequence of having two uORFs in the mRNA leader. On the other hand, the derepression ratio for *GCN4* expression varies considerably with different combinations of uORFs, showing that particular nucleotides associated with these elements are very important in determining the efficiency of *GCN4* translational control.

Different roles for uORFs 1 and 4 in *GCN4* translational control— The importance of sequence differences between the uORFs was confirmed by showing that the 5′- and 3′-proximal uORFs are not functionally interchangeable. Inserting a segment containing uORF1 downstream from uORFs 3–4 resulted in low constitutive *GCN4* expression, indistinguishable from a deletion of uORF1 (Figure 3C).[20] Similarly, replacement of a segment containing uORF1 with a segment of the same length containing uORF4 (producing an allele with two uORF4 sequences) reduced *GCN4* expression to the low constitutive level observed when uORF4 is present alone in the leader (Figure 4). The simplest explanation for these results is that the great majority of 40S subunits cannot advance beyond uORFs 3–4 when these sequences

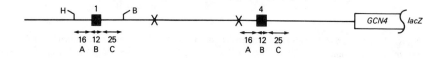

| | Origin of uORF1 Segments | | | Units of Enzyme Activity | | | |
| | | | | Constructs Containing uORF4 | | Constructs Lacking uORF4 | |
Construct	A	B	C	gcn2⁻ (R)	gcd1⁻ (DR)	gcn2⁻ (R)	gcd1⁻ (DR)
pM23	1	1	1	15	200	410	510
pM98	4	4	4	8	20	38	61
pM101	4	1	1	19	230	370	580
pM103	1	4	1	15	160	180	300
pM102	4	4	1	13	106	170	170
pM100	1	1	4	10	56	120	220
pM99	1	4	4	5	22	42	62

Figure 4. Sequences surrounding the stop codons of uORFs 1 and 4 distinguish the functions of these elements in translational control. (A) Schematic of the *GCN4-lacZ* construct with uORFs 1 and 4 shown as solid boxes and "X"s indicating mutations in the ATG codons of uORFs 2 and 3. H and B designate restriction sites employed in cassette mutagenesis of uORF1. Arrows indicate the position and length in bp of the A, B and C segments that were replaced at uORF1 with the corresponding sequences from uORF4. (B) *GCN4-lacZ* fusion enzyme activity was measured in a constitutively repressed *gcn2* mutant (R in parentheses) and in a constitutively derepressed *gcd1* mutant (DR in parentheses). Two sets of analogous constructs were examined that differ only by the presence of the uORF4 ATG codon downstream from the uORF1/uORF4 hybrids.

are present in the mRNA as the 5' proximal initiation sites. By contrast, uORF1 is a leaky translational barrier, reducing the number of scanning 40S subunits that reach downstream start sites by only about one half.[19]

Particular nucleotides at uORF4 that prevent it from functioning as a positive regulatory element were identified by making hybrids between uORF1 and uORF4 (Figure 4). Replacing the 16 bp upstream from uORF1 with the corresponding sequence from uORF4 had no effect on *GCN4* expression, suggesting that uORFs 1 and 4 have very similar initiation efficiencies. (In fact, the same conclusion was reached independently by showing that *lacZ* fusions to uORFs 1 and 4 each express high constitutive levels of fusion enzyme activity in the absence of other uORFs, comparable to that seen for the *GCN4-lacZ* fusion lacking all four uORFs.[21, 22]) By contrast, replacement of the 25 nucleotides immediately following uORF1 with the corresponding sequence from uORF4 significantly lowered *GCN4* expression under derepressing conditions. Additionally replacing the coding sequence of uORF1 with that from uORF4 completely abolished the ability to derepress *GCN4* expression[9] (Figure 4). It was subsequently shown that introduction at uORF4 of only the ten nucleotides immediately 3' to uORF4, plus the rare proline codon immediately preceding the uORF4 stop codon, is sufficient to destroy uORF1 positive regulatory function. The location of inhibitory nucleotides around the uORF4 stop codon suggests that translation terminates differently at uORFs 1 and 4, and the way in which termination occurs at uORF4 is incompatible with the positive regulatory function of uORF1.

Introduction of sequences from the uORF4 termination region at uORF1 also makes the latter a much stronger translational barrier when it occurs singly in the mRNA leader, whereas introduction of nucleotides upstream from uORF4 has no effect on the barrier function of uORF1[9] (Figure 4). Thus, wild type uORF1 blocks translation downstream less than uORF4 does, not because of different initiation efficiencies at the two start sites, but because there is a greater probability for reinitiation following translation at uORF1 vs. uORF4. Presumably, ribosomes dissociate from the mRNA following translation of uORF4 but remain attached and resume scanning after terminating at uORF1.

The fact that reducing the reinitiation potential of uORF1 destroys its ability to stimulate *GCN4* expression when situated upstream from uORF4 could indicate that ribosomes must first translate uORF1 and resume scanning in order to traverse uORF4 sequences. To account for this requirement, it was proposed that ribosomes emerge from translation of uORF1 lacking certain initiation factors that were present during initiation at this site. The absence of such factors upon resumption of scanning, coupled with reduced GCD function under starvation conditions, would alter the behavior of reinitiating ribosomes located downstream from uORF1, allowing a fraction to either scan past the uORF4 AUG codon or to reinitiate more efficiently

Nonstarvation conditions or *gcn2*⁻

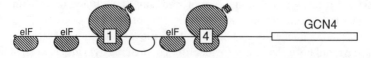

Starvation conditions or *gcd1*⁻

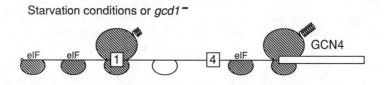

Figure 5. Model for *GCN4* translational control. The leader of *GCN4* mRNA is shown with uORFs 1 and 4 and the beginning of *GCN4* coding sequences indicated as boxes. 40S subunits containing the full complement of initiation factors (eIF) are shown hatched; those lacking certain factors are shown empty. 80S translating ribosomes are also hatched. Ribosomes translate uORF1 and the 40S subunits remain attached to the mRNA and resume scanning. Under repressing conditions (nonstarved wild-type or *gcn2* mutant cells), initiation factors are rapidly reassembled on the 40S subunit and reinitiation occurs efficiently at uORF4. Following uORF4 translation, no reinitiation occurs at *GCN4*, probably because most ribosomes dissociate from the mRNA. Under derepressing conditions (starved wild-type or *gcd1* mutant cells), reassembly of the 40S initiation complex is slower and many 40S subunits are not ready to reinitiate when they reach uORF4; consequently, they continue scanning and most reinitiate at *GCN4* instead. (Figure adapted from ref. 96).

following uORF4 translation.[22] Results described below strongly support the former idea, that ribosomes reach *GCN4* by ignoring the uORF4 start codon. (Figure 5). The important requirements for uORF1 function in this reinitiation model are efficient translation and a high probability for resumed scanning following termination. A modified version of this model was proposed in which reinitiation following uORF1 translation is restricted to starvation conditions by its dependence on positive regulators of *GCN4* expression, such as GCN2.[21] In this version of the model, a reinitiating ribosome generated by translation of uORF1 can advance through uORF4 without additional alterations to the translational apparatus by the *trans*-acting regulatory factors.

The idea that a particular sequence context at the stop codon of uORF1 is required for efficient reinitiation downstream can also explain the fact that mutations that move the uORF1 stop codon upstream or downstream by one or several codons impair the positive regulatory function of uORF1.[9, 22] In addition, the negative effects on *GCN4* expression associated with a variety of single-codon insertions in the middle of uORF1[9] could be explained by proposing that the ribosome persists in a configuration compatible with reinitiation for only the first few elongation steps. Of interest in this connection is the fact that certain antibiotics (e.g., verrucarin A) inhibit elongation only during formation of the first few peptide bonds, suggesting a mechanistic difference between early and advanced elongation steps and a possible overlap between initiation and elongation processes.[24, 25]

Evidence that reinitiating ribosomes reach *GCN4* by ignoring the start codons at uORFs 2-4—The results of recent experiments strongly suggest that ribosomes scanning downstream from uORF1 reach the *GCN4* start site under starvation conditions by scanning past the start codons at uORFsd 2-4 without initiating translation (Figure 5). Ribosomes are thought to physically traverse the uORF4 sequence, rather than "hopping" over it and binding directly to the *GCN4* start codon, because insertions of sequences with the potential to form 8-base-pair stem-loop structures just upstream or downstream from uORF4 almost completely abolishes derepression of *GCN4* expression.[96] A critical piece of evidence that ribosomes traverse the uORF4 sequence by ignoring its initiation codon is that mutations in the uORF4 stop codon that elongate this element from 3 codons to either 46 or 93 codons have virtually no effect on *GCN4* expression.[22, 96] In the latter case, the elongated version of uORF4 actually overlaps the beginning of *GCN4* by 130 nucleotides. Ribosomes that translate this 93-codon-uORF4 would have to scan "backwards" for 130 nucleotides after terminating at uORF4, ignoring four other potential AUG start codons in the overlap region, in order to reach *GCN4*. (It was proven that the normal *GCN4* start codon was used exclusively for GCN4 protein synthesis in the 93-codon-uORF4 construct.[96]) The sequence contexts of the four AUG codons in the overlap region are unlikely to preclude their recognition as start sites, as sequence context has only a modest effect on initiation codon selection in yeast.[97] Therefore, it seems most likely that altering the uORF4 termination site has no effect on *GCN4* expression simply because uORF4 is not translated by the ribosomes that reach *GCN4*.[96] This stands in sharp contrast to the deleterious effects on *GCN4* expression associated with altering the uORF1 termination site, interpreted above to indicate that ribosomes must translate uORF1 and resume scanning in order to reach *GCN4*.

The conclusion that ribosomes which translate *GCN4* have bypassed the uORF4 start codon is also supported by biochemical measurements of uORF4 translation under repressing and derepressing conditions. By making a single base-pair insertion in the uORF4-*GCN4* overlap region of the 93-codon-uORF4 construct described above, the *GCN4-lacZ* coding sequence was placed in-frame with uORF4 to provide a means of measuring uORF4 translation rates. Expression of the resulting uORF4-*lacZ* fusion was found to be about 50% lower in constitutively derepressed *gcd1* mutants than in constitutively repressed *gcn2* mutants, suggesting that roughly 50% of the 40S subunits scanning downstream from uORF1 under derepressing conditions fail to reinitiate at uORF4 and translate *GCN4* instead. As expected if ribosomes are being diverted from translation of uORF4 to allow for increased initiation at *GCN4*, the decrease in uORF4-*lacZ* translation seen in *gcd1* cells is similar in magnitude to the increase in *GCN4-lacZ* expression that occurs in the same situation.[96]

It appears that under starvation conditions, many ribosomes scanning downstream from uORF1 scan past uORF4 without initiating translation because the time required to scan the uORF1-uORF4 interval is insufficient for the reassembly of competent initiation complexes (Figure 5). The evidence for this conclusion is that progressively increasing the distance between uORF1 and uORF4 leads to a progressive decrease in *GCN4* translation under starvation conditions or in a *gcd1* mutant. For example, insertion of 144 nucleotides between uORFs 1 and 4 greatly impairs derepression of *GCN4-lacZ* expression, whereas the same insertion made downstream from uORF4 has virtually no effect. Importantly, the 144 nucleotide-insertion increases the spacing between uORF1 and uORF4 to about the same distance that normally separates uORF1 from *GCN4*. To explain the nonderepressible phenotype of this construct, it was proposed that 40S subunits scanning the expanded uORF1-uORF4 interval now have sufficient time under starvation conditions to reassemble the factors needed for reinitiation before reaching uORF4. Consequently, nearly all subunits reinitiate at uORF4 and are thereby prevented from reaching *GCN4*, presumably because ribosomes dissociate from the mRNA after terminating at uORF4. Supporting this interpretation, the 144 nucleotide-insertion between uORF1 and uORF4 leads to increased translation of the uORF4-*lacZ* fusion under derepressing conditions, eliminating the ca. 50% reduction in uORF4 translation seen with the wild-type uORF1-uORF4 spacing.[96]

Implied in the above explanation is the assumption that under nonstarvation conditions, the normal spacing between uORF1 and uORF4 provides adequate time for nearly all ribosomes scanning downstream from uORF1 to reassemble competent initiation complexes by the time they reach uORF4. Thus, the regulation of *GCN4* expression is seen to arise from the increased scanning time required for reinitiation at downstream start sites following termination at uORF1 in amino acid-starved cells (Figure 5). It was proposed that reinitiation is less efficient under starvation conditions because one or more initiation factors that must associate with 40S subunits to form a competent reinitiation complex is partially impaired in these circumstances by the positive regulators of *GCN4* translation, GCN2 and GCN3.[96] As discussed below, recent studies suggest that initiation factor -2 (eIf-2) is the rate-limiting component for reinitiation antagonized by GCN2 and GCN3 in amino acid-starved cells.

Mutations in the structural genes for initiation factor eIF-2 impair translational control of *GCN4*—Nearly all *gcd* mutations are pleiotropic and lead to conditional lethality at 36°C or unconditional slow growth, irrespective of amino acid levels.[27-29] Synthesis of macromolecules is reduced and the cells arrest early in the G1 phase of the cell cycle when temperature-sensitive *gcd1* and *gcd2* mutants are shifted to the restrictive temperature.[27, 29, 30] Moreover, deletions of *GCD1* and *GCD2* are unconditionally lethal.[30-32] These results suggest that *GCD1* and *GCD2* contribute to an essential metabolic function required for entry into the cell cycle. Given that mutations in *GCD* genes impair *GCN4* translational control, this essential function might be involved with protein synthesis. In this view, GCN2 and GCN3 would function to partially inactivate or modify components of the translational machinery under starvation conditions and thereby alter the behavior of reinitiating ribosomes on *GCN4* mRNA.

Supporting this view of *GCN4* translational control is the recent finding that mutations in the structural genes for two subunits of initiation factor -2 (eIF-2), have a large effect on *GCN4* expression. Mutations in two genes called *SUI2* and *SUI3* were isolated by their ability to allow an in-frame TTG codon to be used as the translational initiation site at *HIS4* in the absence of the normal ATG start codon. The amino acid sequences of the SUI2 and SUI3 proteins show striking homology to the α- and β-subunits, respectively, of human eIF-2. This initiation factor forms a ternary complex with GTP and Met-tRNA$_i^{met}$ and associates with the small ribosomal subunit during the initiation process. Ternary complex formation is defective in extracts prepared from *SUI2* and *SUI3* mutants, supporting the idea that these genes encode subunits of eIF-2 in *S. cerevisiae*.[33, 34] The phenotype of the *SUI* mutations suggests that eIF-2 plays an important role in AUG recognition during the scanning process.

In addition to their effects on translational start site selection, the *SUI* mutations lead to elevated *HIS4* transcription.[35] This secondary phenotype was explained by showing that *GCN4* expression is constitutively elevated in the *SUI* mutants. Importantly, this derepression of *GCN4* requires multiple uORFs in the *GCN4* mRNA leader, since alleles containing only uORF4, or no uORFs at all, are nearly insensitive to the *SUI* mutations. In addition, *SUI2* and *SUI3* mutations overcome the requirement for the positive regulators GCN2 and GCN3 for high-level *GCN4* expression[36] (Figure 6). All of these phenotypes are exhibited by known *gcd* mutants, suggesting that eIF-2 is a GCD factor. If so, eIF-2 activity may be altered by GCN2 and GCN3 in amino acid-starved cells to allow for increased translation of *GCN4* mRNA.

In a related development, it was reported[37] that a temperature-sensitive *gcd1* mutation leads to inefficient binding of the eIF-2·GTP·tRNA$_i^{met}$ ternary complex to the 40S ribosomal subunit after incubation for several hours at the nonpermissive temperature. The same defect was observed transiently in wild type cells when the culture was shifted from medium containing all amino acids (except methionine) to minimal medium. As would be expected from this initiation defect, translation of most mRNAs is greatly reduced by a nutritional shift-down; however, there is a dramatic increase in the rate of *GCN4* protein synthesis. The increased translation of *GCN4* mRNA under these conditions (as revealed by its presence on larger polysomes) is dependent on the presence of uORF1, suggesting that the same translational control mechanism that operates under steady-state starvation conditions is affected transiently during shift-down from amino acid-rich to minimal medium. Based on these findings, it was suggested that the reduced efficiency of an initiation step involving eIF-2 in a *gcd1* mutant, or following a nutritional shift-down in wild type cells, enhances the uORF1-mediated increase in *GCN4* protein synthesis. In the context of the reinitiation model described above, reducing eIF-2 function would allow ribosomes scanning downstream from uORF1 to bypass the AUG codons at uORFs 2-4 and reinitiate at *GCN4* instead.

Interactions between *GCD* factors and the positive regulator *GCN3*— The derepression of *GCN4* expression observed in a nutritional shift-down occurs independently of *GCN2* function, suggesting that a different set of *trans*-acting factors may be involved in this transient phenomenon than that required for steady-state derepression of *GCN4*.[37] However, the same could be said for the derepression that occurs in *GCD* and *SUI* mutants since GCN2 and GCN3 are not required for this response either. For this reason, allele-specific interactions detected between *GCN3* and mutations in *GCD1*, *GCD2* and *SUI2* are important in suggesting that the products of the latter three genes are involved in the same derepression pathway in which GCN3 functions.

Deletion of *GCN3* has no effect on viability under nonstarvation condi-

GCN4-lacZ Enzyme Activity (U)

uORFs present:	1 2 34		4	
	R	DR	R	DR
SUI⁺	13	75	8	21
sui2-1	200	190	23	17
SUI3-2	130	210	18	32
sui2-1 gcn2Δ	240	200	32	32

Figure 6. Mutations in *SUI2* and *SUI3*, structural genes for subunits of eIF-2, lead to constitutive derepression of *GCN4* expression. The presence or absence of uORFs is depicted as in Figure 2. *GCN4-lacZ* fusion enzyme activity was measured under nonstarvation conditions (repressing, R) or under conditions of tryptophan starvation (derepressing, DR). The *SUI3-2* mutation is semi-dominant for derepression of *GCN4*; *gcn2Δ* is a deletion of *GCN2*.

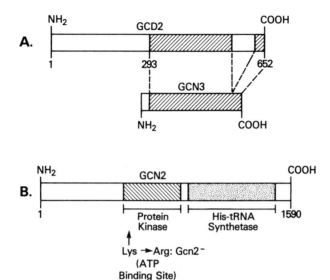

Figure 7. Structural features of *GCN3* and *GCN2*, translational activators of *GCN4*. (A) Amino acid sequence homology between the carboxyl-terminus of *GCD2* and *GCN3*, translational repressors and activators, respectively, of *GCN4* expression. Genetic data suggest that *GCN3* functions as an antagonist of *GCD2* under amino acid starvation conditions. Based on their sequence homology, this interaction may occur through direct competition between *GCN3* and *GCD2* for binding sites or substrates. (B) Sequence homology between *GCN2* and both the catalytic subunit of eukaryotic protein kinases and histidyl-tRNA synthetases from yeast, *E. coli* and humans. Substitution of a highly conserved lysine in the kinase-related domain of *GCN2* destroys its regulatory function *in vivo*, supporting the idea that *GCN2* acts as a protein kinase in stimulating *GCN4* expression. The homology to histidyl tRNA synthetases suggests that *GCN2* may play a direct role in coupling *GCN4* expression to the concentration of uncharged tRNA.

tions in otherwise wild type cells. However, deletion of *GCN3* is lethal in mutants containing a *gcd1-101*[31] or a *sui2-1* mutation.[36] In addition, certain mutations in *GCD2* (first known as *gcd12* alleles) cause derepression of *GCN4* expression and temperature-sensitivity for growth only in the absence of *GCN3* function.[38] These genetic interactions suggest that GCN3 is functionally related to GCD1, GCD2 and SUI2, and contributes to the essential role of these proteins under nonstarvation conditions. In fact, there is significant amino acid sequence similarity between GCN3 and the carboxyl-terminus of GCD2 protein[39] (Figure 7a), prompting the idea that GCN3 and GCD2 are subunits of the same complex that has an important function in translation initiation. Under nonstarvation conditions, GCN3 would either substitute for, or stabilize mutant *gcd2* proteins in the complex, explaining the ability of *GCN3* to suppress the mutant phenotypes associated with the *gcd12* mutations. The activity of complexes containing GCN3 would be altered in amino acid-starved cells, leading to increased translation of *GCN4* mRNA.[39] Presumably, this complex interacts with eIF-2 during the initiation process.

The role of *GCN2* protein kinase in recognition of the amino acid starvation signal—Sequence analysis reveals a strong similarity between a portion of the GCN2 protein and the catalytic domain of eukaryotic protein kinases.[40–42] One of the best characterized subdomains in these kinases contains a lysine residue that is believed to function in the phosphotransfer reaction. Substitutions of this conserved lysine invariably abolish kinase activity.[42] Replacement of the corresponding lysine residue in GCN2 (Lys-559) with arginine or valine abolishes GCN2 regulatory function, whereas substitution of an adjacent lysine residue with valine has no effect on *GCN4* expression[41] (Figure 7b). These results suggest that GCN2 stimulates translation of *GCN4* mRNA by acting as a protein kinase.

The deduced amino acid sequence of the carboxyl-terminus of GCN2 may provide an important clue about how its protein kinase activity is regulated by amino acid availability: 530 residues in the carboxyl-terminal domain of GCN2 are homologous to histidyl-tRNA synthetases (HisRS) from *S. cerevisiae*, humans and *E. coli* (Figure 7b). Several point mutations in the HisRS-related sequences inactivate GCN2 regulatory function, suggesting that this region is required for protein kinase activity under starvation conditions.[41] Given that aminoacyl-tRNA synthetases bind uncharged tRNA as a substrate, and that accumulation of uncharged tRNA is thought to trigger derepression of *GCN4* expression,[43, 44] it was proposed that the HisRS-related domain of GCN2 monitors the concentration of uncharged tRNA in the cell and activates the adjacent protein kinase moiety under starvation conditions when uncharged tRNA accumulates. Given that *GCN4* derepresses in response to starvation for any one of at least ten amino acids, perhaps GCN2 has diverged sufficiently from HisRS that it now lacks the ability to discriminate between different uncharged tRNA species.[41]

No substrates for GCN2 kinase activity have been identified; however, the fact that *gcd* and *sui* mutations restore derepression of *GCN4* in *gcn2* mutants makes SUI2 and the *GCD* factors good candidates. Phosphorylation by GCN2 would partially inactivate one or more of these factors, altering the process of translation initiation in a way that allows reinitiating ribosomes ignore the 3′-proximal uORFs and initiate at the *GCN4* start codon. This hypothesis is attractive given that eIF-2 activity in mammalian cells is regulated by phosphorylation of the α-subunit in certain stress conditions.[45, 46] Alternatively, GCN3 (or another positive factor implicated in *GCN4* control known as GCN1[14]) may be phosphorylated by GCN2 and thereby converted to an antagonist of one or more GCD proteins.

10.2.4. Translational Control by uORFs in Higher Eukaryotes

As many as one tenth of all known mammalian transcripts contain one or more uORFs, and the proportion is even higher for certain interesting classes of mRNAs, such as those encoded by oncogenes.[1] In several cases, it has been shown that such naturally occurring uORFs reduce translation of the downstream coding sequences (reviewed in Reference 47); however, except for the *CPA1* and *GCN4* mRNAs described here, there are no other instances in which uORFs are known to modulate gene expression in response to environmental stimuli. Although translation of *GCN4* mRNA is most efficient under conditions of severe amino acid starvation, *GCN4* is partially derepressed under conditions in which the rate of protein synthesis is only slightly reduced by amino acid limitation.[48] Therefore, similar regulatory mechanisms could operate elsewhere in response to moderate stress.

The possibility that modulation of eIF-2 is involved in regulating *GCN4* expression recalls two other well-studied instances of translational control in response to stress. Heme deprivation in rabbit reticulocytes and certain viral infections in human cells leads to reduced translation initiation due to phosphorylation of the α-subunit of eIF-2.[45, 46] These responses appear to differ from the specific control of *GCN4* expression in that total protein synthesis is regulated by phosphorylation of eIF-2 in mammalian cells. However, partial inhibition of eIF-2 could affect the translation of certain mRNAs more than others. It will be interesting to see whether phosphorylation of eIF-2 or other initiation factors is involved in gene-specific regulatory responses in higher eukaryotes similar to that described for *GCN4* mRNA in yeast.

10.3. A *Cis-Trans* MODEL FOR TRANSLATION CONTROL

10.3.1. Iron Metabolism in Complex Eukaryotic Cells

The introduction of oxygen into the earth's atmosphere created two challenging problems related to the dependency of cells upon elemental iron. First, the oxidation of ferrous iron to the more insoluble ferric form demanded new mechanisms to obtain environmental iron. Second, the ability of ferric iron to catalyze the generation of dangerous hydroxide radicals demanded new mechanisms to control the uptake and detoxification of iron. In humans, the two major, and well-characterized proteins that mediate the uptake and detoxification of iron are the transferrin receptor (TfR) and ferritin, respectively. Iron is delivered to most cells via the binding of diferric-transferrin to its high affinity plasma membrane receptor.[49, 50] The receptor-ligand complex enters via the endocytic pathway where it encounters an acidic endosomal compartment[51, 52] where the iron is released from the transferrin (Tf) and transferred to the cytosol. This transfer process remains an undefined step in cellular iron metabolism. The apotransferrin (apoTf) remains bound to the receptor in an acidic environment, and the TfR-apoTf complex recycles to the cell surface where the apoTf rapidly dissociates, allowing the freed receptor to mediate additional rounds of this Tf cycle.[53, 54] Once in the cytoplasm, the iron is either utilized for the many cellular processes that require this element or it is sequestered into ferritin, a hollow, spherical molecule composed of 24 subunits encoded by two highly homologous genes (H and L).[55] Presumably this sequestration is responsible for the detoxification of unused iron.

The expression of both the TfR and ferritin is highly regulated by the amount of iron available to the cell. Limiting iron results in an increase in the number of TfRs whereas when iron is plentiful, receptor number decreases. Because of the high concentration of transferrin in serum, the uptake of iron by a cell is proportional to the number of cycling TfRs. The level of ferritin is regulated in the opposite direction, increasing in the presence of iron and decreasing in its absence. This regulation allows for adequate sequestration of excess iron and minimizes sequestration when iron is limiting. We are beginning to understand the molecular basis for the coordinate but opposite regulation of these two genes. The common components of these regulatory phenomena were first elucidated in studies of the regulation of ferritin expression.

10.3.2. The Iron Responsive Element: The *Cis*-Acting RNA Element

It had been known since the 1940s that the level of ferritin in tissues varied directly with changes in iron.[56] By the 1970s, indirect evidence suggested that this regulation occurred at the level of translation.[57] Work from Theil and colleagues studying the iron regulated synthesis of amphibian ferritin more clearly demonstrated that the level of this regulation was most probably translational. The molecular cloning of cDNAs corresponding to ferritin subunits[58–61] allowed the direct demonstration of the elements responsible for translational control of ferritin[62–64] and identification of the sequences within the ferritin mRNA responsible for that regulation.[64–66] When cells are exposed to conditions that alter available iron, rapid changes in the rate of synthesis of ferritin are observed. These changes of up to two orders of magnitude in ferritin biosynthesis occurred in the absence of change in the level of cytoplasmic ferritin mRNA.[62–64] The ferritin mRNA within the cytosol was shown to shift from an mRNP fraction to polysomes as the synthesis of ferritin increases.[64, 67, 68] The mRNA for human ferritin H chain (approx 1 kb) contains a 212 nt 5' untranslated region (UTR). When the 5'UTR was deleted, the ferritin protein was made and assembled normally but all acute iron regulation was lost.[65, 66] Further deletion analysis identified a region approximately 35 bases in length that appeared to contain all information required for iron-dependent translational control of ferritin synthesis.[69] That this short sequence is all that is required for this translational control was established by cloning the sequence into the 5'UTR of two different reporter genes. The translation of the resultant chimeric mRNAs were iron-regulated.[65, 69] Examination of all ferritin genes cloned to date (human, rat, mouse, rabbit, chicken, and bullfrog) reveals the presence of this highly conserved sequence in the 5'UTR.[70] All of these sequences are capable of forming moderately stable stem-loop structures. In all instances, the base pairing of the stem is interrupted by an unpaired C that is found six nucleotides (except one example in which it is five) 5' of a six-membered loop. The sequence of the loop is CAGUGN. Because of its function, we have referred to this RNA sequence/structure motif of the ferritin mRNA as an iron responsive element (IRE).[71] There is still no complete definition of exactly what is required for an RNA to be an IRE. It is clear that there must be a stem, although the exact role of the specific nucleotides in this stem has not been definitively elucidated. Some nucleotides have been shown to be absolutely essential to function as an IRE. In particular, the unpaired C on the 5' side of the stem is required for iron-dependent translational control. In addition, if one removes the conserved first C of the loop, resulting in a five-membered loop, there is no iron-dependent translational control observed. This stem-loop structure is reminiscent of the prokaryotic translational control stem-loop element found in the R17 replicase message. Exhaustive mutagenesis of this stem-loop structure by Uhlenbeck and colleagues[72] has demonstrated that the overall structural motif must be conserved, but there is a fair amount of flexibility in the exact nucleotides that make up this structure. However, like the IRE, certain nucleotides are absolutely required in this *cis-trans* prokaryotic model system.

10.3.3. The IRE Binding Protein: The *Trans*-Acting Factor

The location of the IRE in the 5'UTR of ferritin suggested that it functioned to control translation initiation. The precise location of the IRE within the 5'UTR does not determine its function although the distance that it can be displaced from the initiating codon has not been determined. The consequences of placing an IRE within the reading frame have not been examined. Further clues as to its mechanism of action came from observations of the effect of the presence or absence of an IRE on the translatability of mRNA.[69] Removal of the IRE resulted in a constitutively high level of mRNA translation. Addition of an IRE to a 5'UTR depressed the rate of translation unless cells were incubated in the presence of an iron source suggesting that the IRE functions to repress translation if iron is limiting. The presence of an IRE did not yield higher rates of translation than were seen with an identical mRNA lacking an IRE under any iron conditions. The presence of an IRE in the 5'UTR has been shown to repress translation in an *in vitro* translation system.[73] One explanation was that the IRE serves as the target for a repressor molecule. The first evidence of a candidate for such a repressor came with the identification of a cytosolic binding protein that binds to IRE-containing RNAs as determined by an electrophoretic mobility shift assay.[74, 75] The IRE binding protein (IRE-BP) is a 90 kDa cytosolic protein that has been purified to homogeneity by RNA affinity chromatography.[75] In addition, this protein has been purified by conventional chromatographic techniques from rabbit liver.[76] This purification was important in that it resulted in the production of antibodies that were used to demonstrate in an *in vitro* translation system that the purified protein is indeed responsible for translational repression.[76] This protein is the only protein in human cell lysates that is cross-linked to IRE-containing RNAs by UV irradiation. The gene encoding the IRE-BP has been localized to human chromosome 9.[77b] The details of its interactions with IREs and its role in regulating the fate of IRE-containing mRNAs will be discussed below.

10.3.4. Regulating Both Translation and mRNA Stability

As with ferritin, the rapid regulation of the expression of the TfR is the result of altered rates of protein synthesis. However, in contrast to ferritin, this alteration is directly reflected in changes in the levels of cytoplasmic TfR mRNA.[78–80] Initial studies suggested a transcriptional component to this regulation. Indeed the 5' flanking region of the TfR gene can transfer iron-dependent transcriptional regulation to a heterologous reporter gene.[81] However, only a 2- 3-fold effect was observed in contrast to a 20- to 30-fold change in TfR mRNA levels seen when iron availability is manipulated. The major iron-dependent regulation mapped to within the 3'UTR of the TfR message.[81, 82] That this information was present within the RNA had been suggested by the observation that alternative transcripts of a single gene construct was iron regulated only if it contained a specific region of the 3'UTR.[81] Deletion analysis of the 2.5 kb 3'UTR of the TfR mRNA using convenient restriction sites demonstrated that a 680 nt fragment contained all of the information for this regulation.[81, 83] Iron-dependent control of mRNA half-life can be transferred to a heterologous gene by placing the TfR regulatory region in the 3'UTR of the reporter gene.[81, 83] Visual examination and computer-assisted analysis suggested that this 680 nucleotide region is highly structured. In particular, five stem-loop structures (termed A-E) were observed that bore striking resemblance to ferritin IREs although the sequence of the stems differed.[81] We tested whether these TfR elements were functional IREs by two criteria. First we inserted stem-loops B or C from the 3'UTR or the

TfR mRNA into the 5'UTR of an indicator gene and each functioned as a ferritin-like translational control element (i.e., translational repression in response to iron deprivation). Next we asked whether TfR IREs A-E bound the IRE-BP. Each specifically bound the same cytosolic protein as defined for the ferritin IRE although the affinities varied over at least a tenfold range.[84] It should be noted that the IREs in the context of the TfR 3'UTR do not confer translational iron regulation upon the TfR mRNA. In addition, Kuhn and colleagues have reported that multiple IRE-BPs can bind to a single TfR regulatory region simultaneously.[85]

In contrast to the simple regulatory structure (a single IRE) that encodes iron-dependent translational control, the regulatory region required for iron-dependent mRNA stability regulation appears to be much more complex. To understand the function of this region demands both a structural and functional analysis of the 3'UTR regulatory element. Direct structural analysis has not yet been accomplished. However, the availability of TfR sequence from two distinct species (human and chicken) allows prediction of structure based on phylogenetic conservation. Of 160 nt that encode the five IREs, only six changes, all of which are conservative of the IRE sequence/structure motif, have occurred between the two species.[86, 87] This analysis also supports the prediction of other structural motifs within the regulatory region including several small non-IRE stem-loop structures and a long base-paired structure lying between IREs B and C (see Figure 8).

Functional dissection of this region has begun to yield information as to how the TfR regulatory element functions.[88] When the region is split at the central long stem, neither the 5' or 3' half of the region is sufficient for regulation. The 680 nt segment has been trimmed to about 250 nt by removal of the sequences 5' of IRE B, 3' of IRE D and the nucleotides contained within the large central segment which does not contain sequences that are similar between chicken and human. The resultant shorter regulatory fragment, containing only 3' IREs, cannot be distinguished from the full 680 nt region in terms of iron regulation. Two types of deletions within the smaller regulatory fragment can be distinguished: those that specifically affect IRE function and those that affect non-IRE structures. The alteration that best illustrates an IRE mutant results from the removal of the first C residue from each of loops B, C, and D leaving five membered loops. We had previously demonstrated that this mutant IRE is incapable of functioning as a translational control element[70] and that it cannot bind to the IRE-BP with high affinity.[89] The resulting fragment with 3 base changes in 250 nt gives no iron-dependent mRNA stability regulation and no longer binds with high affinity to the IRE-BP. A variety of other deletions, particularly of the small non-IRE stem-loops also eliminate regulation but have no effect on the binding of the region to the IRE-BP.[88] Although both IRE and non-IRE changes can abrogate regulation, a more careful look at their "regulatory phenotypes" yielded a revealing

Ferritin mRNA

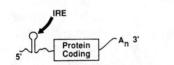

IRE-BP Absent: mRNA Translation

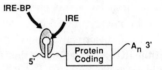

IRE-BP Bound: No mRNA Translation

TfR mRNA

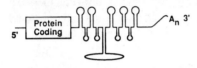

IRE-BP Absent: mRNA Degradation

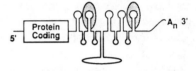

IRE-BP Bound: No mRNA Degradation

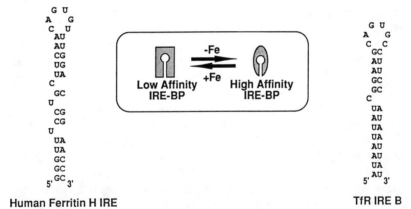

Human Ferritin H IRE

TfR IRE B

Figure 8. Coordinate regulation of the expression of ferritin and the transferring receptor via common *cis*-acting RNA elements and a common *trans*-acting RNA binding protein. The 5'UTR of ferritin mRNA contains a single IRE and the 3'UTR of the TfR mRNA contains five IREs (the latter designated A-E left to right and shown schematically above the horizontal). All of these IREs are capable of interacting with the IRE-BP. The sequence and proposed secondary structure of the human ferritin H chain IRE and one of the TfR IREs are shown to be the preferred binding sites. Treatment of cell with an iron chelator results in high affinity IRE-BP whereas treatment of cells with an iron source results in low affinity IRE-BP (inset). When the IRE-BP is bound to the ferritin IRE, translation of ferritin is inhibited. When the IRE-BP is bound to the TfR IREs, degradation of the TfR mRNA is inhibited. The 3'UTR of the TfR mRNA also contains other RNA structural elements (shown schematically below the horizontal), some of which are critical for iron regulation of TfR expression.

difference. For a given level of transcription, the IRE mutant resulted in very low levels of TfR mRNA. In contrast, the non-IRE mutants resulted in high levels of mRNA, comparable to that seen with the regulated constructs after treatment of cells with the iron chelator desferrioxamine. This difference most probably reflects two distinct functions of the 3' regulatory domain, both of which are required for iron responsive regulation (see below).

10.3.5. A Novel Regulatory Mechanism: The Sulfhydryl Switch

In order to propose that the IRE-BP is the regulatory protein that allows IRE-containing RNAs to respond to cellular iron status, it is necessary to demonstrate some sensitivity of the IRE-BP activity to iron availability. More generally, the mechanism by which any regulatory nucleic acid binding protein responds to physiologic signals represents one of the great unknowns in molecular biology. A small number of examples implicating phosphorylation or ligand binding for DNA binding proteins have been observed.[90–92] Initial studies using either gel shift of UV cross-linking demonstrated that the apparent amount of binding protein increased in lysates of cells that had been starved of iron and decreased when the cells had been loaded with iron.[74, 89] A more quantitative analysis of the interaction between the IRE-BP and its cognate RNA has helped clarify these observations. Scatchard analysis of this interaction revealed a complex binding curve that could be well fit by a two binding site model suggesting that cytosol contains binding activity with two distinct affinities, one K_D = 10 to 30 pM and the other K_D = 2 to 5 nM.[93] That these represent the same protein is supported by the following: both gave identical mobilities by gel shift, a process documented to be quite sensitive to the molecular weight of the protein; both affinities were observed using affinity purified IRE-BP; and variant IREs capable of only binding to the low affinity site could compete (at high concentrations) with both binding sites. The total number of binding sites (high plus low) did not change in response to altered iron status of the cell. What did change was the fraction of total sites displaying picomolar affinity, varying from less than 1% in iron-fed cells to greater than 50% in iron-starved cells. This shift did not require protein synthesis,[77a] suggesting that the same population of IRE-BP had been interconverted by changes in the cell's iron status. The biochemical basis for the affinity change of the IRE-BP appears to be the reversible oxidation-reduction of one or more disulfides in the protein.[77a] The low affinity IRE-BP was converted to the high affinity form *in vitro* by the addition of reducing agents to cell lysates. Conversely, the high affinity form was switched to the low affinity state by treatment with agents that catalyze disulfide formation (diamide or copper orthophenanthroline). The oxidized form appears to involve intramolecular disulfide formation as no change in apparent molecular weight accompanies oxidation. In cells in which the IRE-BP had been maximally activated by treatment with an iron chelator, little, if any IRE-BP isolated from the cell was in an oxidized form whereas in cells treated with an iron source, virtually none of the IRE-BP was in a reduced form. Thus, the iron status of the cell appears to set the redox state of the IRE-BP.[77a, 93] We have termed this novel biochemical regulatory mechanism a sulfhydryl switch. Other examples which may imply sulfhydryl switches in the regulation of biological molecules include the regulation of reductases by light and chloroplasts and the regulation of the hemin-inhibited eIF-2 translation factor kinase.[94] It is not known how the cellular iron status throws this switch. We have, so far, been unable to reproduce the sulfhydryl switch *in vitro* by the addition or chelation of iron. Recently, it has been suggested by Thatch and colleagues[95] that hemin or heme compounds may directly interfere with the binding of the IRE-BP to its cognate RNA. They have demonstrated that hemin, at the 10 to 15 mM concentration, will block the ability of the IRE-BP to repress translation in an *in vitro* system. Our own findings suggest that the interference of the IRE-BP-RNA interaction by heme compounds may be relatively nonspecific and does not seem to mirror the changes in the IRE-BP induced by iron perturbation of cells.

The definition of two affinity states for the IRE-binding protein and the correlation of iron-dependent translational control with the shift between the high and low affinity states have led us to re-examine the role of the sequence/structure of the IRE in light of both the ability to specifically bind to the protein, as well as to be able to sense changes in the affinity state of the protein. We have indeed found that these two phenomena can be dissociated from each other. Thus, if one either removes the bold C from the stem or replaces it with any other nucleotide, one now sees binding to the IRE binding protein that is specific. However, only one binding affinity is seen which corresponds to the low affinity interaction. Regardless of the status of the IRE-binding protein, determined either by the cells in response to iron manipulations or by *in vitro* manipulations of sulfhydryl status, these IREs cannot interact with high affinity with the binding protein. More specifically, they do not apparently sense the two affinity states. Thus, the two affinity model can be viewed both from the protein and the RNA structure.

10.3.6. A Regulatory Model

The above described observations have allowed us to propose a model to explain how the IRE and its binding protein can account for the iron-dependent regulation of both translation and mRNA stability[70, 81] (Figure 8). As cellular iron levels become limiting, a greater fraction of the cell's IRE-BP become recruited into the high affinity binding state through the reduction of one or more critical disulfides in the protein. The high affinity interaction between the protein and an IRE acts as a repressor of translation if an IRE is located within the 5'UTR of a message. Within the regulatory region of the 3'UTR of the TfR this same high affinity interaction represses the degradation of the message. In this way the regulated binding of the IRE-BP can coordinately regulate a decrease in the biosynthesis of ferritin (by translational repression) and an increase in the biosynthesis of the TfR (by repression of mRNA degradation). The TfR regulatory region has two distinct functional elements. One must be an RNA instability determinant which gives this message a short half-life (a requirement for rapid regulation of mRNA levels) and the other is IRE which can regulate the utilization of this instability element.

Much concerning the molecular details of this model remains to be determined. First, we need a clear picture of what sequence/structure defines an IRE. Some progress is now being made toward that end. For example, removal of any nucleotide from the loop renders an IRE nonfunctional. This is not because the IRE-BP cannot recognize the resultant stem-loop but rather because it can bind with only the lower nM affinity and its binding is insensitive to changes in the redox state of the IRE-BP. Only detailed quantitative analysis will allow one to determine the exact contributions of any nucleotide/structure to affinity, specificity and the ability to be responsive to the redox state of the IRE-BP. Second, we need to understand the biochemical basis of the iron-dependent redox shift of the IRE-BP. Does the IRE-BP itself bind iron or does iron indirectly determine the affinity of the protein, perhaps via an oxidoreductase? Only a complete description of the sequence and eventually the structure of the IRE-BP will define the nature of its RNA binding and of its sulfhydryl switch. Third, we need to define how the binding of the IRE-BP actually alters translation initiation and mRNA stability. Does it act merely sterically or by the specific interaction with or modification of components of the translation or degradative machinery? Fourth, we must define much more about the process of TfR and mRNA stability. What is the actual sequence/structure of the instability element and what is the enzymology of the RNA turnover? Finally, we need to determine whether the information gleaned from the study of this RNA regulatory system can guide us to a better understanding of other posttranscriptionally regulated genes.

REFERENCES

1. Kozak M: *Nucleic Acids Res.* 15:8125 (1987).
2. Cigan AM and Donahue TF: *Gene* 59:1 (1987).
3. Sherman F and Stewart JW: In Strathern JN, Jones EW, and Broach JR (Eds): *The Molecular Biology of the Yeast Saccharomyces Metabolism and Gene Expression* Cold Spring Harbor, NY, Cold Spring Harbor Laboratory, 301, 1982.
4. Kozak M: *Nucleic Acids Res.* 12:3873 (1984).

5. Liu C, Simonsen CC, and Levinson AD: *Nature (London)* 309:82 (1984).
6. Johansen H, Schumperli D, and Rosenberg M: *Proc. Natl. Acad. Sci. U.S.A.* 81:7698 (1984).
7. Kozak M: *Cell* 44:283 (1986).
8. Kozak M: *Mol. Cell. Biol.* 7:3438 (1987).
9. Miller PF and Hinnebusch AG: *Genes Dev.* 3:1217 (1989).
10. Messenguy F, Feller A, Crabeel M, and Pierard A: *EMBO J.* 2:1249 (1983).
11. Nyunoya H and Lusty CJ: *J. Biol. Chem.* 259:9790 (1984).
12. Werner M, Feller A, and Pierard A: *Eur. J. Biochem.* 146:371 (1985).
13. Werner M, Feller A, Messenguy F, and Pierard A: *Cell* 49:805 (1987).
14. Hinnebusch AG: *Microbiol. Rev.* 52:248 (1988).
15. Thireos G, Driscoll-Penn M, and Greer H: *Proc. Natl. Acad. Sci. U.S.A.* 81:5096 (1984).
16. Hinnebusch AG: *Proc. Natl. Acad. Sci. U.S.A.* 81:6442 (1984).
17. Hinnebusch AG: *Mol. Cell. Biol.* 5:2349 (1985).
18. Mueller PP, Harashima S, and Hinnebusch AG: *Proc. Natl. Acad. Sci. U.S.A.* 84:2863 (1987).
19. Mueller PP and Hinnebusch AG: *Cell* 45:201 (1986).
20. Williams NP, Mueller PP, and Hinnebusch AG: *Mol. Cell. Biol.* 8:3827 (1988).
21. Tzamarias D and Thireos G: *EMBO J.* 7:3547 (1988).
22. Mueller PP, Jackson BM, Miller PF, and Hinnebusch AG: *Mol. Cell. Biol.* 8:5439 (1988).
23. Hamilton R, Watanabe CK, and DeBoer HA: *Nucleic Acids Res.* 15:3581 (1987).
24. Vazquez D: Inhibitors of protein synthesis. *Springer-Verlag, New York* (1979).
25. Warner JR: In Strathern JN, Jones EW, and Broach JR (Eds): *The Molecular Biology of the Yeast Saccharomyces: Metabolism and Gene Expression,* Cold Spring Harbor, NY, Cold Spring Harbor Laboratory, 529, 1983.
26. Hinnebusch AG, Jackson BM, and Mueller PP: *Proc. Natl. Acad. Sci. U.S.A.* 85:7279 (1988).
27. Wolfner M, Yep D, Messenguy F, and Fink GR: *J. Mol. Biol.* 96:273 (1975).
28. Niederberger P, Aebi M, and Huetter R: *Curr. Genet.* 10:657 (1986).
29. Harashima S and Hinnebusch AG: *Mol. Cell. Biol.* 6:3990 (1986).
30. Hill DE and Struhl K: *Nucleic Acids Res.* 16:9253 (1988).
31. Hannig EM and Hinnebusch AG: *Mol. Cell. Biol.* 8:4808 (1988).
32. Paddon CJ and Hinnebusch AG: *Genetics* 122:543 (1989).
33. Donahue TF, Cigan AM, Pabich EK, and Castilho-Valavicius B: *Cell* 54:621 (1988).
34. Cigan AM, Pabich EK, Feng L, and Donahue TF: *Proc. Natl. Acad. Sci. U.S.A.* 86:2784 (1989).
35. Donahue TF, Cigan AM, De Castilho BA, and Yoon H: in Tuite M (Ed): *Genetics of Translation,* Berlin, Springer-Verlag, 361, 1988.
36. Williams NP, Hinnebusch AG, and Donahue TF: *Proc. Natl. Acad. Sci. U.S.A.* 86:7515-7519 (1989).
37. Tzamarias D, Roussou I, and Thireos G: *Cell* 57:947 (1989).
38. Harashima S, Hannig EM, and Hinnebusch AG: *Genetics* 117:409 (1987).
39. Paddon CJ, Hannig EM, and Hinnebusch AG: *Genetics* 122:551 (1989).
40. Roussou I, Thireos G, and Hauge BM: *Mol. Cell. Biol.* 8:2132 (1988).

41. Wek RC, Jackson BM, and Hinnebusch AG: *Proc. Natl. Acad. Sci. U.S.A.* 86:4579 (1989).
42. Hanks SK, Quinn AM, and Hunter T: *Science* 241:41 (1988).
43. Messenguy F and Delforge J: *Eur. J. Biochem.* 67:335 (1976).
44. Niederberger P, Aebi M, and Huetter R: *J. Gen. Microbiol.* 129:2571 (1983).
45. Safer B: *Cell* 33:7 (1983).
46. Schneider RJ and Shenk T: in Ilan J (Ed): *Translation Regulation of Gene Expression,* New York, Plenum Press, 431 (1987).
47. Kozak M: *J. Cell Biol.* 108:229 (1989).
48. Niederberger P, Miozzari G, and Huetter R: *Mol. Cell. Biol.* 1:584 (1981).
49. Crichton RR and Charloteaux-Waters M: *Eur. J. Biochem.* 164:485 (1987).
50. Harford J, Casey JL, Koeller DM, and Klausner RD: in Steer CJ and Hanover JA (Eds): *Intracellular Trafficking of Proteins,* Cambridge, Cambridge University Press (in press).
51. van Renswoude J et al: *Proc. Natl. Acad. Sci. U.S.A.* 79:6186 (1983).
52. Yamashiro DJ: *Cell* 37:789 (1984).
53. Dautry-Varsat A, Ciechanover A, and Lodish HF: *Proc. Natl. Acad. Sci. U.S.A.* 80:2258 (1983).
54. Klausner RD, Ashwell G, van Renswoude J, Harford JB, and Bridges KR: *Proc. Natl. Acad. Sci. U.S.A.* 80:2263 (1983).
55. Theil EC: *Annu. Rev. Biochem.* 56:287 (1987).
56. Granick S: *J. Biol. Chem.* 164:737 (1946).
57. Munro HN and Linder MC: *Physiol. Rev.* 58:317 (1978).
58. Boyd D: *Proc. Natl. Acad. Sci. U.S.A.* 81:4751 (1984).
59. Boyd D: *J. Biol. Chem.* 260:11755 (1985).
60. Jain SK et al: *J. Biol. Chem.* 260:11762 (1985).
61. Costanzo F et al: *Nucleic Acids Res.* 14:721 (1986).
62. Cairo G, Bardella L, Schiaffonati L, Arosio P, Leui S, and Bernelli-Zazze A: *Biochem. Biophys. Res. Commun.* 133:314 (1985).
63. Rouault TA, Hentze MW, Dancis A, Caughman SW, Harford JB, and Klausner RD: *Proc. Natl. Acad. Sci. U.S.A.* 84:6335 (1987).
64. Aziz N and Munro HN: *Nucleic Acids Res.* 14:915 (1986).
65. Hentze MW, Caughman SW, Rouault TA, Barriocanal JG, Dancis A, Harford JB, and Klausner RD: *Science* 238:1570 (1987).
66. Aziz N and Munro HN: *Proc. Natl. Acad. Sci. U.S.A.* 84:8478 (1987).
67. Zahringer J, Baliga BS, and Munro HN: *Proc. Natl. Acad. Sci. U.S.A.* 73:857 (1976).
68. Rogers J and Munro HN: *Proc. Natl. Acad. Sci. U.S.A.* 84:2227 (1987).
69. Caughman SW, Hentze MW, Rouault TA, Harford JB, and Klausner RD: *J. Biol. Chem.* 263:19048 (1988).
70. Hentze MW, Caughman SW, Casey JL, Koeller DM, Rouault TA, Harford JB, and Klausner RD: *Gene* 72:201 (1988).
71. Hentze MW, Rouault TA, Caughman SW, Dancis A, Harford JB, and Klausner RD: *Proc. Natl. Acad. Sci. U.S.A.* 84:6730 (1987).
72. Uhlenbeck OC, Huey-Nan W, and Sampson JR: in Inouye M and Dudock BS (Eds): *Molecular Biology of RNA: New Perspectives,* San Francisco, Academic Press, 285 (1987).
73. Walden WE et al: *Proc. Natl. Acad. Sci. U.S.A.* 85:9503 (1988).

74. Leibold E and Munro HN: *Proc. Natl. Acad. Sci. U.S.A.* 85:2171 (1988).
75. Rouault TA, Hentze MW, Haile D, Harford JB, and Klausner RD: *Proc. Natl. Acad. Sci. U.S.A.* 86:5768 (1989).
76. Walden WE, Patino MM, and Gaffield L: *J. Biol. Chem.* 264:13765 (1989).
77a. Hentze MW, Rouault TA, Harford JB, and Klausner RD: *Science* 244:357 (1989).
77b. Hentze MW, Suananez HN, O'Brien SJ, Harford JB, and Klausner RD: *Nucleic Acids Res.* 17:6103 (1989).
78. Mattia E, Rao K, Shapiro DS, Sussman HH, and Klausner RD: *J. Biol. Chem.* 259:2689 (1984).
79. Rao KK, Shapiro D, Mattia E, Bridges K, and Klausner RD: *Mol. Cell Biol.* 5:595 (1985).
80. Rao KK, Harford JB, Rouault TA, McClelland A, Ruddle FH, and Klausner RD: *Mol. Cell Biol.* 6:236 (1986).
81. Casey JL, DiJeso B, Rao KK, Klausner RD, and Harford JB: *Proc. Natl. Acad. Sci. U.S.A.* 85:1787 (1988).
82. Owen D and Kuhn LC: *EMBO J.* 6:1287 (1987).
83. Müllner EW and Kühn LC: *Cell* 53:815 (1988).
84. Koeller DM, Casey JL, Hentze MW, Gerhardt EM, Chan LL-N, Klausner RD, and Harford JB: *Proc. Natl. Acad. Sci. U.S.A.* 86:3574 (1989).
85. Mullner EW, Neupert B, and Kuhn LC: *Cell* 58:373 (1989).
86. Casey JL, Hentze MW, Koeller DM, Caughman SW, Rouault TA, Klausner RD, and Harford JB: *Science* 240:924 (1988).
87. Chan LL-N: *Nucleic Acids Res.* 17:3763 (1989).
88. Casey JL, Koeller DM, Ramin V, Klausner RD, and Harford JB: *EMBO J.* 8:3693 (1989).
89. Rouault TA, Hentze MW, Caughman SW, Harford JB, and Klausner RD: *Science* 241:1207 (1988).
90. Yamamoto KK, Gonzalez GA, Biggs WH, and Montminy NR: *Nature* 334:494 (1988).
91. Otwinowski Z, Schevitz RW, Zhang R-G, Lawson CL, Toachimiak A, Marmorstein RQ, Luisi BF, and Sigler PB: *Nature* 335:321 (1988).
92. Magasanik B: *Trends Biochem. Sci.* 13:475 (1988).
93. Haile DJ, Hentze MW, Rouault TA, Harford JB, and Klausner RD: *Mol. Cell. Biol.* 9:5055 (1989).
94. Chan J-J, Yang JM, Petryshyn R, Kosower N, and London IM: *J. Biol. Chem.* 264:9559 (1989).
95. Lin J-J, Daniels-McQueen S, Patino MM, Gaffield L, Walden WE, and Thach RE: *Science* 247:74 (1990).
96. Abastado JP, Miller PF, Jackson BM, Hinnebusch AG: Suppression of ribosomal reinitiation at upstream open reading frames in amino acid-starved cells forms the basis of *GCN4* translational control. *Mol. Cell Biol.* 11:486-496 (1991).
97. Sherman F, Stewart JW: Mutations altering initiation of translation of yeast iso-1-cytochrome *c*; contrasts between the eukaryotic and prokaryotic initiation process. In *The molecular biology of the yeast Saccharomyces metabolism and gene expression*, Strathern JN, Jones EW, and Broach JR (eds), Cold Spring Harbor Laboratory, Cold Spring Harbor, NY, pp 301–334 (1982).

The Translation of Picornaviruses

Karen Meerovitch
*Department of Biochemistry
and
McGill Cancer Center
McGill University
Montreal, Quebec*

Nahum Sonenberg
*Department of Biochemistry
McGill Cancer Center
McGill University
Montreal, Quebec*

Jerry Pelletier
*Department of Cancer Research
Massachusetts Institute of Technology
Cambridge, Massachusetts*

11.1. INTRODUCTION

The Picornaviridae represent one of the largest and most important families of human and agricultural pathogens. For this reason they have figured prominently in the development of modern virology. Their use to study changes which occur in cellular macromolecular synthesis following viral infection has resulted in the description of two important phenomena: (1) a highly discriminatory shut-off mechanism whereby translation of host mRNA is selectively suppressed and (2) a mechanism of internal initiation for translation in which 40S ribosomes bind internally on the viral mRNA template to initiate protein synthesis. Unraveling of the molecular details of these events is important not only for the understanding of virus-host interactions or designing strategies to combat virus infection, but may also shed light on the mechanism governing cellular gene expression at the translational level.

The picornavirus family consists of four genera: (1) enteroviruses (e.g., poliovirus, coxsackievirus, hepatitis A virus), (2) cardioviruses (e.g., encephalomyocarditis (EMC) virus, mengovirus), (3) rhinoviruses, and (4) aphthoviruses (foot-and-mouth disease virus [FMDV]). All members contain a single-stranded RNA molecule of plus sense polarity as their genome (see Reference 1). The RNA contains a poly(A) tail at its 3' end, is blocked at the 5' end by a small viral-encoded peptide (VPg), and encodes a large polyprotein that is posttranslationally cleaved into viral specific polypeptides.

11.2. SHUT-OFF OF HOST PROTEIN SYNTHESIS AFTER POLIOVIRUS INFECTION

11.2.1. Early Studies

The rapid and progressive decline in the rate of host protein synthesis in poliovirus-infected cells was first observed almost 25 years ago, and has been extensively studied ever since. The shut-off of host translation is essentially completed by 2 h after infection, depending on the multiplicity of infection, followed by a burst of protein synthesis representing exclusively the synthesis of viral proteins.[2] Early studies on poliovirus infection have established that cellular polysomes disaggregate and are replaced by virus-specific polysomes.[3] These results suggest that ribosomes remain active during infection since they must translate viral RNA. The defect in translation of cellular RNAs results from a block at the initiation step of protein synthesis.[4-6] Subsequently it was identified that host cellular mRNAs fail to bind to 40S ribosomal subunits.[7, 8] The nontranslated cellular mRNAs, however, remain intact and unmodified, with respect to poly(A) content and the 5'-terminus cap structure,[9, 10] and when extracted from the cytoplasm of infected cells are still capable of directing protein synthesis *in vitro*.[11] Poliovirus also inhibits mRNA translation of other viruses, such as herpes virus,[12] vesicular stomatitis virus,[13] and most adenovirus mRNAs.[14]

The results of studies on poliovirus infection using drugs, or under conditions that restrict poliovirus RNA replication, are consistent with the conclusion that viral gene expression is required for inhibition of cellular protein synthesis. For example, only partial shut-off of host protein synthesis occurs when cells are infected with poliovirus in the presence of guanidine, which prevents viral replication.[15, 16] However, agents which inactivate the infecting genome (e.g., ultraviolet light) or which prevent translation (e.g., cycloheximide) abolish shut-off.[17, 18] Therefore, translation, but not replication, of the viral genome is essential for the shut-off phenomenon.

All of the above studies suggest that poliovirus usurps the host's cellular translational apparatus. A model to explain the mechanism of poliovirus shut-off that has been widely accepted is predicated on the finding that poliovirus RNA is not capped.[19, 20] Consequently, it was presumed that translation of poliovirus RNA would proceed by a cap-independent mechanism. It was proposed that this difference provides the basis for the translational discrimination between capped cellular mRNA and uncapped poliovirus RNA, and that this could be accomplished by the inactivation of a protein that recognizes the cap structure and is required for the translation of host mRNAs. Subsequent efforts were made to identify the component in infected cells that is impaired as a result of poliovirus infection.

11.2.2. CBP Complex Inactivation

Initial experiments with cell-free translation systems demonstrated that initiation factor (IF) preparations from infected cells did not stimulate the translation of capped host cellular mRNAs, but did stimulate poliovirus RNA translation,[11, 21] reflecting the situation *in vivo*. In addition, IF preparations from mock-infected cells could restore the translation of capped mRNAs in extracts from poliovirus-infected cells.[21] Initial attempts to demonstrate that a specific IF is inactivated after poliovirus infection produced conflicting results, suggesting that either eIF-4B[22] or eIF-3[23] was rendered inactive. This discord was resolved when it was shown that a 24 kDa cap-binding protein (CBP; eIF-4E) could be purified from preparations of both eIF-4B and eIF-3.[24] An activity that restored translation of capped mRNAs in extracts from poliovirus-infected cells copurified with eIF-4E,[25] but highly purified eIF-4E by itself had no restoring activity,[26] suggesting that other components may be involved. Subsequently, a multisubunit complex isolated by m[7]GDP-affinity chromatography, termed CBPII or eIF-4F, was shown to have restoring activity.[26-28] eIF-4F is composed of three subunits, a 24 kDa CBP (CBPI, eIF-4E), a 50 kDa protein (eIF-4A), and a 220 kDa polypeptide (p220).[27, 29] In a key experiment Etchison et al.[30] employed the immunoblotting technique to demonstrate that p220 is proteolyzed in poliovirus-infected HeLa cells yielding two to three antigenically related polypeptides of 110 to 130 kDa. These results became the basis for an attractive hypothesis for the mechanism of host cell shut-off, which stated that cleavage of p220 leads to inactivation of eIF-4F and thereby blocks initiation of capped mRNA translation.

The rest of the translational machinery is otherwise unaltered and free to translate uncapped poliovirus mRNA which initiates translation via a distinct cap-independent mechanism (see below).

Several lines of evidence are consistent with the contention that eIF-4F is inactivated as a consequence of poliovirus infection. When purified from infected cells, eIF-4F is inactive in an *in vitro* reconstituted translation system.[31] It has been shown that cross-linking of eIF-4A and eIF-4B is impaired in IF preparations prepared from poliovirus-infected cells,[32-34] although these factors are neither structurally modified nor functionally impaired.[23, 35] These findings are in accord with the finding that cross-linking of eIF-4A and eIF-4B are dependent on the interaction of eIF-4F with the cap structure.[29] Similarly, the cross-linking of eIF-4E is drastically reduced in IF preparations from infected cells.[32, 34] However, a contrary result was obtained by Hansen and Ehrenfeld.[36] In lieu of the observation that cross-linking of eIF-4E to capped mRNA is at least tenfold greater when it exists as part of eIF-4F than when it is free, it is plausible that the eIF-4E subunit in the modified eIF-4F complex in infected cells behaves like uncomplexed eIF-4E with respect to cross-linking efficiency. Indeed, a modified eIF-4F complex has been purified from poliovirus-infected cells consisting of eIF-4E and p220 cleavage products.[33, 37] The above data imply that p220 plays a significant role in mediating the interaction between eIF-4E and mRNA. The possibility that the impairment of eIF-4F was due to modifications of eIF-4E was ruled out, since eIF-4E is not modified after poliovirus infection as judged by two-dimensional SDS-PAGE.[33, 38] Thus, poliovirus infection does not cause any modification to eIF-4E, and inactivation of eIF-4F is a consequence of cleavage of the 220 kDa polypeptide.

What is the proteolytic activity that cleaves p220? Initial experiments ruled out one viral protease, polypeptide 3C as the effector of p220 cleavage, because antiserum to proteinase 3C, which inhibits processing of the poliovirus polyprotein precursor, did not inhibit cleavage of p220 in an *in vitro* assay.[39, 40] A mutant virus displaying a small plaque phenotype and incapable of specific inhibition of host protein synthesis has been described.[41] Significantly, it contains a single codon insertion in the 2A gene and upon infection did not cause p220 proteolysis.[41] In an *in vitro* assay, it was concluded that protease 2A (2A^{pro}) fails to cleave p220.[42] The genetic data taken together with the inability of 2A^{pro} to cleave p220 *in vitro* suggest that 2A^{pro} is required for cleavage, but acts indirectly to mediate this process. In addition, the genetic studies provided the first indication for a causal relationship between cleavage of p220 and a shut-off of host protein synthesis after poliovirus infection. In more recent studies, it was shown that a different insertion mutant in the 2A coding gene, which results in a small plaque phenotype, also prevents cleavage of p220 *in vivo*.[43, 44] It was also shown that anti-2A^{pro} serum inhibits p220 cleavage, if added during *in vitro* translation of 2A^{pro} mRNA, but does not inhibit p220 cleavage if added after translation of 2A^{pro} mRNA. This suggests that activation of a p220 specific protease had occurred.[45] Expression of 2A^{pro} by itself in HeLa cells dramatically inhibits cellular mRNA translation and induces p220 cleavage.[46] Thus, these results provide direct evidence for the role of 2A^{pro} as a mediator of p220 cleavage resulting in inactivation of eIF-4F and shut-off of cellular protein synthesis. Recent experiments aimed at characterizing the cellular protein responsible for p220 cleavage by 2A^{pro} suggested that eIF-3 may be involved.[46a] It was suggested that when eIF-3 interacts with eIF-4F, p220 is presented in a proper conformation to be a substrate for cleavage by 2A^{pro}.

It is puzzling that not all picornaviruses inactivate eIF-4F.[47] A correlation has been documented between the ability of a virus to cleave p220 and the presence of a highly conserved putative thiol protease consensus sequence in 2A^{pro}.[48] This is exemplified by the findings that entero- and rhinoviruses which contain the 2A active site sequence will induce cleavage of p220, whereas two cardioviruses (encephalomyocarditis and Theiler's encephalomyelitis virus), which do not encode this proteinase sequence do not induce cleavage of p220 *in vivo*.[47, 48] The only exception is foot-and-mouth disease virus which induces cleavage of p220, yet does not contain the putative cysteine protease active site in 2A^{pro}. Instead, p220 cleavage activity has been mapped to the amino terminus of the viral polyprotein (the leader protein).[49]

Cleavage of p220 is necessary, but not sufficient to cause complete inhibition of cellular mRNA translation.[16] When cells are infected with poliovirus in the presence of inhibitors of virus replication, such as guanidine or 3-methylquercetin, approximately 30% of host protein synthesis continues, although p220 is completely cleaved.[16] These results raise the possibility that a proportion of cellular mRNAs translate in a cap-independent fashion (analogous to poliovirus) and that a second event, which may require poliovirus replication, is involved in the inhibition of cellular cap-independent translation. This second event could be eIF-2α phosphorylation.

11.2.3. eIF-2α Phosphorylation

Eukaryotic initiation factor-2 (eIF-2) is required for ternary complex formation which is transferred to 40S ribosomes (see Part I, Chapter 8 of this book). eIF-2 can be rendered inactive due to phosphorylation of its alpha subunit by a specific kinase (called p68 kinase, DAI or P1/eIF-2 kinase) which is activated by double stranded RNA subsequent to infection by several viruses (e.g., vesicular stomatitis virus;[50] reovirus[51]). Several viruses have evolved mechanisms to counteract the eIF-2α phosphorylation pathway. For example, adenovirus encodes a small RNA termed virus-associated RNA I (VAI RNA), which complexes to p68 kinase to down-regulate its activity.[52] Influenza virus,[53] EMC virus and vaccinia virus[54] have evolved different mechanisms to ensure translation of viral mRNAs. Recently, it has been shown that in poliovirus-infected cells, p68 kinase becomes autophosphorylated and activated.[55] As a result of p68 activation, the phosphorylation of eIF-2α increased. A significant proportion of p68 kinase is also degraded during poliovirus infection. Proteolysis of p68 kinase may be required to ensure that phosphorylation is controlled thereby enabling poliovirus RNA translation. It would be of interest to determine the phosphorylation state of p68 kinase and eIF-2α in poliovirus-infected cells treated with guanidine. If eIF-2α phosphorylation is the second event required for complete shut-off of host protein synthesis,[16] then in the presence of guanidine, it is predicted that eIF-2α would not be phosphorylated. Phosphorylation of eIF-2α was also shown by O'Neill and Rancaniello,[44] whereas Ransone and Dasgupta[56] have shown that, despite activation of the p68 kinase during poliovirus infection, there was no detectable increase in eIF-2α phosphorylation. Furthermore, Ransone and Dasgupta[57] have characterized an inhibitor or eIF-2α phosphorylation that is present in poliovirus-infected cells. The discrepancies in these results might be due to the different experimental techniques employed. There appears to exist several mechanisms in poliovirus-infected cells which control the extent of eIF-2α phosphorylation. It is intriguing that poliovirus induces the degradation of two proteins which have significant roles in translational regulation: p68 kinase and p220. It is of interest to determine whether similar mechanisms operate to cause the proteolysis of these two proteins. In this regard, it is noteworthy that virus containing mutant 2A protein fails to cleave p220, but still increases phosphorylation of eIF-2α.[44]

The question remains as to whether there exists a differential effect of eIF-2α phosphorylation on viral vs. cellular mRNA translation. Black et al.[55] reported that eIF-2α was maximally phosphorylated at 3 h after poliovirus infection, when poliovirus translation is high but cellular protein synthesis is abrogated. Localized activation of p68 kinase, possibly by double-stranded RNA, may be one mechanism allowing for the specificity observed in the translational control mediated by eIF-2α phosphorylation.[58, 59] In addition, secondary structure in the 5' noncoding region of cellular RNAs could potentially *trans*-activate p68 kinase and cause inhibition of cellular mRNA translation. A precedent for *trans*-inhibition of mRNA translation by mRNA secondary structure has been reported for the HIV-1 TAR sequence element.[60, 61] Clearly, poliovirus is multitalented in ensuring translation of its mRNA.

11.3. INTERNAL INITIATION OF TRANSLATION

Picornavirus mRNAs are unusual in that, unlike eukaryotic cellular mRNAs, they are not capped at their 5' ends but rather terminate by monophosphate residues.[19, 20] In addition, their 5' UTRs contain a high number of AUG codons (e.g., ten for EMC virus; seven or eight for poliovirus, depending on the serotype). Multiple 5' upstream AUGs in cellular mRNAs severely inhibit translation (see Reference 62 for a review), however, they do not restrict picornavirus translation, since EMC virus and mengovirus mRNAs are among the most efficient messengers known.[63–67] This efficient translational property of EMC virus has been shown to be imparted by the 5' UTR.[45, 68] In addition, during initiation of translation ribosomes do not accumulate on the long leader sequence of poliovirus RNA (745 nucleotides for poliovirus type 2) when elongation of protein synthesis is blocked.[69] These observations and the findings that translation of poliovirus RNA is insensitive to cap analog inhibition[70] and occurs *in vivo* under conditions in which cap-mediated translation is impaired (described above) suggested that poliovirus initiation was basically very different from the bulk of cellular mRNAs. These observations led investigators to speculate that initiation on poliovirus mRNA occurs by a mechanism of internal ribosome binding, whereby 43S pre-initiation complexes bind internally on the mRNA template, thus bypassing the majority of the hurdles (i.e., AUG codons) to initiate translation (e.g., Reference 69). This mechanism is different from that proposed to occur on cellular mRNAs, where 43S complexes bind to the 5' end cap structure and then migrate in a linear manner towards the initiator AUG.[71]

Experiments were aimed at defining the structural parameters responsible for the cap-independent translation of poliovirus[72–74] and EMC virus.[75, 76] Capping the poliovirus or EMC virus mRNA 5' end did not stimulate translation, nor did decapping of normally capped VSV mRNAs allow for cap-independent translation in poliovirus-infected extracts.[77] These results were interpreted to indicate that initiation events on the poliovirus template were proceeding in a fashion completely independent of 5' end structure. Consistent with this idea, Shih et al.[78] showed that hybridization of defined cDNA fragments to nucleotides 1 to 338 of the EMC virus 5' UTR had no effect on translation. This is in striking contrast to the inhibitory effects 5' UTR cDNA fragments exert on the translation of cap-dependent mRNAs.[79, 80] In contrast, hybridization of cDNAs to nucleotides 450 to 834 of the EMC virus 5' UTR strongly inhibited translation.

Deletion mutagenesis of the 5' UTRs of poliovirus[72] and EMC virus,[75] followed by *in vitro* and *in vivo* translation studies revealed several interesting points: (1) contrary to what has been found for the majority of cellular mRNAs (see Reference 81 for a review), perturbations (such as inserting or deleting sequences) within the 5' noncoding region of poliovirus or EMC virus can dramatically inhibit translation, and (2) the region responsible for cap-independent initiation of poliovirus type 2 was found to be quite large (300 nucleotides) and resided between nucleotides 320 and 631 of the 5' UTR. This same region was also found to be important for the translation of poliovirus type 1 *in vivo* under conditions of poliovirus infection and in cell-free extracts.[73, 82] For EMC virus, the cap-independent region resided between nucleotides 260 and 848. When fused to heterologous mRNAs, both elements could function in *cis* to render translation cap-independent.

Genetic analysis of insertions and deletions introduced into the 5' UTR of the Mahoney strain[82, 83] or Sabin strain[84] of poliovirus type 1 have confirmed the importance of the cap-independent region for viral translation *in vivo*. Trono et al.[82] analyzed poliovirus insertion mutants at nucleotides 224, 270, and 392 which demonstrated a small plaque and temperature-sensitive phenotype. These mutants made very little protein, did not inhibit host cell translation, but synthesized a significant amount of RNA. These results suggested that the defect was in viral mRNA translation and the region responsible on the mRNA 5′ UTR was called Region P. Other mutations at nucleotides 325, 443, 460, and 499 were lethal demonstrating an absolute involvement of these sequences in viral translation. Consistent with the extended P region, Kuge and Nomoto[84] demonstrated that revertants of a small plaque mutant containing a four nucleotide insertion at nucleotide 220 contained second site mutations at nucleotides 186 and 524 or 186 and 480.

The area defined by Region P (and its EMC counterpart) was shown to direct internal ribosome binding when placed as intercistronic spacer in bi- or tricistronic mRNAs[75, 76, 85] Messenger RNA derived from these constructs contained two ribosome entry points: (1) a 5′ end site mediated through the cap structure, and (2) an internal site mediated by the poliovirus or EMC virus 5′ UTR. Under conditions where 5′ mediated initiation was impaired (i.e., in the presence of cap analog, m⁷GpppG, in poliovirus-infected extracts, or under hypertonic salt conditions), translation was only observed for those cistrons driven by the poliovirus or EMC virus 5′ UTR. The efficiency of initiation was similar whether or not the 5′ UTR was in a mono- or bicistronic context. In some cases, the EMC virus UTR allowed protein product from the cistron under its control to appear before product from the first cistron. Analysis of mRNA integrity by northern blots, primer extension, and direct polysome analysis demonstrated that the results were not due to nucleolytic digestion of the mRNA followed by ''activation'' of spurious translation start sites.

Initiation codons and small cistrons in the 5′ UTRs of cap-dependent mRNAs generally decrease translation from downstream open reading frames (ORFs),[86–89] possibly because not all terminating ribosomes become competent for reinitiation. In an effort to determine if the seven upstream AUGs of poliovirus type 2 Lansing strain encode small peptides with essential function or down-regulate translation initiation at the poliovirus major ORF, they were systematically eliminated by site-directed mutagenesis.[90] Mutagenesis, followed by *in vitro* translation of the mRNA or transfection of infectious RNA into cells, indicated that the AUGs are dispensable for viral biogenesis. Mutagenesis of the A588 nucleotide of AUG-7 surprisingly caused an eightfold decrease in translational efficiency when assayed in poliovirus-infected HeLa cell extracts. When this mutation was reconstructed into the infectious cDNA clone, virus displaying a small plaque phenotype was obtained. Additional studies have demonstrated that mutation of nucleotides U589 and G590 have an inhibitory effect on translation *in vitro* and *in vivo* (Meerovitch,

Nicholson, and Sonenberg, unpublished data). It is possible that AUG-7 exerts a positive effect on translation by serving as a start codon for a short open reading frame, and thus facilitate the transfer of ribosomes to the initiator AUG at position 745. This appears unlikely as it has been found that *in vitro* initiation at AUG-7 is inefficient.[90a]

The possibility that AUG-7 is part of a recognition domain involved in internal ribosome binding is supported by recent experiments from our laboratory. A deletion from the 3' end of poliovirus (type 2, Lansing) extending to nucleotide 585 reduced translation of a reporter gene approximately 3-fold. A further deletion extending to nucleotide 556 completely abrogated translation both *in vitro* and *in vivo*. This latter deletion removes a pyrimidine stretch (nucleotides 560-568: UUUCCUUUU) that is highly conserved among picornaviruses. Mutational analysis of this stretch revealed that nucleotides UUUCC are of critical importance for internal initiation of translation (Nicholson, Meerovitch and Sonenberg, unpublished data). These results are consistent with earlier studies by Nomoto and colleagues who showed that a 3' deletion to nucleotide 564 (type 1, Mahony) resulted in a poliovirus mutant with a small plaque phenotype, but a further deletion to nucleotide 561 completely abrogated viral growth.[84, 90b] Deletion of this pyrimidine stretch in EMC virus and mutations of this stretch in FMDV decreased dramatically the translational efficiency *in vivo* and *in vitro*.[90c, d] Taken together, these results suggest that ribosomes bind upstream of AUG-7 in the vicinity of the pyrimidine stretch and then translocates, possibly by scanning, to the initiator AUG. The placement of initiation codons downstream from AUG_{588}, but upstream from the poliovirus initiator AUG, decreased translational efficiency supposedly by interfering with ribosomes migrating from the internal binding site recognition domain.[91]

A cellular protein (52 kDa) from HeLa cells has been found to bind an RNA fragment (nucleotides 559 to 624) encompassing the pyrimidine stretch and AUG-7.[92] The binding site of this protein and its specificity makes it a likely candidate for a factor involved in internal initiation. p52 does not seem to be one of the previously characterized initiation or elongation factors. Biochemical characterization (is this protein a RNA helicase?) and mRNA specificity (does this factor bind to cellular mRNAs to mediate internal initiation?) are important questions to answer. The importance of the p52 binding region for poliovirus translation has been demonstrated by Bienkowska-Szewczyk and Ehrenfeld,[93] who showed that nucleotides 567 to 627 of poliovirus type 1 (Mahoney strain) were required for efficient translation in a rabbit reticulocyte lysate supplemented with a HeLa extract. Del Angel et al.[94] have also demonstrated specific binding of cytoplasmic factors to a poliovirus sequence from nucleotides 510 to 629. They found that eIF-2 is part of the complex which binds to this region. p52 does not appear to be equivalent to one of the eIF-2 subunits (Meerovitch and Sonenberg, unpublished data).

Specific RNA protein complexes have also been described for EMC virus[90c, 94a] and FMDV 5' UTR.[94b] These groups have identified 57 and 58-kDa proteins in rabbit reticulocyte lysates and Krebs-2 cell extracts, respectively, that crosslink to the viral 5' UTR. These proteins crosslink to regions in the 5' UJTR that are functionally important for translation.

It is probable that more factors are involved in internal initiation than those which have been found to bind to nucleotides 510 to 629 since the ribosome landing pad (RLP) is several hundred nucleotides long. The large nucleotide sequence required for internal ribosome binding is consistent with the model by which this process is dependent upon a particular tertiary structure (as opposed to a strictly sequence-specific signal or secondary structure) for ribosome recognition and binding. This region is functionally conserved among enteroviruses since recombinants containing coxsackievirus sequences substituting nucleotides 66 to 627 of the poliovirus genome have properties similar to those of wild type poliovirus.[95] Several secondary structure models of the poliovirus 5' UTR have been described.[96-98a] These models have been generated by computer-predicted minimal energy folding analysis combined with results from structure mapping studies. The above models subdivide the poliovirus 5' UTR, between nucleotides 240 and 620, into three domains of secondary structure. An interesting possibility is that these separate domains interact to form tertiary structure.

Secondary structure predictions have been made for the EMC virus 5' UTR[99] and do not resemble those of poliovirus. This may imply a difference in requirement for *trans*-acting proteins mediating internal initiation between EMC virus and poliovirus. In this regard, results with EMC virus indicate that following binding, ribosomes are probably transferred directly to the initiator AUG.[100] It has been shown that this is *not* the case with poliovirus, since introduction of secondary structure[74] or AUG codons[91] between the RLP and the initiator AUG inhibits translation.

11.3.1. Internal Initiation on Nonpicornavirus mRNAs

Hybrid-arrest translation experiments demonstrated internal initiation on vesicular stomatitis virus P mRNA,[101] adenovirus DNA polymerase mRNA[102] and the Sendai virus P/C mRNA.[103, 104] These conclusions are based on the findings that DNA complementary to regions upstream of the internal initiation site arrest translation, while not affecting translation of downstream cistrons. Hassin et al.[102] have made the interesting observation that capping adenovirus DNA polymerase mRNA results in increased synthesis of the 5'-end mediated polypeptide while decreasing synthesis of an internally initiated polypeptide. This is consistent with competition between 5'-end mediated and internal initiation thus occluding ribosomes from the internal initiation site by translating ribosomes. Curran and Kolalofsky[104] have inserted AUG codons upstream of the naturally occurring AUGs of the Sendai virus P/C mRNA to intercept and shunt scanning ribosomes before they could reach downstream ORFs on the mRNA. In these constructs, initiation at two downstream AUGs (Y1 and Y2) were shown to be scanning independent *in vitro* and *in vivo*.

A series of elegant genetic and biochemical studies performed by Chang et al.[105] on the nucleocapsid/polymerase gene of duck hepatitis B virus are consistent with internal initiation at the polymerase (pol) cistron. The 5' end of the pol reading frame overlaps with the coding region for the core protein. Translation of the pol cistron was uncoupled from that of the core protein, by creating a stop codon in the core ORF, terminating core translation well upstream of the core-polymerase overlap region. When analyzed *in vivo* in a complementation assay, pol expression was not dependent on translation of the core protein. The results were interpreted as evidence for *de novo* initiation involving direct entry of ribosomes. More recently, Jean-Jean et al.[106] have reached the same conclusion using a different experimental approach.

11.3.2. Molecular Mechanism for Internal Ribosome Binding

The initial event in internal ribosome binding probably involves recognition of the RLP by one (p52) or more cellular factors. This is most likely followed by binding of eIF-4A and -4B to the mRNA (Figure 1). eIF-4A, in the presence of eIF-4B, is capable of unwinding RNA secondary structure using energy generated from the hydrolysis of ATP.[107] A striking feature of this activity is that it functions in a bidirectional manner.[108] The involvement of eIF-4A, in combination with eIF-4B, in internal binding of ribosomes is consistent with the findings that these factors can bind to mRNA (lacking secondary structure) with the same affinity as for capped RNAs in the presence of eIF-4A, -4B, and -4F.[109] These factors scan and unwind the poliovirus 5' UTR until the appropriate AUG codon is encountered where the 43S preinitiation complex then binds, or alternatively the 43S pre-initiation complex binds to the RLP and migrates towards the initiator AUG.

11.4. FUTURE PERSPECTIVES

Considering that poliovirus, EMC virus and other virals RNAs initiate translation via internal ribosome binding, it is likely that other viruses utilize the same mechanism. It is also conceivable that the machinery for internal ribosome binding in eukaryotes exists for the translation of cellular mRNAs. The best candidates are mRNAs which contain very long 5'-noncoding regions. Many of the mRNAs having long 5' noncoding regions code for proteins that function as important regulators of cell growth, differentiation and development. Expression of these proteins could be modulated at the transla-

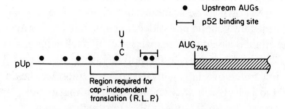

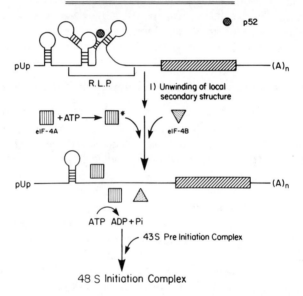

Figure 1. Model for unwinding of poliovirus mRNA 5′ secondary structure prior to ribosome binding (see Section 11.3.2 for details).

tional level by mechanisms involving internal binding of ribosomes. Sarnow and colleagues have provided evidence that the immunoglobulin heavy-chain binding protein (BIP) can be translated in poliovirus-infected cells[109a] by an internal ribosome binding mechanism (Sarnow P., personal communication). More recently *Drosophila* cellular mRNAs encoding the *antennapedia* and *ultrabithorax* proteins have been proposed to initiate translation by internal binding of ribosomes (Sarnow P., personal communication). Identifying more of such mRNAs is an important goal in the immediate future in order to assess how common is this mechanism of translation. It is even possible that most eukaryotic mRNAs can initiate translation to some extent by internal binding of ribosomes, circumventing the highly regulated cap-dependent translation pathway.

Much work has to be accomplished to understand the molecular mechanism of cap-dependent and cap-independent translation initiation in eukaryotes. Clearly, the nature of *cis*-acting elements in the mRNAs 5' noncoding regions that mediate ribosome binding are not fully understood. In particular, the understanding of the involvement of tertiary structure motifs, such as pseudoknots, in translation is in its infancy. Developing research technologies in this area will prove to be instrumental for studying picornavirus translation.

There are surely many *trans*-acting factors that regulate translation initiation, and it is conceivable that p52 is not the only novel factor required for translation of poliovirus. The development of a reconstituted *in vitro* translation system for poliovirus would be of great importance for the identification of these factors.

The future understanding of the host cell shut-off phenomenon will be achieved by current studies on the protease responsible for p220 cleavage and the identification of p220 mutants resistant to cleavage. In addition, in order to assess the importance of eIF-2α phosphorylation an understanding of the kinetics of phosphorylation relative to the shut-off of host protein synthesis is required. It is anticipated that research on poliovirus and other picornaviruses will continue to provide novel insights to basic mechanisms of cellular and viral gene expression in general and translation in particular.

ACKNOWLEDGMENTS

The work cited from the authors' laboratory was supported by the Medical Research Council of Canada. We thank our colleagues who communicated their results prior to publication. K.M. is the recipient of a predoctoral studentship from the Cancer Research Society of Montreal. J.P. is the recipient of a postdoctoral fellowship from the Medical Research Council of Canada and N.S. is a recipient of a Medical Research Council Scientist Award from the Medical Research Council of Canada.

REFERENCES

1. Rueckert RR: *Fields Virology* New York, Raven Press, 705 (1985).
2. Summers DF, Maizel JV, and Darnell JE: Evidence for virus-specific noncapsid proteins in poliovirus infected cells. *Proc. Natl. Acad. Sci. U.S.A.* 54:505 (1965).
3. Penman S, Scherrer K, Becker Y, and Darnell JE: Polyribosomes in normal and poliovirus-infected HeLa cells and their relationship to messenger RNA. *Proc. Natl. Acad. Sci. U.S.A.* 49:654 (1963).

4. Willems M and Penman S: The mechanism of host cell protein synthesis inhibition by poliovirus. *Virology* 30:355 (1966).

5. Summers DF and Maizel JV Jr: Disaggregation of HeLa cell polysomes after infection with poliovirus. *Virology* 31:550 (1967).

6. Leibowitz R and Penman S: Regulation of protein synthesis in HeLa cells. III. Inhibition during poliovirus infection. *J. Virol.* 8:661 (1971).

7. Ehrenfeld E and Manis S: Inhibition of 80S initiation complex formation by infection with poliovirus. *J. Gen. Virol.* 43:441 (1979).

8. Brown B and Ehrenfeld E: Initiation factor preparations from poliovirus-infected cells restrict translation in reticulocyte lysates. *Virology* 103:327 (1980).

9. Koschel K: Poliovirus infection and poly (A) sequences of cytoplasmic cellular RNA. *J. Virol.* 13:1061 (1974).

10. Fernandez-Munoz R and Darnell JE: Structural difference between the 5′ termini of viral and cellular mRNA in poliovirus-infected cells: possible basis for the inhibition of host protein synthesis. *J. Virol.* 18:719 (1976).

11. Kaufmann Y, Goldstein E, and Penman S: Poliovirus-induced inhibition of polypeptide initiation *in vitro* on native polyribosomes. *Proc. Natl. Acad. Sci. U.S.A.* 73:1834 (1976).

12. Saxton RE and Stevens JC: Restriction of herpes simplex virus replication by poliovirus: a selective inhibition of viral translation. *Virology* 48:207 (1972).

13. Ehrenfeld E and Lund H: Untranslated vesicular stomatitis virus messenger RNA after poliovirus infection. *Virology* 80:297 (1977).

14. Bablanian R and Russell WC: Adenovirus polypeptide synthesis in the presence of non-replicating poliovirus. *J. Gen. Virol.* 24:261 (1974).

15. Holland JJ: Inhibition of host cell macromolecular synthesis by high multiplicities of poliovirus under conditions preventing virus synthesis. *J. Mol. Biol.* 8:574 (1964).

16. Bonneau A-M and Sonenberg N: Proteolysis of the p220 component of the cap-binding protein complex is not sufficient for complete inhibition of host cell protein synthesis after poliovirus infection. *J. Virol.* 61:986 (1987).

17. Helentjaris T and Ehrenfeld E: Inhibition of host cell protein synthesis by UV-inactivated poliovirus. *J. Virol.* 21:259 (1977).

18. Penman S and Summers D: Effects on host cell metabolism following synchronous infection with poliovirus. *Virology* 27:614 (1965).

19. Hewlett MJ, Rose JK, and Baltimore D: 5′ Terminal structure of poliovirus polyribosomal RNA is pUp. *Proc. Natl. Acad. Sci. U.S.A.* 73:327 (1976).

20. Nomoto A, Lee YF, and Wimmer E: The 5′ end of poliovirus mRNA is not capped with m⁷G(5′)ppp(5′)Np. *Proc. Natl. Acad. Sci. U.S.A.* 73:375 (1976).

21. Helentjaris T and Ehrenfeld E: Control of protein synthesis in extracts from poliovirus-infected cells. I. mRNA discrimination by crude initiation factors. *J. Virol.* 26:310 (1978).

22. Rose JK, Trachsel H, Leong K, and Baltimore D: Inhibition of translation by poliovirus: inactivation of a specific initiation factor. *Proc. Natl. Acad. Sci. U.S.A.* 75:2732 (1978).

23. Helentjaris T, Ehrenfeld E, Brown-Luedi ML, and Hershey JWB: Alterations in initiation factor activity from poliovirus-infected HeLa cells. *J. Biol. Chem.* 254:10973 (1979).

24. Sonenberg N, Morgan M, Merrick W, and Shatkin AJ: A polypeptide in eukaryotic initiation factors that crosslinks specifically to the 5'-terminal cap in mRNA. *Proc. Natl. Acad. Sci. U.S.A.* 75:4843 (1978).

25. Trachsel H, Sonenberg N, Shatkin AJ, Rose JK, Leong K, Bergman JE, Gordon J, and Baltimore D: Purification of a factor that restores translation of VSV mRNA in extracts from poliovirus-infected HeLa cells. *Proc. Natl. Acad. Sci. U.S.A.* 77:770 (1980).

26. Tahara SM, Morgan MA, and Shatkin AJ: Two forms of purified m^{7}G-cap binding protein with different effects on capped mRNA translation in extracts of uninfected and poliovirus-infected HeLa cells. *J. Biol. Chem.* 256:7691 (1981).

27. Grifo JA, Tahara SM, Morgan MA, Shatkin AJ, and Merrick WC: New initiation factor activity required for globin mRNA translation. *J. Biol. Chem.* 258:5804 (1983).

28. Edery I, Lee KAW, and Sonenberg N: Functional characterization of eukaryotic mRNA cap binding protein complex: effects on translation of capped and naturally uncapped RNAs. *Biochemistry* 23:2546 (1984).

29. Edery I, Humbelin M, Darveau A, Lee KAW, Milburn S, Hershey JWB, Trachsel H, and Sonenberg N: Involvement of eIF-4A in the cap recognition process. *J. Biol. Chem.* 258:11398 (1983).

30. Etchison D, Milburn SC, Edery I, Sonenberg N, and Hershey JWB: Inhibition of HeLa cell protein synthesis following poliovirus infection correlates with the proteolysis of a 220,000 dalton polypeptide associated with eukaryotic initiation factor 3 and a cap binding protein complex. *J. Biol. Chem.* 257:14806 (1982).

31. Etchison D, Hansen J, Ehrenfeld E, Edery I, Sonenberg N, Milburn SC, and Hershey JWB: Demonstration *in vitro* that eucaryotic initiation factor 3 is active but that a cap binding protein complex is inactive in poliovirus-infected HeLa cells. *J. Virol.* 51:832 (1984).

32. Lee KAW and Sonenberg N: Inactivation of cap-binding proteins accompanies the shut-off of host protein synthesis by poliovirus. *Proc. Natl. Acad. Sci. U.S.A.* 79:3447 (1982).

33. Lee KAW and Sonenberg N: Isolation and structural characterization of cap-binding proteins from poliovirus-infected HeLa cells. *J. Virol.* 54:515 (1985).

34. Pelletier J and Sonenberg N: Photochemical cross-linking of cap-binding proteins to eucaryotic mRNAs. *Mol. Cell. Biol.* 5:322 (1985).

35. Duncan R, Etchison E, and Hershey JWB: Protein synthesis initiation factors eIF-4A and eIF-4B are not altered by poliovirus infection of HeLa cells. *J. Biol. Chem.* 258:7236 (1983).

36. Hansen J and Ehrenfeld E: Presence of the cap-binding protein in initiation factor preparations from poliovirus-infected HeLa cells. *J. Virol.* 38:438 (1981).

37. Buckley B and Ehrenfeld E: The cap-binding protein complex in uninfected and poliovirus-infected HeLa cells. *J. Biol. Chem.* 262:13599 (1987).

38. Buckley B and Ehrenfeld E: Two-dimensional gel analyses of the 24 kDa cap-binding protein from poliovirus-infected and uninfected HeLa cells. *Virology* 152:497 (1986).

39. Lloyd RE, Etchison D, and Ehrenfeld E: Poliovirus protease 3C does not mediate the cleavage of the 220,000 dalton component of the cap-binding protein complex. *Proc. Natl. Acad. Sci. U.S.A.* 82:723 (1985).

40. Lee KAW, Edery I, Hanecak R, Wimmer E, and Sonenberg N: Poliovirus protease 3C (P3-7C) does not cleave P220 of the mRNA cap-binding protein complex. *J. Virol.* 55:489 (1985).

41. Bernstein HD, Sonenberg N, and Baltimore D: Poliovirus mutant that does not selectively inhibit host cell protein synthesis. *Mol. Cell. Biol.* 5:2913 (1985).

42. Lloyd RE, Toyoda H, Etchison D, Wimmer E, and Ehrenfeld E: Cleavage of the cap-binding protein complex polypeptide p220 is not effected by the second poliovirus protease 2A. *Virology* 150:299 (1986).

43. Dolph PJ, Racaniello V, Villamarin A, Palladino F, and Schneider RJ: The adenovirus tripartite leader may eliminate the requirement for cap-binding protein complex during translation initiation. *J. Virol.* 62:2059 (1988).

44. O'Neill R and Racaniello VR: Inhibition of translation in cells infected with a poliovirus 2A pro mutation correlates with phosphorylation of the alpha subunit of eucaryotic initiation factor 2. *J. Virol.* 63:5069 (1989).

45. Kraüsslich H-G, Nicklin MJH, Toyoda H, Etchison D, and Wimmer E: Poliovirus proteinase 2A induces cleavage of eukaryotic initiation factor 4F polypeptide p220. *J. Virol.* 61:2711 (1987).

46. Sun S-H and Baltimore D: Human immunodeficiency virus tat-activated expression of poliovirus protein 2A inhibits mRNA translation. *Proc. Natl. Acad. Sci. U.S.A.* 86:2143 (1989).

46a. Wyckoff EE, Hershey JWB, and Ehrenfeld E: Eukaryotic initiation factgor 3 is required for poliovirus 2A protease-induced clevage of the p220 component of eukaryotic initiation factor 4F. *Proc. Natl. Acad. Sci. U.S.A.* 87:9529 (1990).

47. Mosenkis J, Daniels-McQueen S, Janovec S, Duncan R, Hershey JWB, Grifo JA, Merrick WC, and Thach RE: Shut-off of host translation by encephalomyocarditis virus infection does not involve cleavage of the eukaryotic initiation factor 4F polypeptide that accompanies poliovirus infection. *J. Virol.* 54:643 (1985).

48. Lloyd RE, Grubman MJ, and Ehrenfield E: Relationship of p220 cleavage during picornavirus infection to 2A proteinase sequences. *J. Virol.* 62:4216 (1988).

49. Devaney MA, Vakharia VN, Lloyd RE, Ehrenfeld E, and Grubman MJ: Leader protein of foot-and-mouth disease virus is required for cleavage of the p220 component of the cap-binding protein complex. *J. Virol.* 672:4407 (1988).

50. Dratewka-Kos E, Kiss I, Lucas-Lenard J, Mehta HB, Woodley OL, and Wahba AJ: Catalytic utilization of eIF-2 and mRNA binding proteins are limiting in lysates from vesicular stomatitis virus infected L cells. *Biochemistry* 23:6184 (1984).

51. Samuel EC, Duncan R, Knutson GS, and Hershey JWB: Mechanism of interferon action: increased phosphorylation of protein synthesis initiation factor 2 alpha in interferon-treated reovirus infected mouse L-929 fibroblasts *in vitro* and *in vivo*. *J. Biol. Chem.* 259:13451 (1984).

52. Katze MG, DeCorato D, Safer B, Galabru J, and Hovanessian AG: The adenovirus VA1 RNA complexes with the p68 protein kinase to regulate its autophosphorylation and activity. *EMBO J.* 6:689 (1987).

53. Katze MG, Tomita J, Black T, Krug RM, Safer B, and Hovanessian A: Influenza virus regulates protein synthesis during infection by repressing autophosphorylation and activity of the cellular 68,000 Mr protein kinase. *J. Virol.* 62:3710 (1988).

54. Hovanessian AG, Galabru J, Meurs E, Buffet-Janvresse C, Svab J, and Robert N: Rapid decrease in the levels of the double-stranded RNA-dependent protein kinase during virus infection. *Virology* 159:126 (1987).

55. Black TL, Safer B, Hovanessian A, and Katze M: The cellular 68,000 Mr protein kinase is highly phosphorylated and activated yet significantly degraded during poliovirus infection: implications for translational regulation. *J. Virol.* 63:2244 (1989).

56. Ransone LJ and Dasgupta A: Activation of double-stranded RNA-activated protein kinase in HeLa cells after poliovirus infection does not result in increased phosphorylation of eucaryotic initiation factor-2.*J. Virol.* 61:1781 (1987).

57. Ransone LJ and Dasgupta A: A heat sensitive inhibitor in poliovirus-infected cells which selectively blocks phosphorylation of the alpha subunit of eucaryotic initiation factor 2 by the double-stranded RNA-activated protein kinase. *J. Virol.* 62:3551 (1988).

58. deBenedetti A and Baglioni C: Inhibition of mRNA binding to ribosomes by localized activation of dsRNA-dependent protein kinase. *Nature* 311:79 (1984).

59. Murtha P and Kaufman RJ: Translational control mediated by eucaryotic initiation factor-2 is restricted to specific mRNAs in transfected cells. *Mol. Cell. Biol.* 7:1568 (1987).

60. Edery I, Petryshyn R, and Sonenberg N: Activation of double-stranded RNA-dependent kinase (dsI) by the TAR region of HIV-1 mRNA: a novel translational control mechanism. *Cell* 56:303 (1989).

61. Sengupta DN and Silverman RH: Activation of interferon regulated, ds RNA-dependent enzymes by human immunodeficiency virus-1 leader RNA. *Nucleic Acids Res.* 17:969 (1989).

62. Edery I, Pelletier J, and Sonenberg N: *Translational Regulation of Gene Expression,* New York, Plenum Press, 335 (1987).

63. Lawrence C and Thach RE: Encephalomyocarditis virus infection of mouse plasmacytoma cells. I. Inhibition of cellular protein synthesis. *J. Virol.* 14:598 (1974).

64. Golini F, Thach SS, Birge CH, Safer B, Merrick WC, and Thach RE: Competition between cellular and viral mRNAs *in vitro* is regulated by a messenger discriminatory initiation factor. *Proc. Natl. Acad. Sci. U.S.A.* 73:3040 (1976).

65. Hackett PB, Egberts E, and Traub P: Translation of ascites and mengovirus RNA in fractionated cell-free systems from uninfected and mengovirus-infected Ehrlich-asictes-tumor cells. *Eur. J. Biochem.* 83:341 (1978).

66. Hackett PB, Egberts E, and Traub P: Selective translation of mengovirus RNA over host mRNA in homologous, fractionated, cell-free translational systems from Ehrlich-ascites-tumor cells. *Eur. J. Biochem.* 83:353 (1978).

67. Jen G, Birge CH, and Thach RE: Comparison of initiation rates of encephalomyocarditis virus and host protein synthesis in infected cells. *J. Virol.* 27:640 (1978).

68. Parks GD, Duke GM, and Palmenberg AC: Encephalomyocarditis virus 3C protease: efficient cell-free expression from clones which link viral 5' noncoding sequences to the P3 region. *J. Virol.* 60:376 (1986).

69. Perez-Bercoff R: *Protein Biosynthesis in Eukaryotes.* New York, Plenum Press, 245 (1982).

70. Pelletier J, Kaplan G, Racaniello VR, and Sonenberg N: Translational efficiency of poliovirus mRNA: mapping inhibitory *cis*-acting elements within the 5'-noncoding region. *J. Virol.* 62:2219 (1988).

71. Kozak M: Comparison of initiation of protein synthesis in procaryotes, eucaryotes and organelles. *Microbiol. Rev.* 47:1 (1983).

72. Pelletier J, Kaplan G, Racaniello VR, and Sonenberg N: Cap-independent translation of poliovirus mRNA is conferred by sequence elements within the 5' noncoding region. *Mol. Cell. Biol.* 8:1103 (1988).

73. Trono D, Pelletier J, Sonenberg N, and Baltimore D: Translation in mammalian cells of a gene linked to the poliovirus 5' noncoding region. *Science* 241:445 (1988).

74. Pelletier J and Sonenberg N: Internal binding of eukaryotic ribosomes on poliovirus RNA: translation in HeLa cell extracts. *J. Virol.* 63:441 (1989).

75. Jang SK, Kraüsslich H-G, Nicklin MJH, Duke GM, Palmenberg AC, and Wimmer E: A segment of the 5' nontranslated region of encephalomyocarditis virus RNA directs internal entry of ribosomes during *in vitro* translation. *J. Virol.* 62:2636 (1988).

76. Jang SK, Davies MV, Kaufman RJ, and Wimmer E: Initiation of protein synthesis by internal entry of ribosomes into the 5' nontranslated region of encephalomyocarditis virus RNA *in vivo*. *J. Virol.* 63:1651 (1989).

77. Brown D, Jones CL, Brown BA, and Ehrenfeld E: Translation of capped and uncapped VSV mRNAs in the presence of initiation factors from poliovirus-infected cells. *Virology* 123:60 (1982).

78. Shih DS, Park I-W, Evans CL, Jaynes JM, and Palmenberg AC: Effects of cDNA hybridization on translation of encephalomyocarditis virus RNA. *J. Virol.* 61:2033 (1987).

79. Liebhaber SA, Cash FE, and Shakin SH: Translationally associated helix-destabilizing activity in rabbit reticulocyte lysate. *J. Biol. Chem.* 259:15597 (1984).

80. Shakin SH and Liebhaber SA: Destabilization of messenger RNA/complementary DNA duplexes by the elongating 80S ribosome. *J. Biol. Chem.* 261:16018 (1986).

81. Sonenberg N: Cap-binding proteins of eukaryotic messenger RNA: functions in initiation and control of translation. *Prog. Nucl. Acid Res. Mol. Biol.* 35:173 (1988).

82. Trono D, Andino R, and Baltimore D: An RNA sequence of hundreds of nucleotides at the 5' end of poliovirus RNA is involved in allowing viral protein synthesis. *J. Virol.* 62:2291 (1988).

83. Dildine SL and Semler BL: The deletion of 41 proximal nucleotides reverts a poliovirus mutant containing a temperature-sensitive lesion in the 5' noncoding region of genomic RNA. *J. Virol.* 63:847 (1989).

84. Kuge S and Nomoto A: Construction of viable deletion and insertion mutants of the Sabin strain of type I poliovirus: function of the 5' noncoding sequence in viral replication. *J. Virol.* 61:1478 (1987).

85. Pelletier J and Sonenberg N: Internal initiation of translation of eukaryotic mRNA directed by a sequence derived from poliovirus RNA. *Nature* 334:310 (1988).

86. Kaufman RJ, Murtha P, and Davies MV: Translational efficiency of polycistronic mRNAs and their utilization to express heterologous genes in mammalian cells. *EMBO J.* 6:187 (1987).

87. Khalili K, Brady J, and Khoury G: Translational regulation of SV40 early mRNA defines a new viral protein. *Cell* 48:639 (1987).

88. Peabody DS, Subramani S, and Berg P: Effect of upstream reading frames on translation efficiency in simian virus 40 recombinants. *Mol. Cell. Biol.* 6:2704 (1986).

89. Sedman SA and Mertz JE: Mechanisms of synthesis of virion proteins from the functionally bigenic late mRNAs of simian virus 40. *J. Virol.* 62:954 (1988).

90. Pelletier J, Flynn ME, Kaplan G, Racaniello VR, and Sonenberg N: Mutational analysis of upstream AUG codons of poliovirus RNA. *J. Virol.* 62:4486 (1988).

90a. Jackson RJ, Howell MT, and Kaminski A: The novel mechanism of initiation of picornavirus RNA translation. *Trends Biochem. Sci.* 15:477 (1990).

90b. Iizuka N, Kohara M, Hagino-Yamagishi K, Abe S, Komatsu T, Tago K, Arita M, and Nomoto A: Construction of less neurovirulent polioviruses by introducing deletions into the 5' noncoding sequence of the genome. *J. Virol.* 63:5354 (1989).

90c. Jang SK and Wimmer E: Cap-independent translation of encephalomyocarditis virus RNA: structural elements of the internal ribosomal entry site and involvement of a cellular 57kD RNA-binding protein. *Genes Dev.* 4:1560 (1990).

90d. Kuhn R, Luz N, and Beck E: Functional analysis of the internal initiation site of foot-and-mouth disease virus. *J. Virol.* 64:4625 (1990).

91. Kuge S, Kawamura N, and Nomoto A: Genetic variation occurring on the genome of an *in vitro* insertion mutant of poliovirus type I. *J. Virol.* 63:1069 (1989).

92. Meerovitch K, Pelletier J, and Sonenberg N: A cellular protein that binds to the 5'-noncoding region of poliovirus RNA: implications for internal translation initiation. *Genes Dev.* 3:1026 (1989).

93. Bienkowska-Szewczyk K and Ehrenfeld E: An internal 5'-noncoding region required for translation of poliovirus RNA *in vitro*. *J. Virol.* 62:3068 (1988).

94. del Angel RM, Papavassiliou AG, Fernandez-Tomas C, Silverstein SJ, and Racaniello VR: Cell proteins bind to multiple sites within the 5'-untranslated region of poliovirus. *Proc. Natl. Acad. Sci. U.S.A.* 86:8299 (1989).

94a. Borovjagin AV, Evstafieva AG, Ugarova TY, and Shatsky IN: A factor that specifically binds to the 5'-untranslated region of encephalomyocarditis virus RNA. *FEBS Letts.* 261:237 (1990).

94b. Luz N and Beck E: A cellular 57 kDa protein binds to two regions of the internal translation initiation site of foot-and-mouth disease virus. *FEBS Letts.* 269:311 (1990).

95. Johnson VH and Semler BL: Defined recombinants of poliovirus and coxsackievirus: sequence-specific deletions and functional substitutions in the 5'-noncoding regions of viral RNAs. *Virology* 162:47 (1988).
96. Rivera VM, Welsh JD, and Maizel JV Jr: Comparative sequence analysis of the 5' noncoding region of the enteroviruses and rhinoviruses. *Virology* 165:42 (1988).
97. Pilipenko EV, Blinov VM, Romanova LI, Sinyakov AN, Maslova SV, and Agol VI: Conserved structural domains in the 5' untranslated region of picornaviral genomes: an analysis of the segment controlling translation and neurovirulence. *Virology* 168:201 (1989).
98. Skinner MA, Racaniello VR, Dunn G, Cooper J, Minor PD, and Almond JW: New model for the secondary structure of the 5' noncoding RNA of poliovirus is supported by biochemical and genetic data that also show that RNA secondary structure is important in neurovirulence. *J. Mol. Biol.* 207:379 (1989).
98a. Le SY and Zuker M: Common structures of the 5' non-coding RNA in enteroviruses and rhinoviruses. Thermodynamical stability and statistical significance. *J. Mol. Biol.* 216:729 (1990).
99. Pilipenko EV, Blinov VM, Chernov BK, Dmitrieva TM and Agol VI: Conservation of the secondary structure elements of the 5'-untranslated region of cardio- and aphtovirus RNAs. *Nucleic Acids Res.* 75:5701 (1989).
100. Kaminski A, Howell MT, and Jackson RJ: Initiation of encephalomyocarditis virus RNA translation: the authentic initiation site is not selected by a scanning mechanism. *EMBO J.* 9:3753 (1990).
101. Herman RC: Internal initiation of translation on the vesicular stomatitis virus phosphoprotein mRNA yields a second protein. *J. Virol.* 58:797 (1986).
102. Hassin D, Korn R, and Horwitz MS: A major internal initiation site for the *in vitro* translation of the adenovirus DNA polymerase. *Virology* 155:214 (1986).
103. Curran J and Kolakofsky D: Scanning independent ribosomal initiation of the sendai virus X protein. *EMBO J.* 7:2869 (1988).
104. Curran J and Kolakofsky D: Scanning independent ribosomal initiation of the sendai virus Y proteins *in vitro* and *in vivo*. *EMBO J.* 8:521 (1989).
105. Chang L-J, Pryciak P, Ganam D, and Varmus HE: Biosynthesis of the reverse transcriptase of hepatitis B viruses involves *de novo* translational initiation not ribosomal frameshifting. *Nature* 337:364 (1989).
106. Jean-Jean O, Weimer T, De Recondo A-M, Will H, and Rossignol J-M: Internal entry of ribosomes and ribosomal scanning involved in hepatitis B virus P gene expression. *J. Virol.* 63:5451 (1989).
107. Abramson RD, Dever TE, Lawson TG, Ray BK, Thach RE, and Merrick WC: The ATP-dependent interaction of eukaryotic initiation factors with mRNA. *J. Biol. Chem.* 262:3826 (1987).
108. Rozen F, Edery I, Meerovitch K, Dever TE, Merrick WC, and Sonenberg N: Bidirectional RNA helicase activity of eukaryotic translation initiation factors-4A and -4F. *Mol. Cell. Biol.* 10:1134 (1990).
109. Abramson RD, Dever TE, and Merrick WC: Biochemical evidence supporting a mechanism for cap-independent and internal initiation. *J. Biol. Chem.* 263:6016 (1988).
109a. Sarnow P: Translation of glucose-regulated protein 78/ immunoglobulin heavy-chain binding protein mRNA is increased in poliovirus-infected cells at a time when cap-dependent translation of cellular mRNAs is inhibited. *Proc. Natl. Acad. Sci. U.S.A.* 86:5795 (1989).

Adjustment of Translation to Special Physiological Conditions

Virginia M. Pain
School of Biological Sciences
University of Sussex
Falmer, Brighton, United Kingdom

Mike J. Clemens
Department of Cellular and Molecular Sciences
St. George's Hospital Medical School
London, United Kingdom

12.1. INTRODUCTION

All living cells must adapt their functions to changes in their environment and the process of translation is no exception to this rule. Both unicellular organisms and the specialized cells of multicellular organisms are exposed to a variety of extracellular influences which regulate protein synthesis and, in some cases, the ability of the cells to grow and divide.

Among the most notable of such influences are cellular nutrients such as amino acids and glucose; the ionic environment of the cell also has important regulatory effects on translation and calcium in particular is emerging as a major effector. In addition to such essential molecules, there is a large battery of polypeptide growth factors, hormones, and cytokines which regulate many different cellular functions. Several of these molecules are known to play a role in the rapid control of translational activity, some being stimulatory (e.g., insulin) and others inhibitory in certain situations (e.g., interferons). A great deal is now known about how protein synthesis is regulated during the stress response induced by heat shock of cells and there is some evidence that similar mechanisms may be used to control translation during mitosis or during infection by some animal viruses. Finally, several types of regulation exist to allow gene expression to be activated at the translational level during the early stages of development of eggs and embryos. Some of the mechanisms involve overlap with those used in the regulation of somatic cells while others may be specific to the peculiarly dormant state characteristic of unfertilized eggs.

This chapter aims to describe the current state of knowledge concerning several of these aspects of translational control. At the end an attempt is made to summarize this knowledge by defining the stages in the initiation of protein synthesis which are most often found as rate-limiting steps, and which thus constitute key points for regulation. For reviews of earlier work and the background to this field the reader is referred to a number of previous articles.[1-5a]

12.2. REGULATION OF TRANSLATION BY CELLULAR NUTRIENTS

It has been recognized for many years that eukaryotic cells respond to variations in the availability of essential nutrients such as amino acids and glucose in a manner which indicates the existence of specific control mechanisms for translation. In other words, such nutrients exert effects not only as precursors or sources of energy for protein synthesis but because pathways exist for the coupling of translation rates to the extracellular amino acid and glucose concentrations.

12.2.1. Amino Acid Starvation

Although amino acids are obviously utilized as precursors during the elongation stage of protein synthesis, eukaryotic cells respond to deprivation of essential amino acids primarily by decreasing the rate of polypeptide chain initiation.[6-8] Such a phenomenon has been observed in a variety of cell types, including cultured mammalian tumor cells (reviewed in Reference 9), perfused liver[10] and isolated hepatocytes.[11] The lesion in the initiation pathway has been identified as an impairment in the ability of initiation factor eIF-2 to form [eIF-2.GTP.Met-tRNA$_f$] ternary complexes, with a resultant decrease in binding of the Met-tRNA$_i^{Met}$ to 43S initiation complexes.[10-13] Consistent with this, the addition of purified eIF-2 to cell-free protein synthesizing systems from amino acid-starved cells restores 43S initiation complex formation and stimulates protein synthesis to the level seen in fed cell extracts.[13, 14]

These data suggest that inactivation of eIF-2 can largely account for the inhibition of initiation caused by amino acid starvation. However, analysis of native 40S ribosomal subunits on CsCl equilibrium density gradients also indicates that in starved cells some subunits accumulate which are probably deficient in eIF-3; most of these do not carry Met-tRNA$_i^{Met}$[11, 12] and the significance of this change is not yet clear.

Examination of the state of phosphorylation of the α-subunit of eIF-2 in extracts from fed and amino acid-starved Ehrlich ascites tumor cells, using isoelectric focusing and immunoblotting with a monoclonal antibody against the α-subunit, revealed that the inactivation of eIF-2 was accompanied by an increased level of phosphorylation.[15] Thus phosphorylation of this subunit may regulate translation by a mechanism essentially the same as that operating in the reticulocyte lysate system (see Part II, Chapter 8 of this book). As with the latter, a relatively small increase in eIF-2α phosphorylation is associated with inhibition of protein synthesis of more than 80%, although the absolute levels of phosphorylation are higher in Ehrlich cells. As is predicted from our knowledge of the tight binding between eIF-2 (αP) and the guanine nucleotide exchange factor (GEF), the increased phosphorylation of eIF-2α in extracts from starved cells is accompanied by some impairment of GEF activity.[14, 16] However, at least in Ehrlich cells, the effect is not reversed by adding an excess of unphosphorylated eIF-2, suggesting that some additional mechanism for the regulation of GEF during amino acid starvation may exist. This is substantiated by a report from Gross and Rubino[16a] who showed that amino acid depletion of reticulocyte lysates impairs GEF activity, although not enough to inhibit overall initiation.

Although it is clear that eIF-2α becomes more highly phosphorylated during amino acid starvation, the protein kinase responsible and the mechanism by which it may be regulated are not yet known. There is considerable evidence for the involvement of aminoacyl-tRNA synthetases in transduction of the signal for amino acid deficiency to the protein synthetic machinery. When Chinese hamster ovary (CHO) cells which have a temperature-sensitive aminoacyl-tRNA synthetase are incubated at the nonpermissive temperature all of the effects of amino acid starvation on initiation of protein synthesis can be reproduced. These include inhibition of eIF-2 activity, increased phosphorylation of the α-subunit and a decreased rate of GEF-catalyzed guanine nucleotide exchange.[17, 18] These effects are not seen when corresponding temperature shifts (from 34 to 39.5°C) are inflicted on either wild-type cells or single-step revertants of the mutant CHO cells.[19] The latter result establishes beyond a doubt a role for aminoacyl-tRNA synthetases in the mechanism which couples amino acid supply to the control of polypeptide chain initiation. The evidence does not support the earlier idea[20] that tRNA charging itself is important, however, since the control of eIF-2 activity at the nonpermissive temperature is independent of tRNA charging level or the rate of utilization of aminoacyl-tRNA in chain elongation.[19] A similar conclusion was reached in the case of amino acid deficiency.[21, 22]

The mechanism by which an aminoacyl-tRNA synthetase may be able to control the phosphorylation of eIF-2 remains to be established but one exciting possibility is that the synthetases themselves may have some protein kinase activity. Sequence analysis of the GCN2 protein of *Saccharomyces cerevisiae* has recently revealed domains with homologies to both protein kinases and histidyl-tRNA synthetase.[23, 24] In this yeast the GCN2 protein is involved in the stimulation, in response to amino acid starvation, of translation of the GCN4 gene product, a transcriptional activator for enzymes of amino acid biosynthesis (see Part II, Chapter 11 of this book). It has recently been established that eIF-2 also plays a role in the control of GCN4 translation,[25] raising the possibility that the protein kinase activity of GCN2 may directly or indirectly phosphorylate eIF-2. Whether mammalian aminoacyl-tRNA synthetases also have any eIF-2 kinase regulatory activity remains to be determined but these enzymes, like the GCN2 protein, can themselves be phosphorylated.[26,27]

An alternative model by which the synthetases might regulate eIF-2 phosphorylation proposes that, rather than any change occurring in protein kinase or phosphatase activity, the accessibility of eIF-2 as a substrate for phosphorylation may be altered when amino acid supply or synthetase activity is restricted.[19] The presence in eukaryotic cells of large, multi-enzyme complexes containing several synthetase activities (and possibly other protein synthesis factors)[28] raises the question of whether there may be direct protein-protein interactions between these enzymes and eIF-2.

12.2.2. Glucose Starvation

Protein synthesis in mammalian cells responds to glucose deficiency in a way which is very similar to the effects of amino acid deficiency. In Ehrlich ascites tumor cells incubated without glucose or in the presence of 2-deoxy-glucose, initiation is inhibited, 43S initiation complex formation is severely impaired[12] and eIF-2 becomes more extensively phosphorylated on its α-subunit.[15] Again, the biochemical signals linking the metabolism of glucose to the initiation pathway are not clear. One strong candidate for the regulation of early events in the initiation pathway is glucose-6-phosphate which, in gel-filtered reticulocyte lysates, has been shown to be required for initiation activity (see Part II, Chapter 8 of this book). A recent examination of the role of the sugar phosphate in this system suggests that GEF may be directly regulated by glucose-6-phosphate, rather than through changes in the phosphorylation of eIF-2.[29] It is not yet known whether this is the case in other cell types, or indeed whether the effects of glucose starvation are mediated by changes in glucose-6-phosphate concentration. Other metabolic changes result from glucose deficiency, including a decrease in cellular ATP level and energy charge (and hence a decrease in the GTP/GDP ratio).[30] The GTP/GDP ratio has a profound effect on eIF-2 activity since GDP is a potent inhibitor of this initiation factor.[31] ATP is also required at another step in the initiation pathway, namely, eIF-4A-mediated unwinding of 5′ sequences in mRNA, as well as being necessary for aminoacylation of tRNAs and to provide the GTP for polypeptide chain elongation.

12.3. REGULATION OF TRANSLATION BY CHANGES IN IONIC ENVIRONMENT

The overall translational activity of eukaryotic cells is sensitive to the ionic environment in which the cells exist. Changes in intracellular conditions which alter the levels of certain cations (notably calcium) also play an important regulatory role.

12.3.1. Effects of Hypertonic Conditions

It has been known for several years that exposure of cells to high ionic strength (brought about, for example, by raising the concentration of NaCl in the culture medium) inhibits initiation.[32] Detailed examination of the kinetics indicates that this effect is almost instantaneous, and is equally rapidly reversed when isotonicity is restored.[33] The mechanism by which initiation is blocked remains unknown, however. Unlike the situation in nutrient deficiency there is no change in the phosphorylation state of the α-subunit of eIF-2[34] and the activity of the factor remains unimpaired (Clemens, unpublished observations). The phosphorylation of eIF-4B is also unaltered[34] and, although ribosomal protein S6 becomes less highly phosphorylated under hypertonic conditions, the rate at which this occurs is probably too slow to account for the direct effect on translation rate.[33]

There are interesting changes in the relative efficiencies with which different mRNAs are translated under hypertonic conditions;[35] this effect may be interpreted as reflecting the increased stability, under conditions of high ionic strength, of secondary structures in the 5′ untranslated regions of some mRNAs. Such structures have been shown to inhibit translation when placed upstream of initiation codons.[36] Conversely, mRNAs which have largely unstructured 5′ sequences may be preferentially translated under conditions of hypertonic stress,[37, 38] as is the case during heat shock (see below). A further role in the selection of different templates for protein synthesis may be played by changes in the phosphorylation of ribosomal protein S6[39] but this is a hypothesis which needs more experimental verification.

12.3.2. Regulation by Calcium Ions

It is likely that protein synthesis in eukaryotic cells is sensitive to the intracellular concentrations of many ions, but recent evidence has revealed a particularly important role for calcium in the control of both initiation and elongation.[40] When intact cells are depleted of Ca^{2+} by incubation with EGTA, polysomes disaggregate and ribosomes accumulate as 80S monomers.[41, 42] The reversible block in initiation is due to defective formation or stability of 43S complexes[41] and can be reproduced *in vitro* by chelating endogenous calcium with 1,2-*bis*(O-aminophenoxy)-ethane-N,N,N',N'-tetraacetic acid.[42] Unlike the situation with nutrient starvation or serum deprivation, however, this effect is not due to changes in eIF-2 phosphorylation. Rather, the initiation complexes formed on native 40S ribosomal subunits in calcium-depleted conditions are entirely deficient in eIF-3.[42] The significance of this observation remains to be established but the data suggest that the association of eIF-3 with 40S ribosomal subunits is, directly or indirectly, a Ca^{2+}-dependent process and that such an interaction may be required for the stabilization of [40S.eIF-2.GTP.Met-tRNA$_i^{Met}$] initiation complexes. In contrast to the effect on initiation, depletion of cellular calcium does not have any effect on polypeptide chain elongation.[41]

The nature of the calcium pool which regulates translation in intact cells is not yet clear. Treatment of cells with Ca^{2+} ionophores also inhibits protein synthesis to an extent which is dependent on the extracellular concentration of the cation.[43] One interpretation of such findings is that the level of sequestered intracellular calcium, rather than free cytoplasmic calcium, controls the rate of initiation. However, in whole cells it may be difficult to vary these parameters independently. The level of phosphorylation of a 26 kDa ribosome-associated protein decreases in cells treated with the ionophore A23187.[44] This protein is not eIF-4E and its role in initiation is not yet established.

It seems likely that there is a distinct optimum requirement of the protein synthetic machinery for calcium since excess of the cation is also inhibitory. Submillimolar amounts of Ca^{2+} inhibit the reticulocyte lysate system.[45] This may be due at least partly to the phosphorylation of eIF-2α. In addition, calcium can stimulate a strong calmodulin-dependent phosphorylation of elongation factor 2 on one or more threonine residues.[46] This modification almost completely eliminates the ability of the factor to catalyze the translocation of peptidyl-tRNA (Ryazanov[47] *et al.*, 1988 and Part II, Chapter 13 of this book) and may constitute the major way in which elongation rates are modulated *in vivo* during transient rises in the intracellular calcium concentration.

12.3.3. Effects of Heavy Metal Ions

Low concentrations of toxic heavy metal ions, such as cadmium, mercury, and lead, inhibit protein synthesis in the reticulocyte lysate by stimulating the phosphorylation of eIF-2α.[48] This effect appears to be mediated by the activation of the heme-controlled eIF-2 kinase (HCR) (see Part II, Chapter 8 of this book). It is not known if a similar mechanism is operative in nonerythroid cells but protein synthesis is certainly sensitive to inhibition by these metals. It is possible that HCR is activated by interaction between the metal ions and critical sulfhydryl groups in the kinase molecule. A mechanism involving the breaking of disulfide bonds has been implicated in the activation of HCR under conditions of heme deficiency.[49]

12.4. TRANSLATION IN RELATION TO CELLULAR GROWTH CONTROL

Overall rates of protein synthesis are controlled in several ways during cell proliferation. As would be expected, rapidly growing cells maintain high rates of translation in order to provide the necessary increase in protein mass and conversely, when cells enter a state of quiescence or terminal differentiation, rates of protein synthesis fall substantially. These changes occur in response to both stimulatory growth factors and inhibitory cytokines such as the interferons, but the molecular mechanisms involved are only understood in a limited number of cases.

12.4.1. Control by Serum

It has been known for several years that addition of fresh serum to cells in the stationary phase of growth stimulates protein synthesis at the level of initiation. A mechanism involving recruitment of pre-existing mRNA back into polysomes is involved.[50, 51] It seems likely that several regulatory events may occur under these circumstances, resulting in modification of the activity of various initiation factors and perhaps also involving changes in the translatability of "masked" mRNAs (see the section on regulation during embryonic development later in this chapter). In HeLa cells incubated in unreplenished medium (which results in a combination of nutrient starvation and depletion of serum growth factors) the α-subunit of eIF-2 becomes more highly phosphorylated whereas eIF-4B becomes less phosphorylated.[52] These effects are reversed when fresh medium is added to the cells. In addition, there are decreases in the abundance of several initiation factors in starved cells, mainly due to decreased rates of synthesis; this is particularly marked in the case of some subunits of eIF-3.[53] The changes in phosphorylation state of eIF-2α during serum deprivation can be correlated with decreased activity of this initiation factor *in vitro* and, as with nutrient starvation, this has been shown to be due to sequestration of GEF.[16] Serum refeeding reverses these effects within 3 h. In the case of eIF-4B it is not yet clear whether its phosphorylation state influences its activity, but the factor has been shown to be functionally impaired in extracts from serum-deprived neuroblastoma cells.[54]

What is the relationship between changes in initiation factor activity and mRNA recruitment following serum stimulation? Both eIF-2 and eIF-4B may be involved in mRNA recognition and ribosomal scanning. There is also growing evidence that eIF-4E may be regulated by serum factors and other mitogens. In quiescent 3T3 cells exposure to fresh serum or phorbol ester rapidly leads to phosphorylation of this 26 kDa factor; in both cases there is also a shift of previously poorly translated mRNAs (such as that for ribosomal protein L32) into polysomes,[55] suggesting that the cap-binding factor may have a crucial role to play in mRNA recruitment following serum stimulation. A similar mechanism could be responsible for the increased utilization of mRNAs such as that for ornithine decarboxylase in mitogenic lectin-stimulated lymphocytes.[56]

12.4.2. The Role of Individual Growth Factors

The nature of the growth factors in serum which cause the changes in translational activity described above has not been elucidated. It is likely that different factors will be important for different cell types, depending on the receptors present on the cell surface and the nature of the intracellular signaling mechanisms involved in stimulating the translational machinery in different systems. In the case of fibroblastic or epithelial cells, ligands such as epidermal growth factor (EGF) are probably important. For example, EGF has been reported to stabilize some mRNA species in human KB cells.[57]

The most widely studied of the mitogenic or growth-promoting factors which influence protein synthesis is insulin. A very wide range of cell and tissue types show a translational response to insulin, including skeletal muscle,[58-60] cultured myoblasts,[61] adipocytes,[62] 3T3 cells,[63] and chick embryo fibroblasts.[64] It has been shown that the acute stimulation of protein synthesis in fast-twitch skeletal muscle by insulin is due to increased eIF-2 activity and the consequent formation of 43S initiation complexes.[58, 59] Such an effect may be mediated by a change in GEF activity, leading to more efficient recycling of eIF-2.[60] In this particular case, however, a change in the phosphorylation state of eIF-2 may not be involved (Pain, unpublished data), suggesting that a direct effect of insulin on GEF itself may perhaps be responsible. It should be noted that Cox et al.[65] were unable to detect any change in eIF-2 activity in muscle in response to starvation, even though initiation is impaired in this situation and can be stimulated by insulin administration. Interestingly, the translation of different mRNAs is differentially sensitive to insulin, with the synthesis of ribosomal proteins being particularly affected.[61, 66, 67] It is not clear whether such an effect would follow from the modulation of GEF or eIF-2 activity and it is possible that a change in the phosphorylation status of factors involved in mRNA binding may play a role[68] (see below).

The signal transduction pathway by which insulin acts to stimulate protein synthesis has not yet been elucidated. Recent observations indicate that different mechanisms are involved in mediating the effects of insulin on translation and on glucose transport,[62] and that amino acids can enhance the stimulation by insulin of the former but not the latter process.[69] Such synergism between insulin and amino acids has been observed *in vivo* previously,[70, 71] but the basis for it is not yet understood. There is some evidence that insulin may stimulate protein synthesis in 3T3 cells at least partially via phospholipase C-mediated phosphoinositide hydrolysis, with subsequent activation of protein kinase C (PK-C) by the diacylglycerol produced.[63] In light of recent studies on phorbol ester-induced phosphorylation of various initiation factors (discussed below) potential targets for such insulin-mediated effects could include eIF-3, -4B, -4E, or -4F. It should be noted that in the 3T3 cell system the response of protein synthesis to serum stimulation may be mediated by a pathway that is distinct from that affected by insulin.[63]

12.4.3. Protein Kinases Involved in Translational Stimulation by Mitogens

Many of the growth factors which stimulate cell proliferation exert their effects via activation of a range of intracellular protein kinases. The three main classes of kinases which have been intensively studied in recent years are the tyrosine kinases (including both growth factor receptors and cytoplasmic kinases of the *src* family), cyclic AMP-dependent protein kinases and the protein kinase C family. Although the insulin receptor has tyrosine kinase activity no direct phosphorylations of protein synthesis factors on tyrosine residues have been reported. In contrast, the evidence currently favors a protein kinase C-mediated effect on translation, as discussed above. Phorbol esters, which directly activate PK-C, themselves stimulate protein synthesis in some cell types at the level of initiation.[72] This may be a consequence of increased phosphorylation of eIF-3 (170 kDa subunit), eIF-4B or eIF-4F (25 and 220 kDa subunits), all of which can be modified by PK-C *in vitro*[73] or in response to phorbol ester treatment of intact cells[74] (see also Part II, Chapter 9 of this book). However, PK-C may not be the only enzyme able to phosphorylate the eIF-4F 25 kDa component (or free eIF-4E)[75] and in mitogen-activated B cells the evidence points towards the involvement of a different protein kinase in the increased phosphorylation of eIF-4E (Rychlik, personal communication).

Do the cyclic AMP-dependent protein kinases play any role in translational control? Elevation of intracellular cAMP can stimulate protein synthesis in anterior pituitary and pituitary tumor cells,[72] but the site of action on the translational machinery was not identified. In cell-free systems from rabbit reticulocytes high concentrations of cAMP can stimulate protein synthesis either by inactivating the eIF-2α protein kinase HCR[76] or by stimulating elongation via the dephosphorylation of elongation factor-2.[77] However, it is doubtful whether the millimolar amounts necessary to achieve these effects are ever attained *in vivo*. The physiological role, if any, of cAMP-stimulated kinases in controlling translation, and the nature of their targets within the cell, therefore remain to be established.

12.4.4. The Control of Protein Synthesis by Interferons

As well as being controlled by the positive growth factors described above, cell growth and proliferation are influenced by a number of agents which have growth inhibitory and/or differentiation-inducing effects. Of the cytokines which are recognized growth inhibitors, such as transforming growth factor β, the tumor necrosis factors, and the interferons, only the latter group has been studied in detail from the point of view of translational regulation. The mechanisms by which interferons act as antiviral agents frequently result in inhibition of protein synthesis; the two main pathways which have been characterized involve induction of the enzymes 2′5′ oligoadenylate synthetase and a protein kinase variously known as dsI or DAI (for dsRNA-activated inhibitor) or p68 (reviewed in References 78 and 79). Both enzymes require double-stranded RNA (dsRNA) for their activation. The former synthesizes 2′5′ oligoadenylates from ATP, and these molecules activate a ribonuclease (RNase L) which cleaves mRNA and rRNA in ribosomes.[80, 81] The protein kinase phosphorylates the α-subunit of eIF-2, leading to sequestration of GEF activity as described elsewhere in this chapter and in Part II, Chapter 8 of this book. It is clear that these mechanisms play a major role in mediating many of the antiviral effects of the interferons, and both dsRNA-sensitive enzymes have been shown to be activated in interferon-treated, virus-infected cells.[82, 83] However, the evidence for their involvement in the growth inhibition of uninfected cells is less convincing. In the case of the highly sensitive Daudi cell system both 2′5′ oligoadenylate synthetase and DAI are induced within a few hours of interferon treatment, but neither the production of 2′5′ oligoadenylates nor any increase in the phosphorylation of eIF-2 can be demonstrated at a time when cell growth is strongly inhibited.[84, 85] Moreover, although protein synthesis does decline in response to interferon in this system,[86] the mechanism involved does not appear to include any increase in general mRNA turnover or impairment of eIF-2 activity.[85, 87] Although these findings do not rule out a role for either pathway in other systems, in the case of Daudi cells the impairment of protein synthesis may well be a consequence rather than a cause of cell growth inhibition, with specific changes in expression of cellular protooncogenes such as c-*myc* or c-*fgr*[88–90] being more important in blocking proliferation.

A role for the dsRNA-activated protein kinase in the control of growth and differentiation of a strain of 3T3 fibroblasts has been proposed on the basis that the enzyme can be detected in an active form as the cells reach confluence, in the absence of exogenous interferon treatment.[91, 92] It is possible that this activity is due to an autocrine interferon produced in the culture which may serve to limit cell proliferation or to induce differentiation as the cells reach a high density. No correlation was made with possible changes in eIF-2 phosphorylation state or overall protein synthesis under these conditions, but it would seem likely that the latter declines as the cells exit from exponential growth. A possible role for DAI in controlling protein synthesis in uninfected cells raises the question of the nature of the dsRNA which may be present in a form capable of activating the enzyme. A number of reports have indicated that cellular mRNA sequences may have the potential for activation of DAI, presumably because they contain significant amounts of secondary structure.[93–96] On the other hand, there are also several mechanisms by which DAI activation or activity may be inhibited. We know that a number of viruses code for small RNAs or proteins which inhibit DAI activation by dsRNA (see Part II, Chapter 8 of this book) and it is possible that similar cellular factors may exist for this purpose. Dephosphorylation of DAI by a cellular type 1 protein phosphatase[97] may inactivate the enzyme *in vivo* and such a mechanism would potentially be sufficient to limit the extent to which eIF-2 became phosphorylated during the cellular response to endogenous or exogenous interferons.[5]

12.5. REGULATION OF TRANSLATION BY HEAT SHOCK

Heat shock—the state induced by exposing cells to supraphysiological temperatures—is the most extensively studied example of a stress response also observed in a number of other conditions.[98] The onset of this state is characterized by a virtual shut-down of normal cellular protein synthesis and a shift towards high levels of expression of specialized products known as "heat shock proteins". While the major factor in the latter effect is activation of transcription of heat shock genes, the heat-shocked state also raises interesting questions about the translational controls responsible for (a) the profound inhibition of normal cellular translational activity and (b) the ability of the mRNAs for heat shock proteins to escape these inhibitory mechanisms. The inhibition of overall translation is not due to mRNA degradation, since the rate of protein synthesis is rapidly restored when heat-shocked *Drosophila*[98] or mammalian[99] cells are returned to normal physiological temperatures, even in the presence of inhibitors of new RNA synthesis.

The effects of heat shock are particularly pronounced in *Drosophila melanogaster* (reviewed in Reference 98), and it has been demonstrated that the activity of cell-free translation systems prepared from cultured *Drosophila* cells reflects the changes seen *in vivo*.[100] While mRNAs for heat shock proteins could be translated efficiently by extracts from either control or heat-shocked cells, only the normal extracts could utilize cellular mRNAs. Experiments in which components were interchanged between these two types of extract implicated a lesion in the activity of the "ribosomal salt wash" fraction derived from heat shocked cells, suggesting an impairment of one or more initiation factors. Another laboratory[101] located the impaired activity to the postribosomal supernatant fraction, but, as discussed by Lindquist,[98] The apparent discrepancy is probably due to differences in the ionic strength of the extraction buffers used by the two groups.

More detailed examination of initiation factor activity has been carried out with mammalian cells in culture. Severe heat shock (i.e., incubation of the cells at 43 to 45°C) induces polysome disaggregation in both HeLa[102] and Ehrlich ascites tumor cells.[99] In both cases impairment of labeling of 43S initiation complexes with Met-tRNA$_i^{met}$ [99, 102, 103] and increased phosphorylation of eIF-2α[15, 102, 103] have been demonstrated. In the Ehrlich cells this appears to be sufficient to account for the inhibition of GEF activity that is observed.[16] These effects at the level of eIF-2 activity are discussed in more detail in Part II, Chapter 8 of this book. However, heat shock has effects on other initiation factors in mammalian cells, and it could well be that the effects at the level of 43S initiation complex formation are of secondary importance to those at the level of mRNA binding. The inhibition of translation in HeLa cells is accompanied by dephosphorylation of the multiple phosphorylated species of eIF-4B[102] and of the cap-binding polypeptide (eIF-4E) of eIF-4F.[104] While we have no knowledge as yet of the link between the phosphoryaltion status and functional activity of eIF-4B, it is clear that severe heat shock has important effects on the activity of eIF-4F. Duncan and Hershey[102] using a fractionated cell-free system from heat shocked HeLa cells, identified a fraction containing eIF-3 and eIF-4F as having reduced activity. A more detailed study showed that addition of exogenous, purified eIF-4F would rescue translation, particularly of non-heat-shock proteins, in extracts prepared from heat shocked Ehrlich ascites cells.[105] All other factors tried, including eIF-2 and GEF (eIF-2B), failed to stimulate this system. At the same time it was observed that extracts from heat shocked cells had a particularly low K$^+$ optimum for protein synthesis and that the inhibitory effects of heat shock were particularly pronounced at physiological or slightly higher K$^+$ concentrations. Others have pointed out that higher K$^+$ concentrations tend to stabilize secondary structure of mRNA and increase the dependence of initiation on cap binding factors.[38,106] Thus one might expect that a condition that impairs eIF-4F activity would be less effective in inhibiting protein synthesis at low K$^+$ concentrations. Recent work using cell extracts from growing (25°C) or heat-shocked (37°C) *Drosophila* embryos is consistent with this.[107] At K$^+$ concentrations optimal for protein synthesis in the normal lysates translation was severely inhibited in the lysates from heat-shocked embryos. Lowering the K$^+$ concentration lessened this difference and specifically encouraged the translation of normal messages in the heat shocked lysates. Translation in the normal lysates at optimal K$^+$ concentration was potently inhibited by cap analogs, but these were far less effective in inhibiting protein synthesis at lower K$^+$ concentrations or indeed in heat-shock lysates at any K$^+$ concentration.

The hypothesis that a major effect of heat shock on protein synthesis is inactivation of eIF-4F would also go a long way towards explaining the preferential translation of heat shock proteins at a time when overall protein synthesis is inhibited. This effect is particularly pronounced in *Drosophila* cells and can be quite small in mammalian cells,[108] though both show the inhibition of normal, cellular protein synthesis. Heat shock mRNAs have long 5' untranslated regions, which contain little secondary structure. It is known that the 5' untranslated region plays a key role in conferring the ability of these messages to be translated under heat shock conditions.[109,110] One might expect such messages to be less dependent on eIF-4F activity and perhaps even to show enhanced translation when mRNAs that usually compete with them are removed from the translated pool by inactivation of this factor. It is interesting in this context to look at the situation in poliovirus infected cells, in which the activity of eIF-4F is destroyed by the proteolysis of the p220 polypeptide (see Part II, Chapter 10 of this book). Although extracts from these cells are unable to translate most capped cellular mRNAs, they can utilize a natural mRNA, alfalfa mosaic virus 4, with a very unstructured 5' untranslated region.[106] Heat shock mRNAs were found to be more resistant than normal cellular mRNAs to inhibition when HeLa cells recovering from heat shock are subjected to poliovirus infection,[111] and more recently Sarnow[112] has demonstrated enhanced translation of the "heat shock like" protein GRP 78/BiP in poliovirus-infected HeLa cells, at a time when cap-dependent translation of other cellular mRNAs is inhibited.

The evidence discussed so far thus builds up a consistent picture of the mechanisms underlying the effects of severe heat shock on translation. Impairments have been observed in the activity of the eIF-2/GEF system, most likely mediated by eIF-2α phosphorylation, and in the activity of the cap-binding factor, eIF-4F, possibly mediated by the dephosphorylation of the eIF-4E component. The inactivation of eIF-4F could partly or wholly explain the ability of heat shock mRNAs to escape the constraints on translation inflicted on other cellular messages. However, a thought-provoking paper on the effects of milder heat shock (41 to 42°C) on translation in HeLa cells has recently appeared.[108] This paper shows that, while this treatment inhibits translation nearly as much as the more severe conditions, there is no change in the phosphorylation state of the initiation factors eIF-2α, eIF-4B, or eIF-4E. The lack of an effect of these milder conditions on eIF-2α phosphorylation is consistent with an earlier paper showing no change in GEF activity,[113] in contrast to data from severely heat shocked cells.[16] These observations call into doubt the contribution of the changes in initiation factor phosphorylation to the inhibition of protein synthesis in severely heat-shocked cells, and full understanding requires further investigation of the mechanism(s) mediating the effect of mild heat shock.

12.6. TRANSLATIONAL CONTROL DURING MITOSIS

The activity of eIF-4F has also been implicated in the regulation of protein synthesis in mammalian cells undergoing mitosis. It has long been known that the reduced rate of translation in cells in the mitotic phase of the cell cycle is due to impairment of initiation rather than elongation.[114] However, reduced Met-tRNA$_i^{met}$ binding to 43S preinitiation complexes was not found when either intact mitotic cells or extracts prepared from them were incubated with [^{35}S] methionine.[115] More recently, Bonneau and Sonenberg[116] examined the translation patterns when control interphase cells and cells arrested in metaphase with the drug colcemid were subjected to infection with either VSV or poliovirus. The metaphase-arrested cells exhibited reduced translation of capped (VSV) mRNA but undiminished translation of products of the uncapped polio RNA. This suggested that the activity of eIF-4F was decreased during mitosis—a conclusion reinforced by direct measurement of the binding of the 24 kDa cap-binding protein (eIF-4E) to the 5' cap structure of reovirus mRNA in extracts derived from the cells. Furthermore, addition of exogenous eIF-4F was found to stimulate the translation of endogenous mRNA in extracts from the metaphase-arrested cells, but not in extracts from the control, interphase cells. In common with the situation in severely heat shocked cells, the eIF-4E component of endogenous eIF-4F in the metaphase-arrested cells was found to show decreased phosphorylation.

12.7. TRANSLATIONAL CONTROL DURING EARLY DEVELOPMENT

Among the most interesting examples of mechanisms that control translation are those responsible for the changes in protein synthesis that occur during early development of eggs and embryos. Most studied so far have been *Xenopus* oocytes, sea urchin eggs, and the oocytes of the surf clam, *Spisula solidissima*. Clam oocytes are arrested in meiotic prophase, and fertilization triggers germinal vesicle breakdown (i.e., meiotic maturation) followed by embryogenesis. In *Xenopus* one can study meiotic maturation alone, since this can be elicited *in vitro* with the hormone progesterone, whereas the sea urchin egg is already meiotically mature and can be used to study the changes immediately following fertilization. These distinctions partly account for the fact that, while there are profound changes in protein synthesis in all three organisms, there are pronounced differences in emphasis. Fertilization of sea urchin eggs results in a very large increase in overall protein synthesis, with relatively few changes in the pattern of products. In contrast, fertilization of clam oocytes has much less effect on overall translation rate, but produces dramatic changes in the spectrum of proteins made. The latter probably occur during the maturation stage of the process. Maturation of *Xenopus* oocytes results in changes intermediate between these extremes, i.e., a twofold increase in overall protein synthesis together with a few marked changes in the products. Among the proteins whose synthesis is selectively turned on at maturation and/or fertilization in these organisms are cyclins,[117] which are involved in cell cycle regulation and are now known to be capable of inducing maturation when injected into *Xenopus* oocytes,[118,119] the small subunit of ribonucleotide reductase, needed to provide precursors for DNA replication in early embryogenesis[120] and core histones.[121] Concomitantly there is a virtual cessation of synthesis of ribosomal proteins[122] and "housekeeping" proteins like actin and tubulin,[123-125] and the mRNAs for these products are released from polysomes. It is now clear that in all these organisms the responses are complex and involve regulation of protein synthesis at several levels. In unstimulated *Xenopus* oocytes[126] and unfertilized sea urchin eggs[127,128] over 90% of the mRNA and ribosomes are somehow withheld from translation. Quantitative analysis reveals that recruitment of formerly untranslated mRNA is an essential part of the overall increase in protein synthesis during maturation and early embryogenesis.[126,129] However, simple lack of available mRNA cannot be the only limit on protein synthesis, since translation of exogenous mRNA microinjected into *Xenopus* oocytes[130,131] or sea urchin eggs[132] occurs only at the expense of a fall in translation of endogenous mRNAs. Analysis of polysome profiles in sea urchin eggs shows that both initiation and elongation rates are increased following fertilization. For a fuller discussion of this reasoning and of the evidence for a number of proposed mechanisms for the control of translation in early development an excellent review by Rosenthal and Wilt[129] is highly recommended. Here we shall concentrate on recent work on the possible roles of initiation factors in regulating overall rates of protein synthesis and the properties of mRNA molecules that may be responsible for their recruitment for translation.

The potential role of initiation factors has mostly been studied in the sea urchin system, where overall protein synthesis increases about 20-fold after fertilization. The ability to make active cell-free translation systems from unfertilized eggs and from zygotes has been essential to this work, and a number of laboratories have successfully used procedures based on those initially developed by Winkler et al.[133] and further refined by Lopo et al.[134] In particular, the effect of adding exogenous purified initiation factors to cell-free systems from unfertilized eggs has been examined, and there is now a consensus that the factors GEF (eIF-2B) and eIF-4F are stimulatory.[135-137] A low activity of GEF in extracts from unfertilized eggs is suggested in a preliminary report[138] and may explain earlier data showing poor activity in forming [40S.Met-tRNA$_i^{met}$] complexes and stimulatory effects of exogenous eIF-2.[133] However, preliminary evidence that eIF-2α becomes dephosphorylated following fertilization (quoted in Reference 128) has not been confirmed, as the methodology used to examine changes in phosphorylation of the mammalian factor has not proved adequate to detect effects on sea urchin eIF-2α (Gallo, Lopo, and Hershey, personal communication). Work from Jagus' laboratory suggests that GEF activity might be regulated directly by changes in redox state that occur in response to fertilization,[138] but this awaits further evidence. The molecular basis of the stimulation of translation in extracts from unfertilized eggs by addition of eIF-4F also warrants further investigation. One laboratory[136,139] has presented evidence that these extracts contain a potent inhibitor of eIF-4F activity that disappears following fertilization, but the molecular nature of this has not been established. Waltz and Lopo (personal communication), on the other hand, have found increased phosphorylation of the 24 kDa cap-binding polypeptide of eIF-4F (eIF-4E) following fertilization.

Some attention has also been given to the effect of maturation on initiation of protein synthesis in oocytes of *Xenopus laevis*. Cell-free systems from *Xenopus* oocytes show half the activity of those from oocytes subjected to progesterone-induced maturation,[140] reflecting the difference in protein synthesis rate seen in the intact cells.[126] The more active extracts from progesterone-matured oocytes actually showed a lower degree of labeling of 43S initiation complexes with [^{35}S]Met-tRNA$_i^{met}$ suggesting that maturation increases the utilization of these complexes.[140] This would implicate mRNA binding or 60S subunit joining as sites of regulation. Effects of adding exogenous purified initiation factors have been less conclusive in the *Xenopus* system. Audet et al.[141] saw no stimulation of translation in intact oocytes when they microinjected a crude mixture of initiation factors isolated by chromatography on heparin-Sepharose. However, microinjection of purified eIF-4A doubled the rate of protein synthesis, which was not increased further if the oocytes were induced to mature. In our laboratory we have added a number of initiation factors (including eIF-4A) to *Xenopus* oocyte extracts, and failed to observe any effects on protein synthesis (Patrick and Pain, unpublished data).

While increases in the overall capacity for initiation appear to be needed to sustain the higher rates of protein synthesis seen following maturation and fertilization of oocytes and eggs, it is clear that the major factor in the response is the recruitment into polysomes of populations of mRNA that were untranslated prior to stimulation. What are the mechanisms that withhold the bulk of mRNA from translation in unstimulated oocytes? Possibilities that have been investigated are: (a) physical sequestration, e.g., mRNA remaining in the nucleus until germinal vesicle breakdown. With a few exceptions (e.g., Reference 142), the balance of evidence is against this as a widespread mechanism;[121, 143] (b) binding of proteins that "mask" the mRNA and retain it in the untranslated pool—such proteins could either act globally, preventing the translation of all, or many, messages, or act specifically on one or a few species; (c) structural features of the mRNA, such as the 5' cap, the presence or length of the poly(A) tail, or the presence of an inhibitory sequence.

The idea that mRNA molecules can be "masked" by association with specific proteins that bar their accessibility to the translational machinery was first proposed by Spirin,[144] and has been frequently invoked to explain both the failure of unstimulated oocytes to utilize the bulk of their mRNA and the specific withholding of individual mRNA species from translation until a developmental stimulus is received. The possibility that a low rate of translation can be ascribed to global mRNA masking has mainly been investigated in the sea urchin system. A key test is to compare the translational activity of mRNA prepared from unfertilized sea urchin eggs by gentle techniques that preserve its physiological association with proteins (i.e., as mRNP) with that of mRNP prepared similarly from fertilized eggs and with the same mRNA separated from the associated protein by disruptive techniques such as phenol extraction. The majority view, though there has been some dissension,[145] is that the mRNP population from unfertilized eggs does indeed have very low activity, both in overall translatability[146, 147] and in binding to 43S preinitiation complexes.[133, 147] However, it should be recognized that the mRNP fractions analyzed, isolated on the basis of size, are very impure and contain ribosomes and other high molecular weight material.

Work on the role of mRNP proteins in controlling translation in clams and *Xenopus* has focused to a far greater extent on how this may explain changes in the spectrum of proteins made before and after maturation. The classical study of Rosenthal et al.[148] demonstrated that mRNAs encoding a group of translationally regulated proteins (now known to include cyclins[117] and the small subunit of ribonucleotide reductase[120]) were present in unfertilized clam oocytes but were not translated when oocyte extracts containing mRNPs were added to *in vitro* systems. In the *Xenopus* system attempts have been made to identify proteins associated with mRNPs (e.g., Reference 149; see reviews in References 126 and 150). Reconstitution of globin mRNA with mRNP proteins from *Xenopus* oocytes was found to inhibit its translation in wheat germ lysates and following microinjection into *Xenopus* oocytes.[151] One group has localized this type of inhibitory activity to a phosphoprotein present in *Xenopus* RNP preparations.[152] Translatability of globin mRNA associated with this phosphoprotein could be restored by treatment with a protein phosphatase, which destabilized the association between the phosphoprotein and the mRNA. Richter's group have compared proteins bound to specific mRNAs in *Xenopus* oocytes (e.g., Reference 153), and, in the case of one mRNP protein, p56, have identified mRNA sequences to which it binds.[154] Association of this protein with the mRNAs examined diminished concomitantly with the recruitment of the messages into polysomes, suggesting a regulatory function, but this particular example involved gradual recruitment of mRNA species during oogenesis rather than the more rapid mobilization that occurs during maturation.

In parallel with these studies on the protein composition of mRNPs, observations have accumulated indicative of a regulatory role for changes in poly(A) tail length. In such organisms as clams, starfish, and the marine worm *Urechis* several mRNAs that enter the translated pool upon fertilization show a concomitant lengthening of their poly(A) tails, whereas other mRNAs, which cease to be translated at this time, show poly(A) shortening.[125,129] Correlations between polyadenylation and translation were also found in the slime mold, *Dictyostelium*, undergoing differentiation in response to amino acid starvation.[155] Workers using *Xenopus* oocytes have been able to use more interventionist approaches, and Drummond et al.[156] found that the translational efficiencies of *in vitro* transcribed mRNAs following microinjection were greatly enhanced when the transcripts were capped and/or polyadenylated. The observation on the effect of capping has not been followed up, though a claim for a role for cap methylation in activating histone mRNA translation during fertilization of sea urchin eggs[157] has since been contradicted.[158] The effect of polyadenylation on translational efficiency in the work of Drummond et al.[156] was greater than could be explained by the well-known effect on mRNA stability in *Xenopus* oocytes, and has since been confirmed by others.[159] The most exciting recent work in this area has concerned the molecular characterization of sequences in mRNA molecules that may render them susceptible to translational control by changes in polyadenylation. Hyman and Wormington,[122] investigating the cessation of translation of mRNAs encoding ribosomal proteins during maturation in *Xenopus* oocytes, used a microinjected synthetic transcript for ribosomal protein L1 to demonstrate that deadenylation occurred concomitantly with the release of the message from the polysomes to the untranslated pool. Similar behavior could be elicited with a transcript of a fusion gene in which 387 bases at the 3' end of the L1 mRNA sequence was coupled to the 5' end of β-globin mRNA, a message not normally subject to translational regulation during maturation. More precise identification of the sequences required to elicit parallel poly(A) lengthening and translational activation in *Xenopus* oocytes was made by McGrew et al.,[160] studying G10, one of a group of transcripts earlier identified by

Dworkin et al.[161] as being particularly sensitive to recruitment during oocyte maturation. By use of deletion mutants they were able to establish that two key sequences in the 3′ untranslated region were necessary for the poly(A) lengthening/recruitment response, namely, the AAUAAA consensus nuclear polyadenylation signal and a U-rich sequence a few bases upstream of this.

A possible link between poly(A) lengthening and the roles of mRNP proteins in controlling the recruitment of mRNAs for translation is suggested by Richter,[150] who points out that some of the proteins bound specifically to mRNAs in *Xenopus* oocytes appear to be associated with the poly(A) region. However, any feeling that with these recent data a coherent mechanism for the translational activation of mRNAs may at last be emerging could be premature. Examples have been quoted of mRNAs deadenylated during maturation that yet enter polysomes,[162, 163] and Steel and Jacobson[164] were unable to detect changes in poly(A) length in parallel with the cessation of translation of ribosomal protein mRNAs during early development of *Dictyostelium discoideum*. Moreover, there is now some evidence for a role for the 5′ end of mRNA in developmental regulation. Mariottini and Amaldi[164a] find this region to be involved in the recommencement of translation of ribosomal proteins during *Xenopus* embryogenesis, and Lazarus et al.[165] reported that the 5′ untranslated region of murine *c*-myc mRNA, when fused to the coding sequence of the mRNA for the reporter gene, CAT, and microinjected into *Xenopus* oocytes, is able to confer susceptibility to translational regulation. The chimeric message is not translated in unstimulated oocytes, but is recruited into polysomes when they are induced to mature with progesterone.

12.8. CONCLUDING REMARKS

In this chapter we have discussed how a wide variety of physiological and environmental conditions may affect translation. In conclusion it is worth considering the extent to which these effects have features in common that help us to identify key sites of regulation of the overall initiation pathway. It is clear that the most frequently affected steps are the binding of Met-tRNA$_i^{met}$ to 40S ribosomal subunits (controlled by the activity of eIF-2 and/or GEF), and the binding of mRNA to 43S preinitiation complexes. The extent to which changes in phosphorylation of eIF-2α may account for effects on protein synthesis in nonerythroid cells is discussed in Part II, Chapter 8 of this book. Several conditions discussed in this chapter result in changes in activity of the factor eIF-4F, which plays a key role in the binding of capped mRNAs to 43S preinitiation complexes. Increasingly the evidence suggests that the activity of this factor can be modulated by the phosphorylation of the eIF-4E polypeptide. For example, as discussed above, dephosphorylation of eIF-4E accompanies the decreased protein synthesis seen in

HeLa cells in response to heat shock and metaphase-arrest, whereas increased phosphorylation occurs in response to a number of proliferative stimuli in mammalian cells and to fertilization in the case of sea urchin eggs. In considering the possible mechanistic links between eIF-4E phosphorylation and eIF-4F activity one has to question the status of eIF-4F as a permanent association of the 3 polypeptides eIF-4E, eIF-4A and p220. The imbalance between the amounts of these polypeptides in HeLa cells[104] and the ability to isolate active preparations consisting only of eIF-4E and p220[38, 166] suggests that eIF-4F may represent a transient association of independent polypeptides at a particular stage in initiation, probably bound to the 40S subunit at the time of mRNA binding. Rhoads[4] (see Part I, Chapter 5 of this book) has suggested such a mechanism in his model for the mRNA binding step, and his laboratory has demonstrated the presence of eIF-4E in the 48S complexes containing globin mRNA that accumulate in reticulocyte lysates when 60S subunit joining is blocked with one of the inhibitors edeine or guanylate imidodiphosphate.[167] This was determined by adding to the reticulocyte lysate a preparation of eIF-4E synthesized *in vitro* by transcription and translation of a cloned cDNA for this factor. Thus eIF-4E may recognize not only the 5'-terminal cap but also the 43S preinitiation complex and/or p220, whose function may be to aid this process. Present evidence is against an effect of phosphorylation of eIF-4E on cap binding activity,[104, 116, 167] but two recent observations suggest that phosphorylated eIF-4E may be more effective in directing mRNA to the 43S preinitiation complex. First, Joshi-Barve et al.[168] found that the *in vitro*-synthesized product of a mutated eIF-4E sequence, in which the serine residue at the phosphorylation site (Ser-53) was changed to an alanine, did not become associated with 48S complexes in reticulocyte lysates. Moreover, eIF-4E isolated from 48S complexes assembled in the presence of the product of the normal eIF-4E sequence was predominantly in the phosphorylated form. Second, Lamphear and Panniers[168a] have isolated two populations of eIF-4E from Ehrlich ascites cells, one free and one in an association with p220 that survived extensive purification. The eIF-4E in the associated form was much more effective, on a molar basis, in stimulating translation in lysates from heat-shocked cells, and was found to show a considerably higher degree of phosphorylation than the free eIF-4E. Thus phosphorylation appears to modulate the ability of eIF-4E to associate with p220 and with 43S preinitiation complexes. Future investigators should also consider the possible role of eIF-3 in these interactions. This factor is already present on the 43S preinitiation complex at the time of mRNA binding, and both eIF-4E and p220 have been found to be associated with it in cell extracts.[38] However, little is known as yet of its role in the mRNA binding process or, except in the case of Ca^{2+} depletion, its possible regulatory effects during changes in the physiological state of the cell.

Finally, since eIF-2α phosphorylation and eIF-4F activity have both been identified as potential regulatory steps in a number of systems, it is worth ending by mentioning a suggestion that has been made as to how changes in the activity of these two factors could be linked. Edery et al.[169] have hypothesized that impairment of eIF-4F activity may raise the cellular or local concentration of double-stranded structures in mRNA, since, for capped mRNAs, normal eIF-4F action is necessary to initiate the ATP-dependent unwinding of secondary structure in the 5′-untranslated region. An accumulation of double-stranded structures could then activate the double-stranded RNA regulated protein kinase (dsI or DAI) that phosphorylates eIF-2α. Edery et al.[169] suggest that this type of mechanism might operate during heat shock, where it could account for the rather surprising result of Panniers et al.[105] that addition of eIF-4F to lysates of severely heat-shocked cells appeared to stimulate the eIF-2-controlled step of binding Met-tRNA$_i^{met}$ to 40S subunits. However, it should be noted that the experiments upon which Edery et al.[169] based this suggestion were conducted with the reticulocyte lysate, a system known to be particularly sensitive to activation of the dsRNA-dependent kinase, and further evidence is required from nonerythroid systems to evaluate the wider relevance of this mechanism.

ACKNOWLEDGMENTS

The following have been most helpful in discussing their work with us and providing manuscripts prior to publication: C. and M. Brostrom; R. Rhoads, S. Joshi-Barve and W. Rychlik; R. Kaspar; C. Gallo, A. Lopo and J. Hershey; B. Lamphear and R. Panniers. Research in the authors' laboratories is supported by grants from the Wellcome Trust, the Medical Research Council, the Cancer Research Campaign, the Leukaemia Research Fund, and the Gunnar Nilsson Cancer Research Trust.

NOTE ADDED IN PROOF

Since this chapter was prepared, several papers of outstanding importance have appeared on the mechanisms regulating translation during early development. New information has been published on the relationship between translational control and changes in poly(A) tail length during meiotic maturation of *Xenopus* oocytes,[170–175] and a novel approach has been applied to the study of mRNA unmasking activity during early development in the clam, *Spisula solidissima*.[176]

REFERENCES

1. Moldave K: Eukaryotic protein synthesis. *Annu. Rev. Biochem.* 54:1109 (1985).
2. Pain VM: Initiation of protein synthesis in mammalian cells. *Biochem. J.* 235:625 (1986).
3. Proud CG: Guanine nucleotides, protein phosphorylation and the control of translation. *Trends Biochem. Sci.* 11:73 (1986).
4. Rhoads RE: Cap recognition and the entry of mRNA into the protein synthesis initiation cycle. *Trends Biochem. Sci.* 13:52 (1988).
4a. Hershey JWB: Protein phosphorylation controls translation rates. *J. Biol. Chem.* 264:20823 (1989).
5. Clemens MJ: Regulatory mechanisms in translational control. *Curr. Opinion Cell Biol.* 1:1160 (1989).
5a. Clemens MJ: Does protein phosphorylation play a role in translational control by eukaryotic aminoacyl-tRNA synthetases? *Trends Biochem. Sci.* 15:172 (1990).
6. Vaughan MH, Pawlowski PJ, and Forchhammer J: Regulation of protein synthesis initiation in HeLa cells deprived of single essential amino acids. *Proc. Natl. Acad. Sci. U.S.A..* 68:2057 (1971).
7. van Venrooij WJW, Henshaw EC, and Hirsch CA: Nutritional effects on the polyribosome distribution and rate of protein synthesis in Ehrlich ascites tumor cells in culture. *J. Biol. Chem.* 245:5947 (1970).
8. van Venrooij WJW, Henshaw EC, and Hirsch CA: Effects of deprival of glucose or individual amino acids on polyribosome distribution and rate of protein synthesis in cultured mammalian cells. *Biochem. Biophys. Acta* 259:127 (1972).
9. Austin SA and Clemens MJ: The regulation of protein synthesis in mammalian cells by amino acid supply. *Biosci. Rep.* 1:35 (1981).
10. Flaim KE, Peavy DE, Everson WV, and Jefferson LS: The role of amino acids in the regulation of protein synthesis in perfused rat liver. *J. Biol. Chem.* 257:2932 (1982).
11. Everson WV, Flaim KE, Susco DM, Kimball SR, and Jefferson LS: Effect of amino acid deprivation on initiation of protein synthesis in rat hepatocytes. *Am. J. Physiol.* 256:C18 (1989).
12. Pain VM and Henshaw EC: Initiation of protein synthesis in Ehrlich ascites tumour cells. *Eur. J. Biochem.* 57:335 (1975).
13. Pain VM, Lewis JA, Huvos P, Henshaw EC, and Clemens MJ: The effects of amino acid starvation on regulation of polypeptide chain initiation in Ehrlich ascites tumor cells. *J. Biol. Chem.* 255:1486 (1980).
14. Kimball SR, Everson WV, Flaim KE, and Jefferson LS: Initiation of protein synthesis in a cell-free system prepared from rat hepatocytes. *Am. J. Physiol.* 256:C28 (1989).
15. Scorsone KA, Panniers R, Rowlands AG, and Henshaw EC: Phosphorylation of eukaryotic initiation factor 2 during physiological stresses which affect protein synthesis. *J. Biol. Chem.* 262:14538 (1987).
16. Rowlands AG, Montine KS, Henshaw EC, and Panniers R: Physiological stresses inhibit guanine-nucleotide-exchange factor in Ehrlich cells. *Eur. J. Biochem.* 175:93 (1988).

16a. Gross M and Rubino MS: Regulation of eukaryotic initiation factor-2B activity by polyamines and amino acid starvation in rabbit reticulocyte lysate. *J. Biol. Chem.* 264:21879 (1989).

17. Austin SA, Pollard JW, Jagus R, and Clemens MJ: Regulation of polypeptide chain initiation and activity of initiation factor eIF-2 in Chinese-hamster-ovary cell mutants containing temperature-sensitive aminoacyl-tRNA synthetases. *Eur. J. Biochem.* 175:39 (1986).

18. Clemens MJ, Galpine A, Austin SA, Panniers R, Henshaw EC, Duncan R, Hershey JWB, and Pollard JW: Regulation of polypeptide chain initiation in Chinese hamster ovary cells with a temperature-sensitive leucyl-tRNA synthetase. *J. Biol. Chem.* 262:767 (1987).

19. Pollard JW, Galpine AR, and Clemens MJ: A novel role for aminoacyl-tRNA synthetases in the regulation of polypeptide chain initiation. *Eur. J. Biochem.* 182:1 (1989).

20. Vaughan MH and Hansen BS: Control of initiation of protein synthesis in human cells. *J. Biol. Chem.* 248:7087 (1973).

21. Flaim KE, Liao WSL, Peavy DE, Taylor JM, and Jefferson LS: The role of amino acids in the regulation of protein synthesis in perfused rat liver. *J. Biol. Chem.* 257:2939 (1982).

22. Austin SA, Pain VM, Lewis JA, and Clemens MJ: Investigation of the role of uncharged tRNA in the regulation of polypeptide chain initiation by amino acid starvation in cultured mammalian cells; a reappraisal. *Eur. J. Biochem.* 122:519 (1982).

23. Roussou I, Thireos G, and Hauge BM: Transcriptional-translational regulatory circuit in *Saccharomyces cerevisiae* which involves the *GCN4* transcriptional activator and the GCN2 protein kinase. *Mol. Cell Biol.* 8:2132 (1888).

24. Wek RC, Jackson BM, and Hinnebusch AG: Juxtaposition of domains homologous to protein kinases and histidyl-tRNA synthetases in GCN2 protein suggests a mechanism for coupling GCN4 expression to amino acid availability. *Proc. Natl. Acad. Sci. U.S.A.* 86:4579 (1989).

25. Williams NP, Hinnebusch AG, and Donahue TF: Mutations in the structural genes for eukaryotic initiation factors 2-alpha and 2-beta of *Saccharomyces cerevisiae* disrupt translational control of GCN4 mRNA. *Proc. Natl. Acad. Sci. U.S.A..* 86:7515 (1989).

26. Pendergast AM and Traugh JA: Alteration of aminoacyl-tRNA synthetase activities by phosphorylation with casein kinase I. *J. Biol. Chem.* 260:11769 (1985).

27. Pendergast AM, Venema RC, and Traugh JA: Regulation of phosphorylation of aminoacyl-tRNA synthetases in the high molecular weight core complex in reticulocytes. *J. Biol. Chem.* 262:5939 (1987).

28. Mirande M, Le Corre D, and Waller J-P: A complex from cultured Chinese hamster ovary cells containing nine aminoacyl-tRNA synthetases. *Eur. J. Biochem.* 147:281 (1985).

29. Gross M, Rubino MS, and Starn TK: Regulation of protein synthesis in rabbit reticulocyte lysate. Glucose-6-phosphate is required to maintain the activity of eukaryotic initiation factor eIF-2B by a mechanism that is independent of the phosphorylation of eIF-2 alpha. *J. Biol. Chem.* 263:12486 (1988).

30. Live TR and Kaminskas E: Changes in adenylate energy charge in Ehrlich ascites tumor cells deprived of serum, glucose, or amino acids. *J. Biol. Chem.* 250:1786 (1975).

31. Hucul JA, Henshaw EC, and Young DA: Nucleoside diphosphate regulation of overall rates of protein biosynthesis acting at the level of initiation. *J. Biol. Chem.* 260:15585 (1985).

32. Saborio JL, Pong S-S, and Koch G: Selective and reversible inhibition of initiation of protein synthesis in mammalian cells. *J. Mol. Biol.* 85:195 (1974).

33. Kruppa J. and Clemens MJ: Differential kinetics of changes in the state of phosphorylation of ribosomal protein S6 and in the rate of protein synthesis in MPC 11 cells during tonicity shifts. *EMBO J.* 3:95 (1984).

34. Duncan RF and Hershey JWB: Initiation factor protein modifications and inhibition of protein synthesis. *Mol. Cell. Biol.* 7:1293 (1987).

35. Nuss DL and Koch G: Variation in the relative synthesis of immunoglobulin G and non-immunoglobulin G proteins in cultured MPC-11 cells with changes in the overall rate of polypeptide chain initiation and elongation. *J. Mol. Biol.* 102:601 (1976).

36. Kozak M: Leader length and secondary structure modulate mRNA function under conditions of stress. *Mol. Cell. Biol.* 8:2737 (1988).

37. Lee KAW, Guertin D, and Sonenberg N: mRNA secondary structure as a determinant in cap recognition and initiation complex formation. *J. Biol. Chem.* 258:707 (1983).

38. Edery I, Pelletier J, Sonenberg N: Role of eukaryotic messenger RNA cap-binding protein in regulation of translation, in Ilan J (ed): *Translational Regulation of Gene Expression* Plenum Press, New York, 335 (1987).

39. Palen E, and Traugh JA: Phosphorylation of ribosomal protein S6 by cAMP-dependent protein kinase and mitogen-stimulated S6 kinase differentially alters translation of globin mRNA. *J. Biol. Chem.* 262:3518 (1987).

40. Brostrom CO and Brostrom MA: Calcium dependent regulation of protein synthesis in intact mammalian cells. *Annu. Rev. Physiol.* 52:577 (1990).

41. Chin K-V, Cade C, Brostrom CO, Galuska EM, and Brostrom MA: Calcium-dependent regulation of protein synthesis at translational initiation in eukaryotic cells. *J. Biol. Chem.* 262:16509 (1987).

42. Kumar RV, Wolfman A, Panniers R, and Henshaw EC: Mechanism of inhibition of polypeptide chain initiation in calcium-depleted Ehrlich ascites tumor cells. *J. Cell Biol.* 108:2107 (1989).

43. Brostrom CO, Chin KV, Wong WL, Cade C, and Brostrom MA: Inhibition of translational initiation in eukaryotic cells by calcium ionophore. *J. Biol. Chem.* 264:1644 (1989).

44. Fawell EH, Boyer IJ, Brostrom MA, and Brostrom CO: A novel calcium-dependent phosphorylation of a ribosome-associated protein. *J. Biol. Chem.* 264:1650 (1989).

45. de Haro C, de Herreros AG, and Ochoa S: Activation of the heme-stabilized translational inhibitor of reticulocyte lysates by calcium ions and phospholipid. *Proc. Natl. Acad. Sci. U.S.A..* 80:6843 (1983).

46. Nairn AC and Palfrey HC: Identification of the major Mr 100,000 substrate for calmodulin-dependent protein kinase III in mammalian cells as elongation factor-2. *J. Biol. Chem.* 262:17299 (1987).

47. Ryazanov AG, Shestakova EA, and Natapov PG: Phosphorylation of elongation factor 2 by EF-2 kinase affects rate of translation. *Nature* 334:170 (1988).

48. Hurst R, Schatz JR, and Matts RL: Inhibition of rabbit reticulocyte lysate protein synthesis by heavy metal ions involves the phosphorylation of the alpha-subunit of the eukaryotic initiation factor 2. *J. Biol. Chem.* 262:15939 (1987).

49. Chen J-J, Yang JM, Petryshyn R, Kosower N, and London IM: Disulfide bond formation in the regulation of eIF-2α kinase by heme. *J. Biol. Chem.* 264; 9559 (1989).

50. Stanners CP and Becker H: Control of macromolecular synthesis in proliferating and resting Syrian hamster cells in monolayer culture. I. Ribosome function. *J. Cell. Physiol.* 77:31 (1971).

51. Rudland PS, Weil S, and Hunter AR: Changes in RNA metabolism and accumulation of presumptive messenger RNA during transition from the growing to the quiescent state of cultured mouse fibroblasts. *J. Mol. Biol.* 96:745 (1975).

52. Duncan R. and Hershey JWB: Regulation of initiation factors during translational repression caused by serum depletion. Covalent modification. *J. Biol. Chem.* 260:5493 (1985).

53. Duncan R and Hershey JWB: Regulation of initiation factors during translational repression caused by serum depletion. *J. Biol. Chem.* 260:5486 (1985).

54. Salimans MMM, van Heugten HAA, van Steeg H, and Voorma HO: The effect of serum deprivation on the initiation of protein synthesis in mouse neuroblastoma cells. *Biochem. Biophys. Acta* 824:16 (1985).

55. Kaspar RL, Rychlik W, White MW, Rhoads RE, and Morris DR: Simultaneous cytoplasmic redistribution of ribosomal protein L32 mRNA and phosphorylation of eukaryotic initiation factor 4E after mitogenic stimulation of Swiss 3T3 cells. *J. Biol. Chem.* 265:3619 (1990).

56. White MW, Kameji T, Pegg AE, and Morris DR: Increased efficiency of translation of ornithine decarboxylase mRNA in mitogen-activated lymphocytes. *Eur. J. Biochem.* 170:87 (1987).

57. Jinno Y, Merlino GT, and Pastan I: A novel effect of EGF on mRNA stability. *Nucleic Acids Res.* 16:4957 (1988).

58. Harmon CS, Proud CG, and Pain VM: Effects of starvation, diabetes and acute insulin treatment on the regulation of polypeptide-chain initiation in rat skeletal muscle. *Biochem. J.* 223:687 (1984).

59. Kelly FJ and Jefferson LS: Control of peptide-chain initiation in rat skeletal muscle. *J. Biol. Chem.* 260:6677 (1985).

60. Kimball SR and Jefferson LS: Effect of diabetes on guanine nucleotide exchange factor activity in skeletal muscle and heart. *Biochem. Biophys. Res. Commun.* 156:706 (1988).

61. Hammond ML and Bowman LH: Insulin stimulates the translation of ribosomal proteins and the transcription of rDNA in mouse myoblasts. *J. Biol. Chem.* 263;17785 (1988).

62. Marshall S: Kinetics of insulin action on protein synthesis in isolated adipocytes. Ability of glucose to selectively desensitize the glucose transport system without altering insulin stimulation of protein synthesis. *J. Biol. Chem.* 264:2029 (1989).

63. Hesketh JE, Campbell GP, and Reeds PJ: Rapid response of protein synthesis to insulin in 3T3 cells; effects of protein kinase C depletion and differences from the response to serum repletion. *Biosci. Res.* 6:797 (1986).

64. Sato F, Ignotz GG, Ignotz RA, Gansler T, Tsukada K, and Lieberman I: On the mechanism by which insulin stimulates protein synthesis in chick embryo fibroblasts. *Biochemistry* 20:5550 (1981).

65. Cox S, Redpath NT, and Proud CG: Regulation of polypeptide-chain initiation in rat skeletal muscle. *FEBS Lett.* 239:333 (1988).

66. de Philip RM, Chadwick DE, Ignotz RA, Lynch WE, and Lieberman I: Rapid stimulation by insulin of ribosome synthesis in cultured chick embryo fibroblasts. *Biochemistry* 18:4812 (1979).

67. Ashford AJ and Pain VM: Insulin stimulation of growth in diabetic rats. Synthesis and degradation of ribosomes and total tissue protein in skeletal muscle and heart. *J. Biol. Chem.* 261:4066 (1986).

68. Morley SJ and Traugh JA: Insulin and phorbol esters stimulate phosphorylation of eIF-4F. *J. Cell Biol.* 107:109a (1988).

69. Marshall S and Monzon R: Amino acid regulation of insulin action in isolated adipocytes. Selective ability of amino acids to enhance both insulin sensitivity and maximal insulin responsiveness of the protein synthesis system. *J. Biol. Chem.* 264:2037 (1989).

70. Clemens MJ and Pain VM: Hormonal requirements for acute stimulation of rat liver polysome formation by amino acid feeding. *Biochem. Biophsy. Acta* 361:345 (1974).

71. Garlick PJ and Grant I: Amino acid infusion increases the sensitivity of muscle protein synthesis *in vivo* to insulin. *Biochem. J.* 254:579 (1988).

72. Brostrom MA, Chin K-V, Cade C, Gmitter D, and Brostrom CO: Stimulation of protein synthesis in pituitary cells by phorbol esters and cyclic AMP. *J. Biol. Chem.* 262:16515 (1987).

73. Tuazon PT, Merrick WC, and Traugh JA: Comparative analysis of phosphorylation of translational initiation and elongation factors by seven protein kinases. *J. Biol. Chem.* 264:2773 (1989).

74. Morley SJ and Traugh JA: Phorbol esters stimulate phosphorylation of eukaryotic initiation factors 3, 4B, and 4F. *J. Biol. Chem.* 264:2401 (1989).

75. McMullin EL, Haas DW, Abramson RD, Thach RE, Merrick WC, and Hagedorn CH: Identification of a protein kinase activity in rabbit reticulocytes that phosphorylates the mRNA cap binding protein. *Biochem. Biophys. Res. Commun.* 153:340 (1988).

76. Legon S, Brayley A, Hunt T, and Jackson RJ: The effect of cyclic AMP and related compounds on the control of protein synthesis in reticulocyte lysates. *Biochem. Biophys. Res. Commun* 56:745 (1974).

77. Sitikov AS, Simonenko PN, Shestakova EA, Ryazanov AG, and Ovchinnikov LP: cAMP-dependent activation of protein synthesis correlates with dephosphorylation of elongation factor 2. *FEBS Lett.* 228:327 (1988).

78. Clemens MJ and McNurlan MA: Regulation of cell proliferation and differentiation by interferons. *Biochem. J.* 226:345 (1985).

79. Pestka S, Langer JA, Zoon KC, and Samuel CE: Interferons and their actions. *Annu. Rev. Biochem.* 56:727 (1987).

80. Clemens MJ and Williams BRG: Inhibition of cell-free protein synthesis by $pppA^{2'}p^{5'}A^{2'}p^{5'}A$: a novel oligonucleotide synthesized by interferon-treated L cell extracts. *Cell* 13:565 (1978).

81. Wreschner DH, James TC, Silverman RH, and Kerr IM: Ribosomal RNA cleavage, nuclease activation and 2-5A(ppp(A2′p)$_n$A) in interferon-treated cells. *Nucleic Acids Res.* 9:1571 (1981).

82. Samuel CE, Duncan R, Knutson GS, and Hershey JWB: Mechanism of interferon action. Increased phosphorylation of protein synthesis initiation factor eIF-2(alpha) in interferon-treated, reovirus-infected mouse L929 fibroblasts *in vitro* and *in vivo. J. Biol. Chem.* 259:13451 (1984).

83. Rice AP, Duncan R, Hershey JWB, and Kerr IM: Double-stranded RNA-dependent protein kinase and 2-5A system are both activated in interferon-treated, encephalomyocarditis virus-infected HeLa cells. *J. Virol.* 54;894 (1985).

84. Silverman RH, Watling D, Balkwill FR, Trowsdale J, and Kerr IM: The ppp(A2′p)$_n$A and protein kinase systems in wild-type and interferon-resistant Daudi cells. *Eur. J. Biochem.* 126:333 (1982).

85. Clemens MJ and Tilleray VJ: Inhibition of polypeptide chain initiation in Daudi cells by interferons. *Biochem. J.* 237:877 (1986).

86. McNurlan MA and Clemens MJ: Inhibition of cell proliferation by interferons. *Biochem. J.* 237:871 (1986).

87. Clemens MJ, McNurlan MA, Moore G, and Tilleray VJ: Regulation of protein synthesis in lymphoblastoid cells during inhibition of cell proliferation by human interferons. *FEBS Lett.* 171:111 (1984).

88. Einat M, Resnitzky D, and Kimchi A: Close link between reduction of c-myc expression by interferon and G0/G1 arrest. *Nature* 313:597 (1985).

89. Dani C, Mechti N, Piechaczyk M, Lebleu B, Jeanteur P, and Blanchard JM: Increased rate of degradation of c-myc mRNA in interferon-treated Daudi cells. *Proc. Natl. Acad. Sci. U.S.A.* 82:4896 (1985).

90. Sharp NA, Luscombe MJ, and Clemens MJ: Regulation of c-fgr proto-oncogene expression in Burkitt's lymphoma cells: effect of interferon treatment and relationship to EBV status and c-myc mRNA levels. *Oncogene* 4:1043 (1989).

91. Petryshyn R, Chen J-J, London IM: Growth-related expression of a double-stranded RNA-dependent protein kinase in 3T3 cells. *J. Biol. Chem.* 259:14736 (1984).

92. Petryshyn R, Chen J-J, and London IM: Detection of activated double-stranded RNA-dependent protein kinase in 3T3-F442A cells. *Proc. Natl. Acad. Sci. U.S.A..* 85:1427 (1988).

93. Baum EZ and Ernst VG: Inhibition of protein synthesis in reticulocyte lysates by a double-stranded RNA component in HeLa mRNA. *Biochem. Biophys. Res. Commun.* 114:41 (1983).

94. Pratt G, Galpine A, Sharp N, Palmer S, and Clemens MJ: Regulation of *in vitro* translation by double-stranded RNA in mammalian cell mRNA preparations. *Nucleic Acids Res.* 16:3497 (1988).

95. Maran A and Mathews MB: Characterization of the double-stranded RNA implicated in the inhibition of protein synthesis in cells infected with a mutant adenovirus defective for VA RNAI. *Virology* 164:106 (1988).

96. Kaufman RJ, Davies MV, Pathak VK, and Hershey JWB: The phosphorylation state of eucaryotic initiation factor 2 alters translational efficiency of specific mRNAs. *Mol. Cell. Biol.* 9:946 (1989).

97. Szyszka R, Kudlicki W, Kramer G, Hardesty B, Galabru J, and Hovanessian A: A type 1 phosphoprotein phosphatase active with phosphorylated Mr = 68,000 initiation factor-2 kinase. *J. Biol. Chem.* 264:3827 (1989).
98. Lindquist, S: Translational regulation in the heat-shock response of *Drosophila* cells, in Ilan J (Ed): *Translational Regulation of Gene Expression* Plenum Press, New York, p. 187 (1987).
99. Panniers R and Henshaw EC: Mechanism of inhibition of polypeptide chain initiation in heat shocked Ehrlich ascites tumour cells. *Eur. J. Biochem.* 140:209 (1984).
100. Scott MP and Pardue ML: Translational control in lysates of *Drosophila melanogaster* cells. *Proc. Natl. Acad. Sci. U.S.A..* 78:3353 (1981).
101. Sanders MM, Triemer DF, and Olsen AS: Regulation of protein synthesis in heat-shocked *Drosophila* cells. *J. Biol. Chem.* 261:2189 (1986).
102. Duncan R and Hershey JWB: Heat shock-induced translational alterations in HeLa cells. *J. Biol. Chem.* 259:11882 (1984).
103. de Benedetti A and Baglioni C: Activation of hemin-regulated initiation factor-2 kinase in heat-shocked HeLa cells. *J. Biol. Chem.* 261:338 (1986).
104. Duncan R, Milburn SC, and Hershey JWB: Regulated phosphorylation and low abundance of HeLa cell initiation factor eIF-4F suggest a role in translational control. *J. Biol. Chem.* 262:380 (1987).
105. Panniers R, Stewart EB, Merrick WC, and Henshaw EC: Mechanism of inhibition of polypeptide chain initiation in heat-shocked Ehrlich ascites tumor cells involves reduction of eukaryotic initiation factor 4F activity. *J. Biol. Chem.* 260:9648 (1985).
106. Sonenberg N, Guertin D, and Lee KAW: Capped mRNAs with reduced secondary structure can function in extracts from poliovirus-infected cells. *Mol. Cell. Biol.* 2:1633 (1982).
107. Maroto FG and Sierra JM: Translational control in heat-shocked *Drosophila* embryos. *J. Biol. Chem.* 263:15720 (1988).
108. Duncan RF and Hershey JWB: Protein synthesis and protein phosphorylation during heat stress, recovery and adaptation. *J. Cell Biol.* 109:1467 (1989).
109. McGarry TJ and Lindquist S: The preferential translation of *Drosophila* hsp 70 mRNA requires sequences in the untranslated leader. *Cell* 42:903 (1985).
110. Klemenz R, Hultmark D, and Gehring WJ: Selective translation of heat-shock mRNA in *Drosophila melanogaster* depends on sequence information in the leader. *EMBO J.* 4:2053 (1985).
111. Munoz A, Alonso MA, and Carrasco L: Synthesis of heat-shock proteins in HeLa cells: inhibition by viral infection. *Virology* 137:150 (1984).
112. Sarnow P: Translation of glucose-regulated protein 78/immunoglobulin heavy-chain binding protein mRNA is increased in poliovirus-infected cells at a time when cap-dependent translation of cellular mRNAs is inhibited. *Proc. Natl. Acad. Sci. U.S.A..* 86:5795 (1989).
113. Mariano T and Siekierka J: Inhibition of HeLa cell protein synthesis under heat-shock conditions in the absence of initiation factor eIF-2α phosphorylation. *Biochem. Biophys. Res. Commun.* 138:519 (1986).
114. Fan H and Penman S: Regulation of protein synthesis in mammalian cells. II. Inhibition of protein synthesis at the level of initiation during mitosis. *J. Mol. Biol.* 50:655 (1970).

115. Tarnowka MA and Baglioni C: Regulation of protein synthesis in mitotic HeLa cells. *J. Cell. Physiol.* 99:359 (1979).

116. Bonneau A-M and Sonenberg N: Involvement of the 24-kDa cap-binding protein in regulation of protein synthesis in mitosis. *J. Biol. Chem.* 262:11134 (1987).

117. Evans T, Rosenthal ET, Youngbloom J, Distel D, and Hunt T: Cyclin: a protein specified by maternal mRNA in sea urchin eggs that is destroyed at each cleavage division. *Cell* 33:389 (1983).

118. Swenson KI, Farrell KM, and Ruderman JV: The clam embryo protein cyclin A induces entry into M phase and the resumption of meiosis in *Xenopus* oocytes. *Cell* 47:861 (1986).

119. Pines J and Hunt T: Molecular cloning and characterization of the mRNA for cyclin from sea urchin eggs. *EMBO J.* 6:2987 (1987).

120. Standart NM, Bray SJ, George EL, Hunt T, and Ruderman JV: The small subunit of ribonucleotide reductase is encoded by one of the most abundant translationally regulated maternal mRNAs in clam and sea urchin eggs. *J. Cell Biol.* 100:1968 (1985).

121. Adamson ED and Woodland HR: Changes in the rate of histone synthesis during oocyte maturation and very early development of *Xenopus laevis*. *Dev. Biol.* 57:136 (1977).

122. Hyman LE and Wormington WM: Translational inactivation of ribosomal protein mRNAs during *Xenopus* oocyte maturation. *Genes Dev.* 2:598 (1988).

123. Ballantine JEM, Woodland HR, and Sturgess EA: Changes in protein synthesis during the development of *Xenopus laevis*. *J. Embryol. Exp. Morphol.* 51:137 (1979).

124. Grainger JL, Von Brunn A, and Winkler MM: Transient synthesis of a specific set of proteins during the rapid cleavage phase of sea urchin development. *Dev. Biol.* 114:403 (1986).

125. Rosenthal ET and Ruderman JV: Widespread changes in the translation and adenylation of maternal messenger RNAs following fertilization of *Spisula* oocytes. *Dev. Biol.* 121:237 (1987).

126. Richter JD: Molecular mechanisms of translational control during the early development of *Xenopus laevis*, in Ilan, J. (Ed): *Translational Regulation of Gene Expression*, Plenum Press, New York, pp. 111–139 (1987).

127. Goustin AS and Wilt FH: Protein synthesis, polyribosomes, and peptide elongation in early development of *Strongylocentrotus purpuratus*. *Dev. Biol.* 82:32 (1981).

128. Winkler M: Translational regulation in sea urchin eggs: a complex interaction of biochemical and physiological regulatory mechanisms. *BioEssays* 8:157 (1988).

129. Rosenthal ET and Wilt FH: Selective messenger RNA translation in marine invertebrate oocytes, eggs and zygotes, in Ilan J (ed): *Translational Regulation of Gene Expression*, Plenum Press, New York, p. 87 (1987).

130. Laskey RA, Mills AD, Gurdon JB, and Partington GA: Protein synthesis in oocytes of *Xenopus laevis* is not regulated by the supply of messenger RNA. *Cell* 11:345 (1977).

131. Asselbergs FAM, van Venrooij WJ, and Bloemendal H: Messenger RNA competition in living *Xenopus* oocytes. *Eur. J. Biochem.* 94:249 (1979).

132. Colin AM and Hille MB: Injected mRNA does not increase protein synthesis in unfertilized, fertilized, or ammonia-activated sea urchin eggs. *Dev. Biol.* 115:184 (1986).

133. Winkler MM, Nelson EM, Lashbrook C, and Hershey JWB: Multiple levels of regulation of protein synthesis at fertilization in sea urchin eggs. *Dev. Biol.* 107:290 (1985).

134. Lopo AC, Lashbrook CC, and Hershey JWB: Characterization of translation systems *in vitro* from three developmental stages of *Strongylocentrotus purpuratus. Biochem. J.* 258:553 (1989).

135. Colin AM, Brown BD, Dholakia JN, Woodley CL, Wahba AJ, and Hille MB: Evidence for simultaneous derepression of messenger RNA and the guanine nucleotide exchange factor in fertilized sea urchin eggs. *Dev. Biol.* 123:354 (1987).

136. Huang W-I, Hansen LJ, Merrick WC, and Jagus R: Inhibitor of eukaryotic initiation factor 4F activity in unfertilized sea urchin eggs. *Proc. Natl. Acad. Sci. U.S.A.* 84:6359 (1987).

137. Lopo AC, MacMillan S, and Hershey JWB: Translational control in early sea urchin embryogenesis: initiation factor eIF4F stimulates protein synthesis in lysates from unfertilized eggs of *Strongylcentrotus purpuratus. Biochemistry* 27:351 (1988).

138. Akkaraju GR, Hansen LJ, and Jagus R: Fertilization of sea urchin eggs stimulates eIF-2 recycling in cell-free translation systems. *J. Cell Biol.* 107:107a (1988).

139. Hansen LJ, Huang W-I, and Jagus R: Inhibitor of translational initiation in sea urchin eggs prevents mRNA utilization. *J. Biol. Chem.* 262:6114 (1987).

140. Patrick TD, Lewer CE, and Pain VM: Preparation and characterization of cell-free protein synthesis systems from oocytes and eggs of *Xenopus laevis. Development* 106:1 (1989).

141. Audet RG, Goodchild J, and Richter JD: Eukaryotic initiation factor 4A stimulates translation in microinjected *Xenopus* oocytes. *Dev. Biol.* 121:58 (1987).

142. Showman RM, Wells DE, Anstrom J, Hursh DA, and Raff RA: Message-specific sequestration of maternal histone mRNA in the sea urchin egg. *Proc. Natl. Acad. Sci. U.S.A.* 79:5944 (1982).

143. Swenson KI, Borgese N, Pietrini G, and Ruderman JV: Three translationally regulated mRNAs are stored in the cytoplasm of clam oocytes. *Dev. Biol.* 123:10 (1987).

144. Spirin AS: Informosomes. *Eur. J. Biochem.* 10:20 (1969).

145. Moon RT, Danilchik MV, and Hille MB: An assessment of the masked message hypothesis: sea urchin messenger ribonucleoprotein complexes are efficient templates for *in vitro* protein synthesis. *Dev. Biol.* 93:389 (1982).

146. Jenkins NA, Kaumeyer JF, Young EM, and Raff RA: A test for masked message: the template activity of messenger ribonucleoprotein particles isolated from sea urchin eggs. *Dev. Biol.* 63:279 (1978).

147. Grainger JL and Winkler MM: Fertilization triggers unmasking of maternal mRNA in sea urchin eggs. *Mol. Cell. Biol.* 7:3947 (1987).

148. Rosenthal ET, Hunt T, and Ruderman JV: Selective translation of mRNA controls the pattern of protein synthesis during early development of the surf clam, *Spisula solidissima. Cell* 20:487 (1980).

149. Darnbrough CH and Ford PJ: Identification in *Xenopus laevis* of a class of oocyte-specific proteins bound to messenger RNA. *Eur. J. Biochem.* 113:415 (1981).
150. Richter JD: Information relay from gene to protein: the mRNP connection. *Trends Biochem. Sci.* 13:483 (1988).
151. Richter JD and Smith LD: Reversible inhibition of translation by *Xenopus* oocyte-specific proteins. *Nature* 309:378 (1984).
152. Kick D, Barrett P, Cummings A, and Sommerville J: Phosphorylation of a 60 kDa polypeptide from *Xenopus* oocytes blocks messenger RNA translation. *Nucleic Acids Res.* 15:4099 (1987).
153. Swiderski RE and Richter JD: Photocrosslinking of proteins to maternal mRNA in *Xenopus* oocytes. *Dev. Biol.* 128:349 (1988).
154. Crawford DR and Richter JD: An RNA-binding protein from *Xenopus* oocytes is associated with specific message sequences. *Development* 101:741 (1987).
155. Palatnik CM, Wilkins C, and Jacobson A: Translational control during early *Dictyostelium* development: possible involvement of poly(A) sequences. *Cell* 36:1017 (1984).
156. Drummond DR, Armstrong J, and Colman A: The effect of capping and polyadenylation on the stability, movement and translation of synthetic messenger RNAs in *Xenopus* oocytes. *Nucleic Acids Res.* 13:7375 (1985).
157. Caldwell DC and Emerson CP: The role of cap methylation in the translational activation of stored maternal histone mRNA in sea urchin embryos. *Cell* 42:691 (1985).
158. Showman RM, Leaf DS, Anstrom JA, and Raff RA: Translation of maternal histone mRNAs in sea urchin embryos: a test of control by 5′ cap methylation. *Dev. Biol.* 121:284 (1987).
159. Galili G, Kawata EE, Smith LD, and Larkins BA: Role of the 3′-poly(A) sequence in translational regulation of mRNAs in *Xenopus laevis* oocytes. *J. Biol. Chem.* 263:5764 (1988).
160. McGrew LL, Dworkin-Rastl E, Dworkin MB, and Richter JD: Poly(A) elongation during *Xenopus* oocyte maturation is required for translational recruitment and is mediated by a short sequence element. *Genes Dev.* 3:803 (1989).
161. Dworkin MB, Shrutkowski A, and Sworkin-Rastl E: Mobilization of specific maternal RNA species into polysomes after fertilization in *Xenopus laevis*. *Proc. Natl. Acad. Sci. U.S.A.* 82:7636 (1985).
162. Ruderman JV, Woodland HR, and Sturgess EA: Modulation of histone mRNA during the early development of *Xenopus laevis*. *Dev. Biol.* 17:71 (1979).
163. Ballantine JEM and Woodland HR: Polyadenylation of histone mRNA in *Xenopus* oocytes and embryos. *FEBS Lett.* 180:224 (1985).
164. Steel LF and Jacobson A: Translational control of ribosomal protein synthesis during early *Dictyostelium discoideum* development. *Mol. Cell. Biol.* 7:965 (1987).
164a. Mariottini P and Amaldi F: The 5′ untranslated region of mRNA for ribosomal protein S19 is involved in its translational regulation during *Xenopus* development. *Mol. Cell. Biol.* 10:816 (1990).
165. Lazarus P, Parkin N, and Sonenberg N: Developmental regulation of translation by the 5′ noncoding region of murine c-myc mRNA in *Xenopus laevis*. *Oncogene* 3:517 (1988).

166. Etchison D and Milburn S: Separation of protein synthesis initiation factor eIF4A from a p220-associated cap binding complex activity. *Mol. Cell. Biochem.* 76:15 (1987).

167. Hiremath LS, Hiremath ST, Rychlik W, Joshi S, Domier LL, and Rhoads RE: *In vitro* synthesis, phosphorylation, and localization on 48S initiation complexes of human protein synthesis initiation factor 4E. *J. Biol. Chem.* 264:1132 (1989).

168. Joshi-Barve S, Rychlik W, and Rhoads RE: Alteration of the major phosphorylation site of eukaryotic protein synthesis initiation factor 4E prevents its association with the 48S initiation complex. *J. Biol. Chem.* 265:2979 (1990).

168a. Lamphear B and Panniers R: Cap binding protein complex that restores protein synthesis in heat-shocked Ehrlich cell lysates contains highly phosphorylated eIF-4E. *J. Biol. Chem.* 265:5333 (1990).

169. Edery I, Petryshyn R, and Sonenberg N: Activation of double-stranded RNA-dependent kinase (dsI) by the TAR region of HIV-1 mRNA: a novel translational control mechanism. *Cell* 56:303 (1989).

170. McGrew LL and Richter JD: Translational control by cytoplasmic polyadenylation during *Xenopus* oocyte maturation: characterization of *cis* and *trans* elements and regulation by cyclin/MPF. *EMBO J.* 9:3743 (1990).

171. Paris J and Richter JD: Maturation-specific polyadenylation and translational control: diversity of cytoplasmic polyadenylation elements, influence of poly(A) tail size and formation of stable polyadenylation complexes. *Mol. Cell. Biol.* 10:5634 (1990).

172. Varnum SM and Wormington WM: Deadenylation of maternal mRNAs during *Xenopus* oocyte maturation does not require specific *cis*-sequences: a default mechanism for translational control. *Genes Dev.* 4:2278 (1990).

173. Fox CA and Wickens M: Poly(A) removal during oocyte maturation: a default mechanism selectively prevented by a specific sequence in the 3' untranslated region of certain maternal mRNAs. *Genes Dev.* 4:2287 (1990).

174. Wickens M: In the beginning is the end: regulation of poly(A) addition and removal during early development. *Trends. Biochem. Sci.* 15:320 (1990).

175. Jackson RJ and Standard N: Do the poly(A) tail and 3' untranslated region control mRNA translation? *Cell* 62:15 (1990).

176. Standard N, Dale M, Stewart E, and Hunt T: Maternal mRNA from clam oocytes can be specifically unmasked *in vitro* by antisense RNA complementary to the 3' untranslated region. *Genes Dev.* 4:2157 (1990).

Regulation of Elongation Rate

Alexander S. Spirin and Alexey G. Ryazanov
Institute of Protein Research
Academy of Sciences of the USSR
Pushchino, Moscow Region, USSR

13.1. INTRODUCTION: REGULATION AT ELONGATION STAGE IS IMPORTANT IN EUKARYOTIC TRANSLATION

Since the overwhelming amount of information on translational control concerns the regulation at the initiation stage, little attention has been paid until recently to the elongation stage as a possible target for regulatory influences. At the same time, facts demonstrating variations and changes of elongation rate in eukaryotic cells are accumulating. Now it becomes clear that elongation can be regulated both totally and differentially and that this regulation has a great significance for overall translational control in eukaryotes. The facts and the mechanisms will be considered below.

During the elongation stage each codon of mRNA is read in a cycle (*elongation cycle*) consisting of three main steps, namely, aminoacyl-tRNA binding, transpeptidation and translocation (Figure 1 and Part I, Chapter 6 of this book). In principle, the speed (rate) of each step may be regulated. If the regulation affects a given step of all cycles more or less uniformly the *cycle frequency* will be altered and, hence, the general elongation rate will change. On the other hand, one or another step of specific elongation cycles along mRNA can be intervened; e.g., aminoacyl-tRNA binding to a specific codon can be slowed down, or translocation through a specific site of mRNA can be hindered. In this case the elongation becomes uneven (discontinuous),

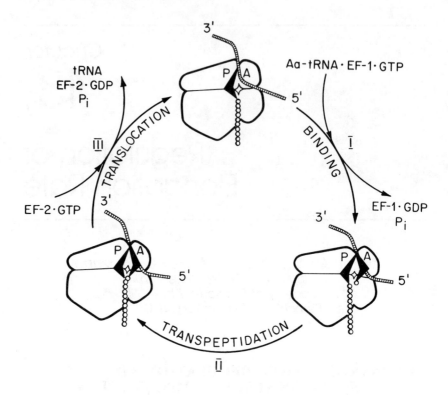

Figure 1. The elongation cycle.

and *translational pauses* may arise, resulting in a change in average elongation rate. Regulation of the cycle frequency and the translational pausing seem to be two main phenomena in the control of elongation rate.

13.2. ELONGATION RATE CHANGES CAN BE REVEALED FROM TRANSIT TIME DETERMINATION AND POLYRIBOSOME PROFILE ANALYSES

13.2.1. Transit Time

The time during which a growing nascent peptide remains attached to the translating ribosome, i.e., the time of its elongation plus termination, is called transit time (see also Part I, Chapter 6 of this book). If the termination time is neglected, the number of amino acids in a given protein divided by its transit time is the average rate of elongation on a corresponding mRNA. Hence, the transit time determinations for proteins of known size may give an information about the elongation rate.

When a radioactive amino acid is given to a cell, the radioactivity will soon appear in the polyribosome fraction in the form of growing nascent peptides attached to ribosomes. The radioactivity of the polyribosome fraction will increase as the nascent peptides on ribosomes will be elongated by the labeled amino acids (Figure 2). When all the growing peptides become fully labeled the plateau of radioactivity of the polyribosome fraction will be established. The time of reaching the plateau is the transit time.

In practice, however, it is more convenient and accurate to determine the transit time from the kinetics of radioactivity incorporation into the ribosome-free fraction of the soluble (completed) proteins as compared with that of the total incorporation into polypeptides. The incorporation into soluble protein will increase first exponentially and then, after reaching the plateau in the polyribosome fraction, linearly at the expense of the release of fully labeled polypeptides. At the same time it is evident that the total incorporation of the radioactive amino acid into polypeptides should be linear almost from the beginning (provided the elongation rate is constant during the experiment). Since after reaching the radioactivity plateau in polyribosomes the increment of the radioactivity both in total peptides and in soluble protein fraction is determined by the same process of releasing labeled polypeptides from ribosomes the two linear plots should be parallel (Figure 3). The distance between them along the abscissa corresponds to the time required for the completion of the synthesis of nascent peptides in polyribosomes, i.e., to half transit time. (A reminder should be that the *average* length of nascent peptides in a polyribosome is *half* of the full length of completed polypeptides, i.e., at each given moment a polyribosome contains half-completed polypeptides on average.)

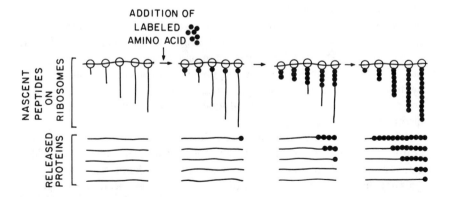

Figure 2. Schematic representation of the time-course of radioactivity accumulation in the fraction of growing nascent peptides (polyribosome fraction) and in the fraction of completed released proteins (supernatant fraction) after addition of radioactive amino acid. Note that after all the growing peptides are fully labeled the radioactivity in the polyribosome fraction becomes constant.

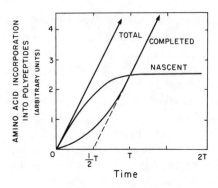

Figure 3. Theoretical kinetic curves of incorporation of radioactive amino acid into total polypeptides, completed released protein, and nascent polypeptides attached to ribosomes. (Modified from Wu RS and Warner JR: *J. Cell Biol.* 51:643 (1971).

In practice the procedure of transit time determination is sensitive to several experimental variables. Most common difficulties and ways of overcoming them can be found in the survey of Nielsen and McConkey.[1]

13.2.2. Polyribosome Profile

Absolute rates of elongation are not necessarily required for studies of regulation at the elongation stage. Often it is sufficient to know just relative changes of elongation rates. This can be done by recording a change in the profile of polyribosome distribution upon sucrose gradient centrifugation (Figure 4).

The polyribosome profile depends on the rates of initiation, elongation and termination. Generally, slowing down the movement of ribosomes along mRNA at a constant initiation rate will result in an increase of heavy polyribosome fraction (i.e., increase of the density of ribosomes on mRNA). It is obvious, however, that the increase of initiation rate at a constant elongation rate will give the same result. Hence, the analysis of polyribosome profiles by itself does not always permit judgement about changes in elongation rate, and the information about the rate of total translation (the rate of amino acid incorporation) is required.

Three clear cases can be considered: (1) the "light shift" in polyribosome profile (decrease of the number of ribosomes per mRNA) indicates the elongation rate rise, if the amino acid incorporation (translation rate) increases or does not change, (2) the "heavy shift" in polyribosomes means the elongation rate drop, if the amino acid incorporation goes down or does not change, and (3) the absence of shifts in polyribosome profile with simultaneous increase or decrease of the amino acid incorporation will give evidence that the elongation rate increases or decreases, respectively. The last case can be illustrated by the following example. The stimulation of protein synthesis in HeLa cells by serum did not result in the change of polyribosome profile.[1] It was concluded that the addition of serum enhanced the elongation rate (and the initiation rate as well). The transit time measurement supported the conclusion about elongation.

13.2.3. Uncertainty Because of Termination Pause

It should be mentioned that the changes in termination rate will affect the polyribosome profile, as well as the transit time, in the same way as the elongation rate changes. Therefore, strictly speaking, the transit time determination and the polyribosome profile do not permit to conclude unambiguously about elongation rate changes. Nevertheless, since no direct facts were known concerning the regulation of termination, and no indication existed that termination is a limiting stage of translation, the possibility of termination control used to be neglected.

At the same time, it was demonstrated that in the case of globin synthesis in rabbit reticulocytes termination required much more time than a single elongation cycle.[2] In the case of cell-free translation of bovine preprolactin mRNA one of the four ribosomal pausing sites was found at the termination codon.[3] The above-said means that in the condition when the rates of initiation and elongation are sufficiently high, termination could become rate limiting. In support of this, it was reported that under heat shock conditions in the rabbit reticulocyte cell-free translation system the inhibition of translation of non-heat-shock mRNAs was due to the block of termination.[4] Thus it is not

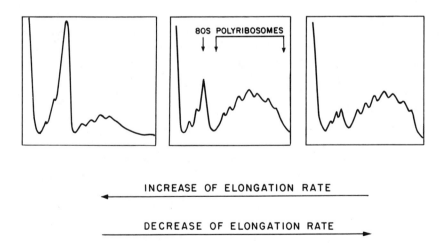

Figure 4. Diagrams of mono- and polyribosome distribution profiles upon sucrose gradient centrifugation.

excluded that in some special cases termination can be a rate-limiting stage, and therefore transit time measurements and polyribosome profile analyses cannot be directly used to test the elongation rate.

13.3. ELONGATION STAGE OF TRANSLATION IS SHOWN TO BE REGULATED

13.3.1. Overall Elongation Rate in the Cell Can Be Stimulated or Slowed Down

Changes in the rate of polypeptide chain elongation, i.e., in the speed of ribosome movement, on all cellular mRNAs at the same time were reported as a result of different stimulatory effects, cell differentiation events, hormone treatment, viral infection, egg fertilization, etc.

One of the examples is the case mentioned above when elongation rate measured for total mRNA of HeLa cells responded to serum stimulation.[1] A significant change in the overall rate of polypeptide elongation was reported for human placental explants during normal placental development and in diabetic pregnancies.[5]

The injection of estradiol into cockerels was demonstrated to induce a drop of the elongation rate on almost all mRNAs in liver cells, from the average value of seven amino acids per second per ribosome at the beginning to five amino acids per second after 1 day, three amino acids per second after 2 days, and two amino acids per second after 3 days.[6] The infection of mouse L cells with encephalomyocarditis virus led to drastic slowing down of polypeptide elongation on host mRNAs, from nine to ten amino acids per second per ribosome to two amino acids per second in 5 h after infection.[7] The fertilization of sea urchin eggs was accompanied by the twofold increase of the total elongation rate.[8, 9] During yeast-to-hyphae morphogenesis in the fungus *Mucor racemosus* the fourfold rise of the elongation rate was observed.[10] Elongation rate on most cellular mRNAs was shown to be reduced in *Drosophila* tissue culture cells upon heat shock,[11] as well as in HeLa cells treated with amino acid analogs.[12] Intraperitoneal glucagon administration produced temporary inhibition of the elongation rate in rat liver.[13]

It should be mentioned that the elongation rate in general can also be regulated by nonspecific factors, such as ionic conditions or temperature. In particular, the concentration of K^+ was found to be important specifically for elongation: reduced K^+ concentration inhibited predominantly the elongation stage but not the initiation in reticulocyte cells and cell extracts.[14] In all cases the elongation strongly depends on the temperature.[15] The uniform temperature dependence was found for the elongation rate in the case of mouse cells with an apparent activation energy of 16 kCal/mol.[16] In rabbit reticulocytes, however, elongation exhibited two different apparent activation energies; 35 kCal/mol at 10 to 25°C, and 12 kCal/mol at 25 to 40°C.[17]

13.3.2. Elongation Rates Can Be Differentially Affected on Different mRNAs

The elongation rate is not necessarily the same on different mRNAs. Moreover, elongation on specific mRNAs can be regulated selectively. For example, in the cockerel liver one day after estradiol injection the elongation rate for vitellogenin was reported to be about nine amino acids per ribosome per second while that for serum albumin was just about two amino acids per second, as compared with the elongation rate of average cellular polypeptides of about five amino acids per second.[18] A very selective effect was observed in cultured hepatoma cells: their exposure to dibutyryl cyclic AMP resulted in no change in the elongation rate of total protein (about two amino acids per ribosome per second) and at the same time a significant stimulation of the elongation rate for tyrosine aminotransferase (to ten amino acids per second).[19] In rat liver the elongation rate for ornithine aminotransferase was shown to be stimulated by the administration of a high-protein diet, while the elongation rate for total protein was somewhat reduced.[20] During heat shock of chicken reticulocytes the synthesis of the 70 kDa heat shock protein (hSP 70), but not globin, increased several times due to specific increase of the elongation rate of the hSP 70 nascent peptide.[21]

Thus, differential effects on reading different mRNAs can be observed, and specific mRNAs can be regulated selectively. In some cases the effects can be opposite for different mRNAs of the same cell.

13.3.3. Regulatable Pauses Occur During Elongation

The elongation rate may also not be uniform along the same mRNA chain, not only on different mRNAs. In other words, the ribosome can move with different speed at different sites within mRNA. The first indications of the nonconstancy of the elongation rate were observations of discontinuities or pauses during translation in various eukaryotic cells: the size distributions of growing nascent peptides analyzed by gel electrophoresis or gel filtration were found to be not smoothly continuous but more or less discrete[22-28] (see also the review in Reference 29).

It seems that two classes of pauses can be distinguished: short transient retardations of ribosomes at some rare codons (modulating codons) or other local elements of mRNA structure, and longer arrests of ribosomes at several specific sites of mRNA (structural barriers). An example of the latter is the translational barrier in encephalomyocarditis virus RNA causing a significant delay of expression of the coding sequence after that of the polypeptide preA (precursor of capsid protein) and before that of preF and preC, i.e., somewhere in the middle of the RNA.[30] Two relatively long pauses in elongation were found during translation of bovine preprolactin mRNA: one occurred after 75 amino acids had been polymerized, and the other at 160 amino acids.[3] In the last case an elegant new method was proposed to detect pausing of ribosomes on mRNA (Figure 5).

The average elongation rate on a mRNA can be regulated through changes in the number and duration of ribosomal pauses. In the case of pausing on modulating codons the alterations of minor isoacceptor tRNA concentrations may be such a regulating influence (see below). Pauses during translation of globin mRNA and tobacco mosaic virus RNA in wheat germ cell-free system could be reduced by increased spermidine concentration, thus indicating that the duration of pauses can be regulated by polyamines.[27] The translational barrier in encephalomyocarditis virus RNA was found to be overcome by increased concentration of elongation factor 2.[30] The pause on preprolactin mRNA at the 75th amino acid is enhanced by the signal recognition particle (SRP), and the arrest is released upon interaction with microsomal membrane.[3] The inhibitory effect of low glucose on the synthesis of insulin in rat pancreatic islets was demonstrated to be accompanied by increased ribosome pausing after approximately 70 amino acids of preproinsulin had been synthesized; this suggested the inhibition of translation to be at the level of elongation and SRP to play a role in the regulatory event.[31]

13.4. OVERALL ELONGATION RATE CAN BE MODULATED THROUGH CHANGES IN THE ACTIVITY OF THE ELONGATION FACTORS

In principle, overall elongation rate can be regulated through changes in the concentration of aminoacyl-tRNA, or modifications of ribosomes and elongation factors, as well as through changes in GTP concentration.

13.4.1. Do Aminoacyl-tRNA and GTP Concentrations Control the Elongation Rate?

Theoretically amino acids starvation or inhibition of aminoacyl-tRNA synthetase activities should reduce concentrations of corresponding aminoacyl-tRNAs and thus result in slowing down the elongation. Indeed there is evidence that starvation for certain amino acids inhibits elongation.[32] Also the phenomenon of tRNA adaptation (adaptation of relative tRNAs abundances to the frequence of usage of respective codons in particular cell or tissue) is known[33-35] indicating that definite levels of the concentrations of different aminoacyl-tRNAs may be important for maintaining a high elongation rate. At the same time, however, it has been reported that the decrease in the extent of aminoacylation of certain elongator tRNA causes the inhibition of initiation rather than elongation. For example, when the degree of charging of tRNA^{His} or tRNA^{Ile} in HeLa cells was reduced by 20% with two specific

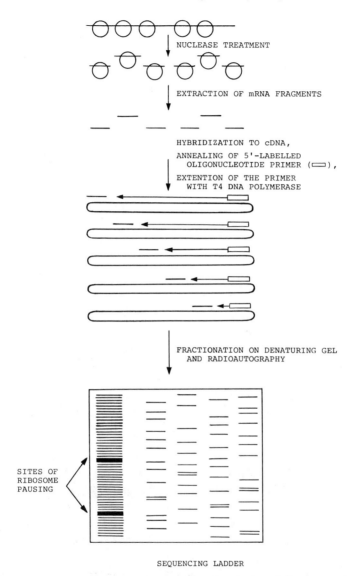

NUCLEASE TREATMENT

EXTRACTION OF mRNA FRAGMENTS

HYBRIDIZATION TO cDNA,

ANNEALING OF 5'-LABELLED
OLIGONUCLEOTIDE PRIMER (▭),

EXTENTION OF THE PRIMER
WITH T4 DNA POLYMERASE

FRACTIONATION ON DENATURING GEL
AND RADIOAUTOGRAPHY

SITES OF
RIBOSOME
PAUSING

SEQUENCING LADDER

Figure 5. Schematic representation of the method for determination of ribosome pausing developed by Wolin and Walter (Modified from Wolin SL and Walter P: *Embo. J.* 7:3559 (1988).) Polyribosomes are treated with a ribonuclease to digest those mRNA sequences which are not shielded by ribosomes; mRNA section covered by ribosomes are protected. As pause sites are statistically more likely to contain a ribosome than nonpausing sites, there will be relatively more identical mRNA fragments corresponding to pause sites. The fragments are purified and hybridized to DNA complementary to the original mRNA (antisense cDNA). A labeled oligonucleotide primer is also annealed to the cDNA, upstream from the bound mRNA fragment. T4 DNA polymerase is used to extend the primer and generate a transcript stretching to the 5'-end of the bound mRNA fragment. Gel electrophoresis fractionation reveals the labeled products of the primer extension; enhanced bands correspond to the pause sites. By using the same primer in a sequencing procedure, the 5'-ends of the ribosome-protected fragments can be resolved with single nucleotide precision.

inhibitors of amino acid activation, histidinol and *O*-methylthreonine, a four-fold reduction in the rate of initiation was observed, while the elongation rate was reduced only by a factor of 1.5 to 2.[36] Similarly, the omission of any of the 13 essential amino acids from the medium upon incubation of Ehrlich ascites tumor cells caused stronger inhibition of initiation than elongation.[37] When mutants of Chinese hamster ovary cells that are temperature-sensitive for certain aminoacyl-tRNA synthetases were incubated at the nonpermissive temperature inhibition of initiation was observed due to increased phosphorylation of eIF-2 (see Part II, Chapter 12 of this book). Thus changes in the ratio of aminoacylated tRNA to deacylated tRNA does not ultimatively serve for regulation of the elongation rate.

GTP to GDP ratio is important for maintaining a high rate of elongation, but changes of this ratio in the cell also affects initiation much stronger than elongation.[38]

13.4.2. Is the Ribosome a Target for Physiological Regulation of Elongation?

The ribosome seems to be a good candidate for a target of regulatory effects on elongation. Many antibiotics inhibiting eukaryotic translation act by binding to ribosomes and blocking one of the steps of the elongation cycle (see Part III, Chapter 15 of this book). Plant toxins, such as ricin, abrin, modeccin, and bacterial toxin sarcin, are known to inhibit elongation by a local damage of ribosomal RNA. There are no indications, however, that modifications of the ribosome activity are involved in physiological mechanisms of the elongation rate regulation.

At the same time, natural modifications of ribosomes, e.g., methylation of ribosomal RNA, and methylation and phosphorylation of ribosomal proteins have been reported (see Part I, Chapter 1 of this book). Unfortunately, the effects of most of these modifications on the activity of ribosomes in elongation were not investigated. In the case of the ribosomal protein S6 which is the most prominent phosphorylatable ribosomal protein no effect on poly(U)-directed translation was found as a result of phosphorylation of this protein with cyclic AMP-dependent protein kinase (Reference 39 and Part II, Chapter 9 of this book). Several other protein kinases have been identified which also can phosphorylate protein S6 at different sites; it is not excluded that phosphorylation of protein S6 by some of them may affect ribosome activity in elongation.

Thus, while aminoacyl-tRNAs, ribosomes and GTP can be in principle involved in the regulation of the elongation rate there is no convincing evidence of the role of these components in the modulation of the elongation rate.

13.4.3. Elongation Factors Activities Can Be Regulated

The situation is found to be quite different for the two elongation factors, EF-1 and EF-2. In most cases where components responsible for the regulation of the elongation rate were examined, the elongation factors, either EF-1 or EF-2, were detected.

EF-1—The separation of ribosomes from elongation factors with high speed centrifugation and the subsequent analysis of the activity of the two fractions in the poly(U)-directed system of poly(Phe) synthesis has demonstrated that EF-1 activity can vary depending on cultivating conditions, hormone action, and aging. The changes in the overall protein synthesis rate have been demonstrated to correlate with the EF-1 activity. This was the case in toad-fish liver during cold acclimation,[40, 41] in cultivated mammalian cells depending on the presence of serum in the medium,[42, 43] in rat liver after regeneration[44] or thyroidectomy,[41, 45] in sea urchin eggs as a result of fertilization,[46] in fungus *Mucor racemosus* during the course of spore germination,[47] in rat spleen during the immune response,[48] as well as during aging in rat liver,[49] in *Drosophila*,[50] and in human cell lines.[51] Despite all these indications of the involvement of EF-1 in regulation of the elongation rate the nature of EF-1 modification was determined in no case.

At the same time several kinds of modifications of EF-1 are known. First of all, EF-1 can be isolated from eukaryotic cells in various forms: "light" form consisting of only 50 kDa polypeptide (EF-1α), and two types of "heavy" form—one consisting of three polypeptides of 50, 48, and 32 kDa (EF-1α, EF-1γ and EF-1β, see Part I, Chapter 6 of this book) and the other which is the complex of the same three polypeptides with valyl-tRNA synthetase.[52] The relative proportion of the "light" and "heavy" forms in the cell is different depending on the type of cells and is affected by cultivating conditions, hormones, and other factors. A correlation between changes in the rate of protein synthesis and conversion of one EF-1 form to the other was observed in some cases, such as aging of nematoda *Turbatrix*,[53] development of brain in chicken,[54] or in rat,[55] during germination of wheat seeds,[56] during development of *Artemia salina*,[57] upon cold acclimation in toad-fish liver.[41] The mechanism controlling the interconversion of various forms of EF-1 is unknown; posttranslational covalent modifications can be involved.

EF-1α can be methylated at several lysine residues, which was demonstrated in fungus *Mucor racemosus*,[58] in mouse 3T3B cells,[59] and in brine shrimp *Artemia salina*.[60] In the case of *Mucor* the rise of EF-1α methylation correlated with the increase in the overall elongation rate upon yeast-to-hyphae transition,[10] and the reduction of EF-1α methylation correlated with the decrease in EF-1 activity during spore germination.[47]

Possibilities of phosphorylation of all three subunits of EF-1 have been demonstrated. Phosphorylation of EF-1α *in vitro* in polyribosome fraction of rabbit reticulocytes was reported[61] but the effect of this phosphorylation on the activity was not studied. A significant portion of EF-1β purified from *Artemia* was shown to be present in a phosphorylated form.[62] In addition, EF-1β was capable of being phosphorylated *in vitro* by endogenous protein kinase present in highly purified EF-1.[62] Similarly, EF-1 preparation from wheat embryos had intrinsic protein kinase activity which could phosphorylate β-subunit.[63] The effect of EF-1β phosphorylation on EF-1 activity is unclear; both stimulation[63] and inhibition[62] were reported. Also, EF-1γ was found to be a substrate for maturation promoting factor protein kinase in *Xenopus* oocytes.[64]

Finally, EF-1α from various mammalian cells and tissues has been shown to be ethanolaminated at two specific glutamic acid residues.[65] The role of this modification in the EF-1 functioning in protein synthesis is obscure.

Thus, while there is evidence that EF-1 activity can be involved in the regulation of the elongation rate, and several covalent modifications of EF-1 are demonstrated, no convincing information clarifying the interrelationship between these two groups of facts is available.

EF-2—Concerning the other elongation factor, EF-2, several covalent modifications of the factor were also reported, at least some of them being important for the regulation of the elongation rate. The most known is the modification of histidine 714 into diphthamide. Diphthamide is a unique derivative of histidine which is specifically present in EF-2 from all eukaryotic organisms. At least five enzymes participate in this modification (for detailed description of diphthamide structure and biosynthesis see Reference 66). Mutations in some of these enzymes have no effect on cell viability, suggesting that the transformation of histidine 714 into diphthamide is not strictly compulsory for the EF-2 function. These mutations, however, affected late steps of diphthamide biosynthesis and it is not excluded that at least a partial modification of His 714 is important for EF-2 functioning. In any case, the substitutions of other amino acids for histidine 714 by means of site-directed mutagenesis abolished EF-2 activity.[67]

EF-2 can be ADP-ribosylated at the diphthamide residue by diphtheria and *Pseudomonas* toxins. ADP-ribosylated EF-2 is inactive in translation. An intriguing question is if a cellular enzyme which can regulate EF-2 activity through ADP-ribosylation exists. Such an enzymatic activity in animal cells has been reported.[68-70] It is not certain, however, that in these reports ^{32}P-incorporation into EF-2 catalyzed by cellular extracts was due to ADP-ribosylation but not due to phosphorylation by a protein kinase (see below).

Phosphorylation represents another covalent modification of EF-2. In various mammalian tissue extracts, e.g., extracts from rabbit reticulocytes and rat liver, EF-2 is the major phosphorylatable protein[71] (see Figure 6). Phosphorylation of EF-2 is catalyzed by a specific Ca^{2+}/calmodulin-dependent protein kinase III[72,73] called also EF-2 kinase.[74] The EF-2 kinase was found in virtually all mammalian protein-synthesizing cells and tissues. The only identified substrate of the kinase was EF-2. The kinase phosphorylates threonine residues located in the N-terminal part of EF-2.[75]

Phosphorylation of EF-2 makes it inactive in protein synthesis. The rate of poly(Phe) synthesis in poly(U)-directed cell-free translation system was found to be dependent on the fraction of nonphosphorylated EF-2[73, 76, 77] (Figure 7). cAMP-dependent activation of globin mRNA translation in rabbit reticulocyte cell-free translation system was shown to correlate with dephosphorylation of EF-2.[78] Inhibition of globin mRNA translation in the same system by okadaic acid was found to correlate with the increase of the phosphorylated fraction of EF-2.[79] It has been demonstrated that phosphorylated EF-2 can form complexes with GTP and ribosomes but is unable to catalyze translocation of peptidyl-tRNA from the ribosomal A-site to the P-site.[80]

Thus, it is likely that changes of EF-2 phosphorylation *in vivo* under

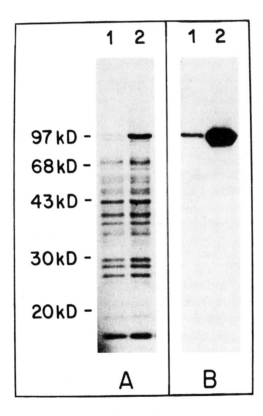

Figure 6. Phosphorylation of EF-2 in the ribosome-free extract from rat liver. (A) Coomassie blue-stained gel after electrophoresis of proteins of the ribosome-free extract incubated with [γ-^{32}P]ATP without (1) or with (2) the addition of purified rabbit reticulocyte EF-2. (B) Radioautograph of the same gel. (From Ryazanov AG: *FEBS Lett.* 214:331 (1987). With permission.)

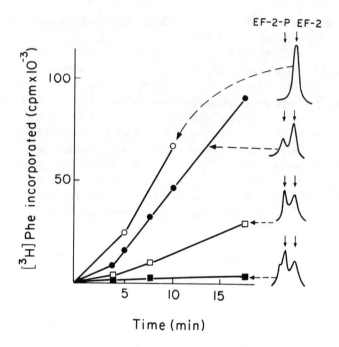

Figure 7. Dependence of polyphenylalanine synthesis kinetics in the poly(U)-directed translation system on the addition of the EF-2 preparations with different proportions of the phosphorylated and nonphosphorylated forms of EF-2. (○) 0:100, (●) 30:70, (□) 50:50, (■) 70:30). The results of scanning the stained gels after isoelectrofocusing of the EF-2 preparations used in the given experiments are shown at the right. (From Ryazanov AG et al., *Nature* 334:170 (1988). With permission.)

different conditions and effectors provide a direct mechanism for the regulation of the elongation rate. As the EF-2 kinase is a Ca^{2+}/calmodulin-dependent enzyme, all hormones and events which increase Ca^{2+} concentration in the cytoplasm must lead to the increase in EF-2 phosphorylation. For example, growth-arrested human fibroblasts respond to mitogenic stimulation with a rapid transient increase in free Ca^{2+} concentration in the cytoplasm as well as with transient increase in the phosphorylation of EF-2.[81] Similar transient increase in EF-2 phosphorylation was demonstrated in the case of thrombin and histamine treatment of the human umbilical vein endothelial cells.[82]

Transient decrease in the EF-2 phosphorylation was observed during differentiation of rat pheochromocytoma cells (PC-12 cells) induced by nerve growth factor.[83–86] The decrease was caused by inactivation of the EF-2 kinase.[84, 85] This example indicates that phosphorylation of EF-2 can be important for the regulation of cell differentiation.

When changes in the level of phosphorylated EF-2 was investigated throughout the cell cycle of transformed human amnion cells (AMA), a dramatic increase in the level of phosphorylated EF-2 was found specifically in mitosis.[87] The increase in EF-2 phosphorylation correlated with the decline in the rate of protein synthesis during cell division.

Hence, phosphorylation of the EF-2 represents the mechanism where the dependence on different effectors and second messengers and the direct effect on translation have been demonstrated.

13.5. SPECIFIC ELONGATION RATES ON DIFFERENT mRNAs MAY BE UNDERLAIN BY INTERCYTOPLASMIC COMPARTMENTATIONS AND RIBOSOME PAUSING

In the previous sections several examples were presented where elongation rate on different mRNAs was different in the same cell at the same time. Moreover, various factors were reported to affect elongation rate differentially on different mRNAs. The mechanisms of the differential regulation of elongation are unknown. Several possible mechanisms, however, can be considered. Experimental evidence supporting some of them is available.

One possibility is that different mRNAs and/or some elongation factors are distributed nonuniformly in the cell. This is in fact the case. It was shown that mRNAs for cytoskeleton proteins in fibroblasts[88] and several mRNAs in *Xenopus* oocytes[89] are distributed in a nonrandom way in the cytoplasm. There is also evidence of nonuniform distribution of elongation factors and aminoacyl-tRNA synthetases in the cytoplasm of various eukaryotic cells (see Reference 90). These proteins possess an affinity for polyribosomes. This dynamic compartmentation of mRNAs, elongation factors, and aminoacyl-tRNA synthetases suggests the possibility that some polyribosomes, together with associated translational machinery proteins, could be physically separated from other polyribosomes in the same cell, providing a mechanism for differential regulation of elongation rate on various mRNAs.

Another possibility which can explain differences in the elongation rate between mRNAs based on the fact that the elongation rate is nonuniform along mRNA. The number and the time of duration of ribosomal pauses are different in various mRNAs and some of these pauses can be more sensitive to regulation than others. This can be a good explanation to the observed differences in the elongation rate between mRNAs.

13.6. MECHANISMS OF RIBOSOME PAUSING AND ITS MODULATION CAN BE DIFFERENT

13.6.1. Ribosomes Can Be Retarded at Modulating Codons

The most often considered explanation of the ribosome pausing is the limitation of certain aminoacyl-tRNAs which can retard ribosome movement at the corresponding codons. There are several facts in favor of this idea. First of all, some synonymous codons are used much more frequently than others. As a rule, highly expressed genes use almost exclusively frequent codons. For example, in the yeast *Saccharomyces cerevisiae* more than 96% of amino acids in highly expressed genes are encoded by a set of 25 codons.[91] A strong correlation exists between codon usage of highly expressed genes and the relative abundance of corresponding tRNAs.[35, 92] Similar adaptation of tRNAs abundancy to the frequency of codon usage was also observed in highly specialized cells, producing predominantly one type of proteins, e.g., in silk glands of *Bombyx mori* producing fibroin,[93, 94] in mammary glands of cows,[95] in mammalian reticulocytes producing hemoglobin.[96] It was suggested from these facts that the rate of ribosome movement along mRNA at different codons can be different depending on the concentration (availability) of corresponding aminoacyl-tRNA. The slowest rate of ribosome movement should be at codons corresponding to minor tRNA species. These codons were called modulating codons.

More experimental evidence in favor of ribosome pausing at modulating codons exists for prokaryotic cells. In the yeast cells it was demonstrated that the replacement of a number of major codons with synonymous minor ones in the coding sequence of phosphoglycerate kinase caused dramatic decline of the expression level.[97] In this case, however, the relationship between the expression level and ribosomal pausing is not clear. The only example where the effect of tRNA set on ribosome pausing was demonstrated is silk fibroin synthesis in heterologous cell-free translation systems: full-length product can be observed only when tRNA from silk gland of *Bombyx mori,* rather than that from heterologous sources, was used.[24, 25] Since in the silk gland there is a proportionality between the intracellular level of different tRNAs and the amino acid composition of fibroin,[93] as well as proportionality between the distribution of isoacceptor tRNAs and the frequencies of corresponding synonymous codons in fibroin mRNA,[94] the limitation of certain aminoacyl-tRNAs when the heterologous tRNA was used is the most evident cause of ribosomal pausing.

13.6.2. Structural Barriers May Exist Along mRNA

Ribosomal pausing in the central region of encephalomyocarditis virus RNA can be overcome by increasing the EF-2 concentration[30] which means that pauses can be caused also by some other reasons. In the case mentioned, the pause occurred despite the saturating amount of EF-2 to catalyze translocation was present. It was suggested that the additional amount of EF-2 was required to read through a peculiar structured region of mRNA.

In the case of globin mRNA translation the distribution of pauses was found to be independent of the tRNA concentrations, as well as of the concentration of the fraction of initiation and elongation factors.[23] Essentially the same pattern of pauses on globin mRNA was found in intact rabbit reticulocytes, reticulocyte cell-free translation system and wheat germ cell-free translation system.[22, 23] It was suggested that mRNA secondary structure induces ribosomal pausing.

It should be mentioned that, in principle, elongation rate can be affected not only by mRNA secondary structure per se, e.g., double-helical hairpins, but also by mRNA-associated proteins which may hinder the elongation at specific sites either directly, or by stabilizing a local secondary or tertiary structure of mRNA.

13.6.3. Structure of Nascent Peptide May Serve as a Signal for the Elongation Pause

During translation of bovine preprolactin mRNA in both wheat germ and rabbit reticulocyte cell-free translation systems two elongational pauses were observed.[3] One pause occurred at the 75th amino acid and another at the 160th amino acid. Both amino acids were found to be glycines coded by GGC codons. Interestingly, two other GGC codons are in preprolactin mRNA, but ribosomes passed these codons without pausing. This ruled out the possibility that the pauses were induced by limitations in the aminoacyl-tRNA. It is important that the addition of SRP increased ribosome pausing at glycine 75 but not at glycine 160[3, 98] which means that ribosome pausing can be affected through nascent signal peptide. This raises an intriguing possibility that discontinuities in elongation may be generally caused also by nascent peptide structure and modulated by a peptide-recognizing protein or particle.

13.7. CAN THE REGULATION AT THE ELONGATION STAGE RESULT IN TRANSLATIONAL DISCRIMINATION OF DIFFERENT mRNAs?

It is clear from the phenomena described in this chapter that the rate of elongation on all or specific cellular mRNAs can be stimulated or inhibited depending on different conditions and effectors. The physiological role of these changes of the elongation rate, however, is not obvious. Certainly, in some cases, e.g., tyrosine aminotransferase induction by cAMP,[19] changes in the elongation rate serve for increasing or decreasing the protein production. But it is difficult to imagine that just 1.5 to 2-fold changes in the elongation rate described in most cases are needed for this purpose.

It is known that in many cases where changes in the elongation rate were reported qualitative changes in the spectrum of synthesized proteins were observed. Some facts and considerations suggest that changes in the overall elongation rate may differentially affect the availability of different mRNAs for translation.

In the simplest case when several mRNAs are translated simultaneously and the rate-limiting stage is initiation, the rate of protein synthesis on these mRNA will be determined by respective initiation rates. When the elongation stage becomes rate limiting, the efficiency of translation of each mRNA will be determined by the relative amounts of mRNAs. This was demonstrated in the case of differential translation of α- and β-globin mRNAs. β-globin mRNA has higher initiation rate than α-globin mRNA. Slowing down the elongation by antibiotics had a stronger inhibitory effect on β-globin mRNA translation than on α-globin mRNA translation, thus resulting in the increase of the α- to β-chain ratio.[99] In this case the reduction of the polypeptide chain elongation led to overcoming the translational discrimination and allowed the protein synthesis to be expressed more as a function of relative mRNA concentrations and less as a function of initiation rate constants.

A more common situation can be encountered when translation is limited by availability of a certain initiation factor, which has different affinities to different mRNAs. In this case it was demonstrated that even moderate reduction of the elongation rate, which makes elongation rate limiting only on ''strong'' mRNAs (mRNAs with high initiation rate), can stimulate translation of ''weak'' mRNAs (mRNAs with low initiation rate) by mobilizing the limiting initiation factor from ''strong'' mRNAs and thus making it available for ''weak'' mRNAs.[100–103] For example, when protein synthesis in mouse fibroblasts was partially inhibited by the elongation inhibitor, cycloheximide, the synthesis of a number of polypeptides was stimulated.[104, 105] The paradoxical effect of cycloheximide can be explained by the fact that under conditions when the elongation rate is reduced ''strong'' messengers can no longer utilize initiation factor molecules so rapidly, leaving more of them available for use by ''weak'' messengers.

The same mechanism can explain the role of the increased EF-2 phosphorylation and decreased rate of protein synthesis during mitosis.[87] During mitosis the cell needs few special proteins which are necessary for cell division and does not need many constitutive proteins which are important for other phases of the cell cycle. Reduction of the elongation rate through EF-2 phosphorylation can serve for the inhibition of translation of most cellular mRNAs for constitutive proteins and thus makes some limiting translational components available for mRNAs specific for mitosis-related proteins.

The modulatory effect of elongation rate on the mRNA discrimination could be important in the case of activation of certain cells and tissues by hormones or as a result of fertilization, when a large amount of new mRNAs are mobilized for translation. If these new mRNAs have very high rate constants for initiation, the sudden demand for their translation could limit the availability of initiation factors and thus favor the synthesis of the new proteins at the expense of constitutive protein synthesis. The estradiol injection into cockerels was demonstrated to cause a rapid increase in the amount of mRNA in the liver, together with the decrease in the overall elongation rate.[6] The most evident explanation for the role of the reduction of the elongation rate in this case is that it could minimize translational discrimination and hence prevent interruption of the syntheses of constitutive proteins when many egg-related proteins begin to be synthesized.

The inhibition of elongation can activate translation of some stored messenger ribonucleic acid protein (mRNP) particles by the same or similar mechanism. Indeed, it was shown that together with relatively stable mRNP particles which are inactive in translation due to the presence of tightly bound repressor, there are other mRNP particles in which mRNA can be mobilized into polyribosomes by treatment of cells with low doses of cycloheximide.[106, 107]

Interestingly, mRNA for EF-1α is relatively weak and the inhibition of elongation by cycloheximide stimulates translation of EF-1α polypeptide.[108] This means that slowing down the elongation rate can automatically stimulate the synthesis of the elongation factor, thus providing a mechanism for autogenously controlled maintenance of a constant concentration of EF-1 in the cytoplasm.

Thus, it is clear that changes in the overall elongation rate can differentially affect translation of different mRNAs, which will provide synthesis of a unique set of proteins in a cell. Since the cell phenotype is determined by the composition of synthesized proteins, regulation of the elongation rate seems to be important for such processes as differentiation, proliferation, and division of eukaryotic cells.

REFERENCES

1. Nielsen PJ and McConkey EH: Evidence for control of protein synthesis in HeLa cells via the elongation rate. *J. Cell. Physiol.* 104:269 (1980).
2. Lodish HF and Jacobsen M: Regulation of hemoglobin synthesis. Equal rates of translation and termination of α- and β-globin chains. *J. Biol. Chem.* 247:3622 (1972).
3. Wolin SL and Walter P: Ribosome pausing and stacking during translation of a eukaryotic mRNA. *EMBO J.* 7:3559 (1988).
4. Denisenko ON: Protein synthesis regulation in the cell-free system at heat shock. *Dokl. Akad. Nauk SSSR* 297:725 (1987).

5. Ilan J, Pierce DR, Hochberg AA, Folman R, and Gyves MT: Increased rates of polypeptide chain elongation in placental explants from human diabetics. *Proc. Natl. Acad. Sci. U.S.A..* 81:1366 (1984).

6. Gehrke L, Bast RE, and Ilan J: An analysis of rates of polypeptide chain elongation in avian liver explants following *in vivo* estrogen treatment. I. Determination of average rates of polypeptide chain elongation. *J. Biol. Chem.* 256:2514 (1981).

7. Ramabhadran TV and Thach RE: Translation elongation rate changes in encephalomyocarditis virus-infected and interferon-treated cells. *J. Virol.* 39:573 (1981).

8. Brandis JW and Raff RA: Elevation of protein synthesis is a complex response to fertilisation. *Nature* 278:467 (1979).

9. Hille MB and Albers AA: Efficiency of protein synthesis after fertilisation of sea urchin eggs. *Nature* 278:469 (1979).

10. Orlowski M. and Sypherd PS: Regulation of translation rate during morphogenesis in the fungus *Mucor. Biochemistry* 17:569 (1978).

11. Ballinger DG and Pardue ML: The control of protein synthesis during heat shock in Drosophila cells involves altered polypeptide elongation rates. *Cell* 33:103 (1983).

12. Thomas GP and Mathews MB: Control of polypeptide chain elongation in the stress response: a novel translation control, in Schlesinger, MJ, Ashburner M and Tissiéres A (Eds): *Heat Shock from Bacteria to Man* Cold Spring Harbor, NY, Cold Spring Laboratory, p. 207 (1982).

13. Ayuso-Parrilla MS, Martín-Requero A, Pérez-Díaz J, and Parrilla R: Role of glucagon on the control of hepatic protein synthesis and degradation in the rat *in vivo. J. Biol. Chem.* 251:7785 (1976).

14. Cahn F and Lubin M: Inhibition of elongation steps of protein synthesis at reduced potassium concentrations in reticulocytes and reticulocyte lysate. *J. Biol. Chem.* 253:7798 (1978).

15. Hunt T, Hunter T, and Munro A: Control of haemoglobin synthesis: rate of translation of the messenger RNA for the α and β chains. *J. Mol. Biol.* 43:123 (1969).

16. Craig N: Effect of reduced temperatures on protein synthesis in mouse L cells. *Cell* 4:329 (1975).

17. Craig N: Regulation of translation in rabbit reticulocytes and mouse L-cells: comparison of the effects of temperature. *J. Cell. Physiol.* 87:157 (1975).

18. Gehrke L, Bast RE, and Ilan J: An analysis of rates of polypeptide chain elongation in avian liver explants following *in vivo estrogen treatment*. II. Determination of the specific rates of elongation of serum albumin and vitellogenin nascent chains. *J. Biol. Chem.* 256:2522 (1981).

19. Roper MD and Wicks WD: Evidence for acceleration of the rate of elongation of tyrosine aminotransferase nascent chains by dibutyryl cyclic AMP. *Proc. Natl. Acad. Sci. U.S.A..* 75:140 (1978).

20. Mueckler MM, Merrill MJ, and Pitot HC: Translational and pretranslational control of ornithine aminotransferase synthesis in rat liver. *J. Biol. Chem.* 258:6109 (1983).

21. Theodorakis NG, Banerji SS, and Morimoto RI: HSP70 mRNA translation in chicken reticulocytes is regulated at the level of elongation. *J. Biol. Chem.* 263:14579 (1988).

22. Protzel A and Morris AJ: Gel chromatographic analysis of nascent globin chains. Evidence of nonuniform size distribution. *J. Biol. Chem.* 249:4594 (1974).

23. Chaney WG and Morris AJ: Nonuniform size distribution of nascent peptides: the role of messenger RNA. *Arch. Biochem. Biophys.* 191:734 (1978).

24. Lizardi PM, Mahdavi V, Shields D, and Candelas G: Discontinuous translation of silk fibroin in a reticulocyte cell-free system and in intact silk gland cells. *Proc. Natl. Acad. Sci. U.S.A..* 67:6211 (1979).

25. Chavancy G, Marbaix G, Huez G, and Cleuter Y: Effect of tRNA pool balanace on rate and uniformity of elongation during translation of fibroin mRNA in a reticulocyte cell-free system. *Biochimie* 63:611 (1981).

26. Wieringa B, van der Zwaag-Gerritsen J, Mulder J, Ab G, and Gruber M: Translation *in vivo* and *in vitro* of mRNAs coding for vitellogenin, serum albumin and very-low-density lipoprotein II from chicken liver. A difference in translational efficiency. *Eur. J. Biochem.* 114:635 (1981).

27. Abraham AK and Pihl A: Variable rate of polypeptide chain elongation *in vitro*. Effect of spermidine. *Eur. J. Biochem.* 106:257 (1980).

28. Candelas G, Candelas T, Ortiz A, and Rodriguez O: Translational pauses during a spider fibroin synthesis. *Biochem. Biophys. Res. Commun* 116:1033 (1983).

29. Candelas GC, Carrasco CE, Dompenciel RE, Arroyo G, and Candelas TM: Strategies of fibroin production. in Ilan J (ed): *Translational Regulation of Gene Expression*, New York, Plenum Press, p. 209 (1987).

30. Svitkin Yu V and Agol VI: Translational barrier in central region of encephalomyocarditis virus genome. Modulation by elongation factor 2 (eEF-2). *Eur. J. Biochem.* 133:145 (1983).

31. Wolin SL and Walter P: Translational regulation of secretory protein synthesis. Abstracts of papers presented at the 1989 Meeting of Translational Control. September 13–17, 1989, Cold Spring Harbor, NY, Cold Spring Harbor Laboratory, 92.

32. Ogilvie A, Huschka U, and Kersten W: Control of protein synthesis in mammalian cells by aminoacylation of transfer ribonucleic acid. *Biochim. Biophys. Acta* 565:293 (1979).

33. Garel J-P: Functional adaptation of tRNA population. *J. Theor. Biol.* 43:211 (1974).

34. Chavancy G, Chevallier A, Fournier A, and Garel J-P: Adaptation of iso-tRNA concentration to mRNA codon frequency in the eukaryote cell. *Biochimie* 61:71 (1979).

35. Ikemura T: Correlation between the abundance of yeast transfer RNAs and the occurrence of the respective codons in protein genes. Differences in synonymous codon choice patterns of yeast and *Escherichia coli* with reference to the abundance of isoaccepting transfer RNAs. *J. Mol. Biol.* 158:573 (1982).

36. Vaughan MH and Hansen BS: Control of initiation of protein synthesis in human cells. Evidence for a role of uncharged transfer ribonucleic acid. *J. Biol. Chem.* 248:7087 (1973).

37. van Venrooij WJW, Henshaw EC, and Hirsch A: Effects of deprival of glucose or individual amino acid on polyribosome distribution and of protein synthesis in cultured mammalian cells. *Biochim. Biophys. Acta* 259:127 (1972).
38. Hucul JA, Henshaw EC, and Young DA: Nucleoside diphosphate regulation of overall rates of protein biosynthesis acting at the level of initiation. *J. Biol. Chem.* 260:15585 (1985).
39. Eil C and Wool IG: Function of phosphorylated ribosomes. The activity of ribosomal subunits phosphorylated *in vitro* by protein kinase. *J. Biol. Chem.* 248:5130 (1973).
40. Haschemeyer AEV: Rates of polypeptide chain assembly in liver *in vivo:* relation to the mechanism of temperature acclimation in *Opsanus tau. Proc. Natl. Acad. Sci. U.S.A..* 62:128 (1969).
41. Nielsen JBK, Plant PW, and Haschemeyer AEV: Polypeptide elongation factor 1 and the control of elongation rate in rate in rat liver *in vivo. Nature* 264:804 (1976).
42. Hassell JA and Engelhardt DL: The regulation of protein synthesis in animal cells by serum factors. *Biochemistry* 15:1375 (1976).
43. Fischer I, Arfin SM, Moldave K: Regulation of translation analysis of intermediary reactions in protein synthesis in exponentially growing and stationary phase Chinese hamster ovary cells in culture. *Biochemistry* 19:1417 (1980).
44. Nicholls DM, Carey J, and Sendecki W: Growth of liver accompanied by an increased binding of aminoacyl-transfer ribonucleic acids. *Biochem. J.* 166:463 (1977).
45. Mathews RW, Oronsky A, and Haschemeyer AEV: Effect of thyroid hormone on polypeptide chain assembly kinetics in liver protein synthesis *in vivo. J. Biol. Chem.* 248:1329 (1973).
46. Felicetti L, Metafora S, Gambino R, and Di Matteo G: Characterization and activity of the elongation factors T1 and T2 in the unfertilized egg and in the early development of sea urchin. *Cell Different.* 1:265 (1972).
47. Fonzi WA, Katayama C, Leathers T, and Sypherd PS: Regulation of protein synthesis factor EF-1α in *Mucor racemosus. Mol. Cell Biol.* 5:1100 (1985).
48. Willis DB and Starr JL: Protein biosynthesis in the spleen. III. Aminoacyltransferase I as a translational regulatory factor during the immune response. *J. Biol. Chem.* 246:2828 (1971).
49. Moldave K, Harris J, Sabo W, and Sadnik I: Protein synthesis and aging: studies with cell-free mammalian systems. *Fed. Proc.* 38:1979 (1979).
50. Webster GC and Webster SL: Specific disappearance of translatable messenger RNA for elongation factor 1 in ageing *Drosophila melanogaster. Mech. Ageing Dev.* 24:335 (1984).
51. Rattan SIS, Cavallius J, and Clark BFC: Heat shock-related decline in activity and amounts of active elongation factor 1α in ageing and immortal human cells. *Biochem. Biophys. Res. Commun.* 152:169 (1988).
52. Motorin Yu A, Wolfson AD, Orlovsky AF, and Gladilin KL: Mammalian valyl-tRNA synthetase forms a complex with the first elongation factor. *FEBS Lett.* 238:262 (1988).
53. Bolla R and Brot N: Age dependent changes in enzymes involved in macromolecular synthesis in *Turbatrix aceti. Arch. Biochem. Biophys.* 169:227 (1975).

54. Murakami K and Miyamoto K: Polypeptide elongation factors of the developing chick brain. *J. Neurochem.* 38:1315 (1982).
55. Vargas R and Castaneda M: Role of elongation factor 1 in the translation control of rodent brain protein synthesis. *J. Neurochem.* 37:687 (1981).
56. Sacchi GA, Zocchi G, and Cocucci S: Changes in form of elongation factor 1 during germination of wheat seeds. *Eu. J. Biochem.* 139:1 (1984).
57. Slobin LI and Möller W: Changes in form of elongation factor during development of *Artemia salina*. *Nature* 258:452 (1975).
58. Hiatt WR, Garcia R, Merrick WC, and Sypherd PS: Methylation of elongation factor 1α from the fungus *Mucor*. *Proc. Natl. Acad. Sci. U.S.A.* 79:3433 (1982).
59. Coppard NJ, Clark BFC, and Cramer F: Methylation of elongation factor 1α in mouse 3T3B and 3T3B/SV40 cells. *FEBS Lett.* 164:330 (1983).
60. Amons R, Pluijms W, Roobol K, and Moller W: Sequence homology between EF-1α, the α-chain of elongation factor 1 from *Artemia salina* and elongation factor EF-Tu from *Escherichia coli*. *FEBS Lett.* 153:37 (1983).
61. Davydova EK, Sitikov AS, Ovchinnikov LP: Phosphorylation of elongation factor 1 in polyribosome fraction of rabbit reticulocytes. *FEBS Lett.* 176:401 (1984).
62. Janssen GMC, Maessen GDF, Amons R, and Moller W: Phosphorylation of elongation factor 1β by an endogenous kinase affects its catalytic nucleotide exchange activity. *J. Biol. Chem.* 263:11063 (1988).
63. Ejiri S and Honda H: Effect of cyclic AMP and cyclic GMP on the autophosphorylation of elongation factor 1 from wheat embryos. *Biochem. Biophsy. Res. Commun.* 128:53 (1985).
64. Bellé R, Derancourt J, Poulhe R, Capony J-P, Ozon R, and Mulner-Lorillon O: A purified complex from *Xenopus* oocytes contains a p47 protein, an *in vivo* substrate of MPF, and a p30 protein respectively homologous to elongation factors EF-1γ and EF-1β. *FEBS Lett.* 255:101 (1989).
65. Whiteheart SW, Shenbagamurthi P, Chen L, Cotter RJ, and Hart GW: Murine elongation factor 1α (EF-1α) is posttranslationally modified by novel amide-linked ethanolamine-phosphoglycerol moieties. Addition of ethanolamine-phosphoglycerol to specific glutamic acid residues on EF-1α. *J. Biol. Chem.* 264:14334 (1989).
66. Bodley JW and Veldman SA: Biosynthesis of diphthamide: The ADP-Ribose acceptor for diphtheria toxin. In Moss J, Vaughan M (eds): ADP-Ribosilating Toxins and G-Proteins. ASM, Washington, D.C. 21–30 (1990).
67. Omura F, Konho K, and Uchida T: The histidine residue of codon 715 is essential for function of elongation factor 2. *Eur. J. Biochem.* 180:1 (1989).
68. Lee H and Iglewski WJ: Cellular ADP-ribosyltransferase with the same mechanism of action as diphtheria toxin and *Pseudomonas* toxin A. *Proc. Natl. Acad. Sci. U.S.A.* 81:2703 (1984).
69. Sitikov AS, Davydova EK, Ovchinnikov LP: Endogenous ADP-ribosylation of elongation factor 2 in polyribosome fraction of rabbit reticulocytes. *FEBS Lett.* 176:261 (1984).
70. Fendrick JL and Iglewski WJ: Endogenous ADP-ribosylation of elongation factor 2 in polyoma virus-transformed baby hamster kidney cells. *Proc. Natl. Acad. Sci. U.S.A.* 86:554 (1989).

71. Ryazanov AG: Ca^{2+}/calmodulin-dependent phosphorylation of elongation factor 2. *FEBS Lett.* 214:331 (1987).

72. Nairn AC, Bhagat B, and Palfrey HC: Identification of calmodulin-dependent protein kinase III and its major M_r 100,000 substrate in mammalian tissues. *Proc. Natl. Acad. Sci. U.S.A.* 82:7939 (1985).

73. Nairn AC and Palfrey HC: Identification of the major M_r 100,000 substrate for calmodulin-dependent protein kinase III in mammalian cells as elongation factor-2. *J. Biol. Chem.* 262:17299 (1987).

74. Ryazanov AG, Natapov PG, Shestakova EA, Severin FF, and Spirin AS: Phosphorylation of the elongation factor 2: the fifth Ca^{2+}/calmodulin-dependent system of protein phosphorylation. *Biochimie* 70:619 (1988).

75. Ovchinnikov LP, Motuz LP, Natapov PG, Averbuch LJ, Wettenhall REH, Szyzka R, Kramer G, and Hardesty B: Three phosphorylation sites in elongation factor 2. *FEBS Lett.* 275:209 (1990).

76. Shestakova EA and Ryazanov AG: Influence of elongation factor 2 phosphorylation on its activity in the cell-free translation system. *Dokl. Akad. Nauk SSSR* 297:1495 (1987).

77. Ryazanov AG, Shestakova EA, and Natapov PG: Phosphorylation of elongation factor 2 by EF-2 kinase affects rate of translation. *Nature* 334:170 (1988).

78. Sitikov AS, Simonenko PN, Shestakova EA, Ryazanov AG, and Ovchinnikov LP: cAMP-dependent activation of protein synthesis correlates with dephosphorytion of elongation factor 2. *FEBS Lett.* 228:327 (1988).

79. Redpath NT and Proud CG: The tumour promoter okadaic acid inhibits reticulocyte-lysate protein synthesis by increasing the net phosphorylation of elongation factor 2. *Biochem. J.* 262:69 (1989).

80. Ryazanov AG and Davydova EK: Mechanism of elongation factor 2 (EF-2) inactivation upon phosphorylation. Phosphorylated EF-2 is unable to catalyze translocation. *FEBS Lett.* 251:187 (1989).

81. Palfrey HC, Nairn AC, Muldoon LL, and Villereal ML: Rapid activation of calmodulin-dependent protein kinase III in mitogen-stimulated human fibroblasts. Correlation with intracellular Ca^{2+} transients. *J. Biol. Chem.* 262:9785 (1987).

82. Mackie KP, Nairn AC, Hampel G, Lam G, and Jaffe EA: Thrombin and histamine stimulate the phosphorylation of elongation factor 2 in human umbilical vein endothelial cells. *J. Biol. Chem.* 264:1748 (1989).

83. End D, Hanson M, Hashimoto S, and Guroff G: Inhibition of the phosphorylation of a 100,000-dalton soluble protein in whole cells and cell-free extracts of PC12 pheochromocytoma cells following treatment with nerve growth factor. *J. Biol. Chem.* 257:9223 (1982).

84. Togari A and Guroff G: Partial purification and characterization of a nerve growth factor-sensitive kinase and its substrate from PC12 cells. *J. Biol. Chem.* 260:3804 (1985).

85. Nairn AC, Nicholls RA, Brady MJ, and Palfrey HC: Nerve growth factor treatment or cAMP elevation reduces Ca^{2+}/calmodulin-dependent protein kinase III activity in PC12 cells. *J. Biol. Chem.* 262:14265 (1987).

86. Koizumi S, Ryazanov A, Hama T, Chen H-C, Guroff G: Identification of Nsp 100 as elongation factor 2 (EF-2). *FEBS Lett.* 253:55 (1989).

87. Celis JE, Madsen P, Ryazanov AG: Increased phosphorylation variants of elongation factor 2 during mitosis in transformed human amnion cells correlates with a decreased rate of protein synthesis. *Proc. Natl. Acad. Sci. U.S.A.* 87:4231 (1990).

88. Lawrence JB and Singer RH: Intracellular localization of messenger RNAs for cytoskeletal proteins. *Cell* 45:407 (1986).

89. Rebagliati MR, Weeks DL, Harvey RP, and Melton DA: Identification and cloning of localized maternal RNAs from *Xenopus* eggs. *Cell* 42:769 (1985).

90. Ryazanov AG, Ovchinnikov LP, and Spirin AS: Development of structural organization of protein-synthesizing machinery from prokaryotes to eukaryotes. *Biosystems* 20:275 (1987).

91. Bennetzen JL and Hall BD: Codon selection in yeast. *J. Biol. Chem.* 257:3026 (1982).

92. De Boer HA and Kastelein RA: Biased codon usage: an exploration of its role in optimization of translation, in Davis JJ, Reznikoff B, and Gold L (Eds): *From Gene to Protein: Steps Dictating the Maximal Level of Gene Expression*, New York, Butterworth.

93. Garel J-P, Mandel P, Chavancy G, and Daillie J: Functional adaptation of t-RNAs to protein biosynthesis in a highly differentiated cellular system. V. Fractionation by countercurrent distribution of t-RNAs of sericigen gland from *Bombyx mori L. Biochimie* 53:1195 (1971).

94. Garel J-P: Quantitative adaptation of isoacceptor tRNAs to mRNA codons of alanine, glycine and serine. *Nature* 260:805 (1976).

95. Elska A, Matsuka G, Matiash U, Nasarenko I, and Semenova N: tRNA and aminoacyl-tRNA synthetases during differentiation and various functional states of the mammary gland. *Biochim. Biophys. Acta* 247:430 (1971).

96. Smith DWE: Reticulocyte transfer RNA and hemoglobin synthesis. *Science* 190:529 (1975).

97. Hoekema A, Kastelein RA, Vasser M, and de Boer HA: Codon replacement in the PGK1 gene of *Saccharomyces cerevisiae:* experimental approach to study the role of biased codon usage in gene expression. *Mol. Cell. Biol.* 7:2914 (1987).

98. Wolin SL and Walter P: Signal recognition particle mediates a transient elongation arrest of preprolactin in reticulocyte lysate. *J. Cell Biol.* 109:2617 (1989b).

99. Lodish HF: Alpha and beta globin messenger ribonucleic acid. Different amounts and rates of initiation of translation. *J. Biol. Chem.* 246:7131 (1971).

100. Walden WE, Godefroy-Colburn T, and Thach RE: The role of mRNA competition in regulating translation. I. Demonstration of competition *in vivo*. *J. Biol. Chem.* 256:11739 (1981).

101. Brendler T, Godefroy-Colburn T, Carlill RD, and Thach RE: The role of mRNA competition in regulating translation. II. Development of a quantitative *in vitro* assay. *J. Biol. Chem.* 256:11747 (1981).

102. Brendler T, Godefroy-Colburn T, Yu S, and Thach RE: The role of mRNA competition in regulating translation. III. Comparison of *in vitro* and *in vivo* results. *J. Biol. Chem.* 256:11755 (1981).

103. Godefroy-Colburn T and Thach RE: The role of mRNA competition in regulating translation. IV. Kinetic model. *J. Biol. Chem.* 256:11762 (1981).

104. Ray A, Walden WE, Brendler T, Zenger VE, and Thach RE: Effect of medium hypertonicity on reovirus translation rates. An application of kinetic modelling *in vivo. Biochemistry* 24:7525 (1985).

105. Walden WE and Thach RE: Translational control of gene expression in a normal fibroblast. Characterization of a subclass of mRNAs with unusual kinetic properties. *Biochemistry* 25:2033 (1986).

106. Lee GT-Y and Engelhardt DL: Peptide coding capacity of polysomal and non-polysomal messenger RNA during growth of animal cells. *J. Mol. Biol.* 129:221 (1979).

107. Bergmann IE, Cereghini S, Georghegan T, and Brawerman G: Functional characteristics of untranslated messenger ribonucleoprotein particles from mouse sarcoma ascites cells. Possible relation to the control of messenger RNA utilization *J. Mol. Biol.* 156:567 (1982).

108. Slobin LI and Jordan P: Translational repression of mRNA for eucaryotic elongation factors in friend erythroleukemia cells. *Eur. J. Biochem.* 145:143 (1984).

109. Wu RS and Warner JR: Cytoplasmic synthesis of nuclear proteins. Kinetics of accumulation of radioactive proteins in various cell fractions after brief pulses. *J. Cell Biol.* 51:643 (1971).

Appendix

Comparison of Prokaryotic and Eukaryotic Translation

John W.B. Hershey
Department of Biological Chemistry
School of Medicine
University of California
Davis, California

14.1. INTRODUCTION

Studies of the process of protein synthesis began in the 1950s with the identification of a number of key components. Ribosomes, whereas observed earlier as particles in a variety of cells, were first shown to function in protein synthesis in rat liver homogenates by Zamecnik's group.[1] Discoveries of aminoacyl-tRNA synthetases[2] and tRNAs[3] were reported soon after. The first soluble factors implicated in translation were identified in rat liver by Fessender and Moldave[4] and in rabbit reticulocytes by Bishop and Schweet.[5] Thus the earliest pioneering work in the translation field involved primarily mammalian cells. In the beginning of the 1960s, three major events caused investigators to regard bacteria as a suitable organism for studying translation. The concept of mRNA in bacteria was clearly enunciated by Jacob and Monod,[6] who demonstrated the power of bacterial genetics for addressing problems of gene expression. Nirenberg and Matthaei[7] showed that *E. coli* lysates synthesize poly-phenylalanine when programmed with the synthetic oligonucleotide, poly(U). Capecchi and Gussin[8] then found that bacterial lysates can synthesize *in vitro* a "natural" protein from a viral RNA template. Interest in solving the genetic code, and the realization that protein and RNA components of protein synthesis could be purified from bacterial cells as demonstrated in Lipmann's laboratory[9] stimulated studies of the mechanism of translation in prokaryotes. Only later in the 1970s did the major focus of the field shift back to mammalian cells, in large part due to expectations of finding translational control mechanisms in eukaryotic cells.

As details of the process of translation in prokaryotic and eukaryotic organisms emerged, striking similarities between the two systems became apparent. Identical genetic codes are employed, the machinery is comparable though not identical, and the pathways are remarkably the same, at least at a superficial level. Only the process of initiation of protein synthesis differs significantly, involving a more complex set of components and a fundamentally different strategy. Since a detailed understanding of protein synthesis at the molecular level remains elusive in either system, a comparison of these similar yet slightly different systems may provide insights in both mechanisms. It is this idea that motivates drawing the comparisons described in this chapter.

14.2. PATHWAY OF PROTEIN SYNTHESIS IN BACTERIA

14.2.1. Overview

Protein synthesis in prokaryotic organisms proceeds along the same general pathway as that in eukaryotes: synthesis initiates at a precise site in the mRNA sequence, elongation follows, and a termination step specified by the mRNA releases the completed peptide chain. Ribosomes initiate as subunits, are found in polysomes during their active phase, and terminate protein synthesis as 70S particles in dynamic equilibrium with their subunits. The three phases of translation are promoted by three distinct sets of soluble factors. One of the major differences involves the informational content of mRNAs. In prokaryotes many mRNAs are polycistronic, possessing a number of translational start and stop signals whereas in eukaryotes, most appear to direct the synthesis of a single protein product. The strategy and mechanism of initiation of protein synthesis are very different, whereas those for elongation and termination are quite comparable for the two systems. We shall review briefly the pathway for bacteria, then compare macromolecular components, pathway and mechanisms with those in eukaryotic cells. A working model of the pathway of protein synthesis in bacteria is shown in Figure 1. The reader should consult available reviews of translation in prokaryotes for further details.[10-12]

14.2.2. Initiation

The process of initiation selects a mRNA for translation by the ribosome, recognizes an initiator codon amongst all other similar codons found in-frame or out-of-frame, and precisely phases translation of the mRNA. To accomplish

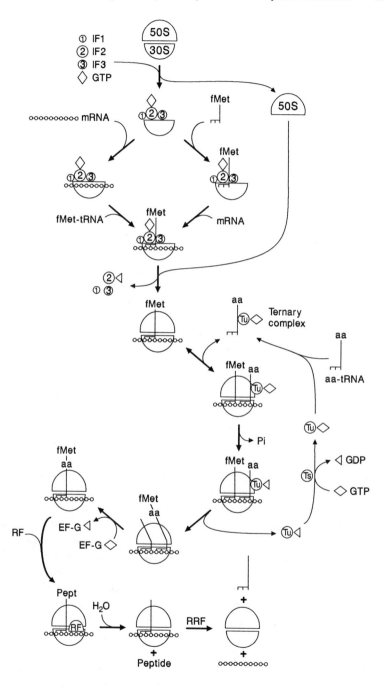

Figure 1. Pathway of protein synthesis in *E. coli*.

this, ribosomes dissociate into subunits and a pre-initiation complex forms containing the 30S ribosomal subunit, a mRNA and the initiator tRNA, fMet-tRNA$_f$. Junction of the 50S subunit to form a 70S initiation complex results in GTP hydrolysis and the ejection of initiation factors. The most frequently used initiator codon is AUG, although GUG is used for ca. 10% of initiation events, and a number of other codons function in rarer instances. To distinguish the proper AUG, additional information is needed. Most important are accessibility to the mRNA binding site (lack of secondary structure) and an interaction between the 16S ribosomal RNA (rRNA) in the 30S ribosomal subunit and a purine-rich sequence upstream from the initiator codon, called the Shine/Dalgarno (S/D) sequence. Evidence suggests that other sequences in a 30 b region of the mRNA surrounding the AUG also may contribute to the recognition process. RNA-RNA interactions between the mRNA and both 16S rRNA and the anticodon of the bound fMet-tRNA$_f$ are sufficient in most cases to select and phase the mRNA.[13]

Because the major difference between prokaryotic and eukaryotic protein synthesis concerns the initiation phase, a more detailed development of the various sub-reactions of initiation is provided (see Figure 1). The initiation pathway begins with the dissociation of 70S particles into 30S and 50S ribosomal subunits. A dynamic equilibrium of 70S particles and subunits exists which is sensitive to the Mg^{++} concentration. At physiological concentrations (near 5 mM), pure ribosomes are primarily in the associated state, whereas below 1 mM Mg^{++}, they are mostly dissociated. Initiation factor IF1 enhances the rates of both dissociation and association, but does not significantly alter the position of equilibrium. In contrast, IF3 when bound to the 30S subunit greatly decreases the rate of association, leading to dissociation of 70S ribosomes. Bacteria growing in exponential phase in rich medium contain more ribosomal subunits than free 70S particles (however, note that >85% of total ribosomes are found as functioning 70S particles in polysomes). The dissociation of 70S ribosomes into subunits occurs at every round of protein synthesis and is an obligatory step for initiation.

Formation of 30S preinitiation complexes is promoted by three initiation factors: IF1 (8 kDa); IF2 (97 kDa); and IF3 (20 kDa) (see Table 1 and sections below). All three initiation factors bind to 30S ribosomal subunits to form a stable tri-factor complex, which likely is an intermediate in the pathway.[14] The 30S preinitiation complex contains the initiation factors, fMet-tRNA$_f$, GTP, and mRNA, but the order of addition of these components is not clear. Binary complexes of either mRNA or fMet-tRNA$_f$ with 30S subunits are not stable and cannot be isolated easily. However, evidence has been obtained for mRNA-30S complexes that can be chased into stable 70S initiation complexes,[15] whereas other experiments suggest that fMet-tRNA$_f$-30S complexes form before mRNA binding.[16] Kinetic studies led Gualerzi et al.[17] to propose a random-order mechanism, whereas Ellis and Conway[18] claim that fMet-

TABLE 14–1.
Soluble Factors in *Escherichia coli*

Factor	Mass	No. of Amino Acids	Gene	Cell Level: Factors per Ribosome	Function
				Initiation Factors	
IF1	8,119	71	*infA*	0.18	Promotes IF2 and IF3 functions
IF2α	97,300	889	*infB*	0.10	Binds fMET-tRNA$_f$ and
IF2β	79,700	732	*infB*	0.05	GTPase
IF3	20,668	181	*infC*	0.14	Ribosome dissociation; mRNA binding
				Elongation Factors	
EF-Tu	43,195	393	*tufA*	2-7	Bind aminoacyl-tRNA
	43,225	393	*tufB*	2-7	to A site; GTPase
EF-Ts	30,257	282	*tsf*	1	GDP-GTP exchange on EF-Tu
EF-G	77,444	703	*fus*	1	Translocation; GTPase
				Termination Factors	
RF-1	35,911	323	*prfA*	0.02	Recognizes UAA, UAG
RF-2	38,404	339	*prfB*	0.02	Recognizes UAA, UGA
RF-3	46,000				GTPase; stimulates RF-1 and RF-2
RRF	23,500		*frr*		Ejection of tRNA and mRNA

tRNA$_f$ binds first. Thus this important mechanistic problem remains unresolved. The actual pathway utilized may depend on the specific mRNA involved.

What features of a bacterial mRNA confer efficient translation? Examination of mRNA sequences in the region of ribosome binding sites suggests that most of the ca. 30 nucleotide residues in contact with the 30S subunit may influence the efficiency of binding.[19] Besides the initiator codon, the most important region is the Shine/Dalgarno sequence, a purine-rich sequence five to seven nucleotides upstream from the AUG that is complementary to a short sequence at the 3'-terminus of 16S rRNA. A rather loose correlation of extent of complementarity and mRNA strength exists, and mutants that increase or decrease the complementarity in the mRNA lead to corresponding changes in translation rates. Proof that the interaction between mRNA and rRNA actually occurs was obtained by constructing specialized ribosomes with altered 3'-terminal sequences in the 16S rRNA.[20, 21] Thus the binding of mRNA involves two RNA-RNA interactions: the initiator codon with the anticodon of fMet-tRNA$_f$ and the S/D sequence with the 16S rRNA. These interactions confer stability to the 30S preinitiation complex to an extent likely proportional to the number of complementary base pairs that form. Computer-assisted analyses of ribosome binding sites indicate that A's and U's are preferred between the S/D sequence and the AUG, that an A is preferred at position -3, and that the residues between -21 and $+13$ are nonrandom as well (reviewed in References 19 and 22). How these other regions contribute to mRNA binding is unclear, but an interaction of mRNA with either rRNA or ribosomal proteins is implied.

The information for the binding and precise phasing of the mRNA resides primarily in the S/D and initiator codon sequences and in their interaction with the 30S ribosomal subunit and fMet-tRNA$_f$. This is elegantly demonstrated by the toeprint assay of Gold and co-workers,[13] who found that the 30S ribosome is precisely placed on a mRNA in a binding reaction that requires only the anticodon arm of tRNA$_f$, but not initiation factors. It appears that the S/D region helps to confer binding stability to the complex, placing the initiator codon near the initiator tRNA, whereas fine tuning of mRNA phasing occurs by the codon-anticodon interaction. The role of IF2 is to assure that fMet-tRNA$_f$ rather than some other tRNA occupies the peptidyl (P) site of the 30S subunit. The role of IF1 and IF3 are less certain, although some evidence suggests that IF3 monitors the codon-anticodon interaction of fMet-tRNA$_f$, disfavoring interactions with non-initiator tRNAs. Another possible function of the initiation factors is to increase the rates of binding of fMet-tRNA$_f$ and mRNA to the 30S ribosome, possibly by promoting a favorable conformation of the ribosome. Finally, a specific role for ribosomal protein S1 in mRNA translation is indicated.[23] S1 binds to pyrimidine-rich, unstructured RNAs, and may contribute to a stabilization of mRNA-30S subunit interactions. S1 and IF3 together promote the formation of 30S preinitiation complexes.[24] Perhaps S1, which binds relatively weakly to 30S ribosomes, should be regarded as a factor rather than a ribosomal protein.

The 30S preinitiation complex contains fMet-tRNA$_f$, mRNA, GTP, IF1, and IF2. The presence of bound IF3 is controversial, as some experiments suggest that it is ejected from the complex upon fMet-tRNA$_f$ binding.[25] The 50S subunit then joins the 30S preinitiation complex, labilizing IF1 binding but stabilizing bound fMet-tRNA$_f$ and mRNA. The GTPase activity of IF2 is activated by formation of the 70S complex, resulting in the hydrolysis of a molecule of GTP. The resulting IF2-GDP binary complex binds less avidly to the 70S initiation complex and is ejected. Thus the role of GTP hydrolysis in initiation appears to be to enhance the rate of dissociation of IF2 from 70S initiation complexes, thereby allowing the elongation phase to begin more rapidly.

14.2.3. Elongation

The elongation phase of protein synthesis is promoted by three elongation factors (Table 1) and is conveniently divided into three subreactions: the binding of aminoacyl-tRNA (aa-tRNA); peptide bond formation; and translocation (see Figure 1). Mechanistic details are provided in a number of reviews;[26, 27, 40] a brief description follows. In the first reaction, a ternary complex of aa-tRNA, EF-Tu, and GTP is an obligatory precursor. Ternary complexes representing aa-tRNA species whose anticodons complement the triplet mRNA codon in the acceptor (A) site bind to the ribosomal A site about ten times faster than noncognate ternary complexes, but the latter are much more rapidly released. Before a true equilibrium is established between the ternary complex and the ribosome, ribosome-bound EF-Tu catalyzes the hydrolysis of GTP to GDP. Dissociation of incorrectly matched aa-tRNAs again occurs more rapidly than those whose anticodons complement the codon, leading to higher accuracy of aa-tRNA selection. Bound EF-Tu-GDP prevents the aminoacyl group from entering the peptidyl transferase center, thereby blocking peptide bond formation and allowing sufficient time for dissociation of noncognate aa-tRNAs from the ribosome. The subsequent release of EF-Tu-GDP may be the rate-limiting step in the aa-tRNA binding reactions. In order for EF-Tu to promote another round of aa-tRNA binding, bound GDP must be replaced by GTP. The rate of guanine nucleotide exchange is catalyzed by EF-Ts, which displaces GDP to form a complex with EF-Tu, and which in turn is displaced by GTP. EF-Tu binds GDP about two orders of magnitude more avidly than GTP. The exchange reaction is driven towards the EF-Tu-GTP binary complex by the high energy-charge of the cell and by removal of EF-Tu-GTP through formation of ternary complexes.

Peptide bond formation occurs very rapidly following the release of EF-Tu-GDP. The reaction constitutes an $O \rightarrow N$ acyl shift whereby the peptidyl portion of peptidyl-tRNA bound in the ribosomal P site migrates from the hydroxyl of the tRNA to the α-amino group of the aa-tRNA bound in the A site. This reaction is energetically favorable and does not require the hydrolysis of GTP. It is catalyzed by elements of the 50S ribosomal subunit, called the peptidyl transferase center, but the precise catalytic entities are not yet identified. The resulting ribosomal complex has peptidyl-tRNA in the A site and stripped tRNA in the P site.

Translocation is the step whereby peptidyl-tRNA and mRNA move relative to the ribosome from the A site to the P site. This reaction is catalyzed by elongation factor EF-G and involves the hydrolysis of GTP to GDP and ejection of the stripped tRNA. Recent structural studies[28] support the view that translocation occurs in two steps: in the first, the peptidyl moiety and the acceptor arm of its tRNA shift to the P site but the anticodon arm complexed to its cognate codon in the mRNA remains in the A site; in the second, the mRNA and tRNA anticodon arm shift into the P site (see Figure 1). The ribosome then contains an open A site ready to accept the binding of another ternary complex and begin another cycle of elongation. When a nonhydrolyzable GTP analog is used, EF-G binding to ribosomes is stabilized, but binding of the next aa-tRNA is blocked.[29] Since this peptidyl-tRNA reacts with puromycin, its peptidyl moiety must already be in the P site. Therefore, the binding of EF-G-GTP and GTP hydrolysis promote translocation of the anticodon arm and the mRNA; the "energy" for translocation of the peptidyl end of the tRNA comes from the relative binding energies of peptidyl-tRNA in the P˙and A sites, and therefore ultimately derives from the high energy ester bond of aa-tRNA. The process of elongation proceeds at about 15 cycles per second and consumes at least two high energy phosphate bonds. Since aa-tRNA charging consumes at least another two high energy bonds of ATP, the incorporation of one amino acid into protein costs at least four high energy bonds.

14.2.4. Termination

When a termination (nonsense) codon appears in the A site following translocation, no aa-tRNA is present to recognize the codon. Instead, one of two release factors, RF-1 and RF-2, binds and promotes the hydrolysis of peptidyl-tRNA. RF-1 recognizes the codons UAA and UAG; RF-2 recognizes UAA and UGA. Catalysis of peptidyl-tRNA hydrolysis is thought to be provided by the peptidyl transferase center of the ribosome, and a third factor, RF-3, may be involved. Following release of the hydrolyzed peptide, another protein called ribosome release factor (RRF) is involved in ejecting the stripped tRNA and in dissociation of the ribosome from the mRNA.[30]

14.2.5. Reinitiation

Many bacterial mRNAs are polycistronic and in some cases the translation of a downstream cistron is tightly coupled to translation of the upstream cistron. It is not clear whether initiation of the downstream cistron involves the same ribosome that just terminated upstream or whether a new ribosome is used. It is possible that some mechanistic features of reinitiation differ. Two possibilities exist. (1) Initiation at the downstream cistron occurs by the same mechanism as described above for initiation with 30S ribosomal subunits. In this case, coupling might occur by the presence of mRNA secondary structure which masks the ribosomal binding site of the downstream cistron but which is transiently destroyed by ribosomes translating the upstream cistron. (2) Initiation at the downstream cistron occurs by the same ribosome that just completed translating of the upstream cistron. Upon termination, the ribosome (possibly as the 70S particle) moves either upstream or downstream along the mRNA, initiating protein synthesis upon reaching an initiator codon.

In many cases of coupled translation, the terminator and initiator codons are very close to each other and may even overlap. Neither model is proven. The latter mechanism resembles formally some aspects of the scanning model for initiation in eukaryotes.

14.3. COMPARISON OF THE TRANSLATIONAL MACHINERY

Because the overall process of protein synthesis is quite comparable in prokaryotic and eukaryotic cells, it is possible to compare the macromolecular components responsible for individual steps or phases of the pathway. Here we describe the extent of conservation of structures of components carrying out similar functions, as well as our knowledge of how the components act.

14.3.1. mRNA

Bacterial mRNAs are transcribed from DNA by a DNA-dependent RNA polymerase which recognizes specific start (promoter) and stop (terminator) sequences in the DNA. The unit of transcription, called an operon, may encode one (monocistronic) or many (polycistronic) different proteins. It follows that the polycistronic mRNAs possess more than one translational start and stop signal. The cistrons usually are nonoverlapping, although there are examples in bacteriophages where two different reading frames are used to translate a mRNA into two different proteins. Most mRNAs are thought to function as unprocessed primary transcripts which therefore possess a triphosphate at the 5'-terminus. However, posttranscriptional processing of mRNAs is known to occur (reviewed in Reference 31). mRNAs may be altered by endonucleolytic cleavage by RNase III, by 3'-exonucleolytic shortening by polynucleotide phosphorylase or RNase II, and/or by polyadenylation to generate short poly(A) tails. Most mRNAs exhibit short half-lives of 0.5 to 2 min, although some bacterial mRNAs are quite stable, with half-lives greater than a generation time (15 to 20 min). Because bacterial genomes are not separated by a membrane from the translational machinery in the cytoplasm, translation of nascent mRNAs begins as soon as a translational start signal emerges from the polymerase and well before transcription terminates. Thus transcription and translation are coupled both physically and temporally. Because transcription is very rapid (ca. 50 bases/sec, or about 20 sec for an average-sized cistron), completely transcribed mRNAs are able to accummulate, presumably as polysomes, prior to their degradation. Detection of discrete, full-length mRNAs by Northern blot analysis reinforces this view. Ribosomes in nascent polysomes prevent or retard the formation of secondary structures in the coding sequence of a mRNA that might otherwise form were the mRNA completely formed before its translation begins.

The structure and metabolism of eukaryotic mRNAs differ substantially. Eukaryotic mRNAs are capped and usually polyadenylated, and generally encode a single protein (monocistronic). Their transcription and processing are temporally separated from translation by the nuclear membrane. As a result, mature, completely formed mRNAs enter the cytoplasm for presentation to the translational machinery. Such mRNAs appear to possess considerable secondary structure and are complexed with a discrete set of proteins to form a messenger ribonucleoprotein particle (mRNP). Thus mRNAs are present in the cytoplasm in two forms: as nonactive mRNPs; and as actively translated polysomes. To the best of our knowledge, bacteria generally do not store mRNAs as nonactive mRNPs, although certain long-lived mRNAs (e.g., bacteriophage T4 gene 32 mRNA) may assume such a form when strongly repressed.

Because of the rapid turnover of most bacterial mRNAs, it is difficult to isolate and purify natural mRNAs from prokaryotes. This stands in contrast to eukaryotic cells, from which either mixtures or individual species of mRNAs can be purified and translated *in vitro*. Historically, investigators of prokaryotic translation have used the genomes of the RNA bacteriophages as sources of "natural" mRNAs. With the development of *in vitro* SP6 and T7 polymerase transcription systems, prokaryotic cellular mRNAs are now readily available.

A further difference between prokaryotic and eukaryotic mRNAs concerns the codons used to specify the various amino acids. Because most amino acids are specified by more than one code "word", it is possible for cells to exhibit a bias in the selection of synonymous codons. Codon choice was tabulated for 83 *E. coli* genes[32] and a bias was found for one or two codons in each degenerate codon family. Highly expressed genes showed a much stronger bias than those encoding low abundance proteins. Codon biases similar to *E. coli* are found in other Enterobacteriaceae, but differ as the species become more distant evolutionarily. A strong codon bias is seen in the eukaryote, *Saccaromyces cerevisiae*,[33] although the codons favored in yeast differ from those used most frequently in bacteria. In mammalian cells, a still different codon bias is used, although analysis is complicated by differences due to cellular differentiation.[34] Highly expressed genes encoding different proteins in different tissues of the same organism may use strikingly different codon biases.

14.3.2. tRNAs and Aminoacyl-tRNA Synthetases

About 15% of *E. coli* RNA is tRNA, which corresponds to ca. 400,000 molecules per cell or 10 tRNAs per ribosome. There are about 65 to 75 individual tRNA species, most of which differ in primary sequence as well as posttranscriptional modifications. These species are grouped into 20 isoacceptor families according to the amino acid that is attached. About half the tRNAs or their genes have been sequenced.[35, 36] The number of different tRNA molecules in prokaryotic and eukaryotic cells is comparable, although a precise comparison is not possible since all species in either cell type have not yet been identified and sequenced. The molecular weights and tertiary structures also are very similar. Details of tRNA structures are beyond the scope of this chapter, but are provided in many excellent reviews.[37, 38] The cellular content of *E. coli* isoacceptor tRNAs was measured for about 35 molecular species, which were found to differ tenfold in concentration.[39] In general, tRNAs that recognize frequently used codons (due to the amino acid content in proteins and to codon bias) are more abundant than those that recognize rarely used codons.

There are 20 different aminoacyl-tRNA synthetases (aa-RSs) in bacterial cells, one for each isoacceptor tRNA family. All of the *E. coli* synthetases have been purified and characterized and their subunit structures determined (reviewed in Reference 10). The genes for ten of these have been cloned and sequenced as well (reviewed in Reference 40). In the eukaryotic cytoplasm there are also 20 different aa-RSs, as well as an additional set in eukaryotic organelles such as mitochondria and chloroplasts. Both prokaryotic and eukaryotic aa-RSs carry out comparable charging reactions on similar tRNAs, and their mechanisms of action are thought to be very similar. The aa-RS binds to the acceptor arm of the tRNA, and usually appears to interact with the anticodon region as well. This has been elegantly shown by X-ray crystallographic analysis of glutaminyl-tRNA synthetase in a complex with its tRNA.[41] The aa-RSs from *E. coli* can aminoacylate many but not all eukaryotic tRNAs. For example, Met-RS recognizes mammalian $tRNA_i^{met}$, but not $tRNA_m^{met}$. The structures of the various aa-RSs differ within a single species; thus species to species comparisons even of homologous enzymes are difficult to evaluate. The most striking difference is that many mammalian aa-RSs are found in large complexes with eight or more different aa-RSs.[42] How this contributes to their function in eukaryotic cells is unclear.

The ability of the *E. coli* initiator tRNA to function in a mammalian system has been tested.[43] The unformylated Met-$tRNA_f$ from *E. coli* forms a ternary complex with eIF-2 and GTP and binds to 40S ribosomes, but reacts poorly on the 80S ribosome in peptide bond formation. When formylated, ternary complex function is somewhat inhibited but 40S binding is stimulated. However, the $fMet$-$tRNA_f$ does not participate in peptide bond (fMet-puromycin) formation, in contrast to the mammalian species, whose reactivity in this reaction is stimulated by formylation. In conclusion, the *E. coli* initiator species shares certain structural features that promote its specific activity in early steps of the initiation pathway, but lacks those elements that enable the tRNA to react efficiently in the peptidyl transferase step.

14.3.3. Ribosomes

The *E. coli* 70S ribosome comprises a large (50S) and small (30S) subunit, each consisting of rRNAs and ribosomal proteins. As in eukaryotes, the proteins are present in one copy each per particle, except for the 50S proteins L7/L12, which together are present in four copies. There are fewer 70S proteins than 80S proteins (53 vs. ca. 75), and the weight average molecular weights are smaller (21,000 vs. 33,000). Even though the eukaryotic rRNAs are about 50% larger than their prokaryotic equivalents, the RNAs comprise only about 50% of the particle mass, compared with 67% in bacteria.

The primary structures of all of the protein and RNA components of bacterial ribosomes are known. The *E. coli* rRNA genes have been cloned and sequenced and their expression has been studied in detail (reviewed in Reference 44). All three rRNAs are transcribed together on a single transcript from seven different loci in the *E. coli* genome. The precursor transcript is then processed by endonucleolytic and exonucleolytic cleavages to generate mature 5S, 16S, and 23S rRNAs.[31] In this respect, rRNA synthesis and processing are quite similar in bacteria and eukaryotes. The 52 genes encoding ribosomal proteins have been cloned, mapped and sequenced. The synthesis of ribosomal proteins occurs from many different mRNAs, some of which are monocistronic whereas others are polycistronic with strong translational coupling between cistrons.[44] Coordination of ribosomal protein and rRNA expression is accomplished by autogenous regulation of ribosomal protein synthesis in large part at the level of translation. Certain proteins, when present in excess over rRNA, bind to their mRNAs and inhibit translation of all the cistrons of the mRNA. The control of overall ribosome biogenesis occurs by regulating the rRNA promoters. The rRNA promoters vary in efficiency over time, thus responding to the need for more ribosomes or the presence of excess ribosomes, possibly by RNA polymerase sensing the metabolic signal molecules, ppGpp and pppGpp.

The three-dimensional structure of bacterial ribosomes is being studied intensively and considerable progress has been made in recent yeast.[45–47] Early work focused on the arrangement of the ribosomal proteins in the particle using primarily immunoelectron microscopy, neutron scattering, and protein cross-linking techniques. The tertiary structure of the rRNAs was elucidated by distinguishing regions of single-stranded, double-stranded, and long-range interactions by differential chemical and enzymic reactivities, phylogenetic comparisons, and cross-linking. The protein and RNA structural models have been merged by placing the ribosomal proteins within the structural framework of the rRNAs to create a comprehensive model.[48, 49] The overall shape and surface contours of the two subunits are being refined by X-ray crystallographic analyses and by electron microscope tomography. Together these analyses have a resolution of about 20 Å, sufficient to identify the relative positioning of the proteins and RNA, but insufficient to describe the precise course of a peptide or RNA strand. Nevertheless, these models now contribute to the formulation of experiments that relate precise structural elements to function. For example, the site of binding of the 3'-end of aminoacyl-tRNA to ribosomes has been localized to specific conserved nucleotides in 23S rRNA, depending on whether the aa-tRNA binds in the A, P, or exit (E) sites.[50] Progress in defining the structure of bacterial ribosomes is reviewed every 3 years at an international workshop, whose proceedings are published.[51] The most recent meeting held in 1989 in Montana provides an up-to-date description of our current knowledge.[52]

Mammalian ribosomes are not yet so well defined but share many general characteristics of shape. Thus both types of small subunit exhibit a cleft, thought to be where mRNA binds and is decoded. Both larger subunits possess a number of protuberances, one of which comprises the acidic proteins L7/L12 (prokaryotic) or P1/P2 (eukaryotic). Whereas the sequences of the rRNAs are known, the primary structures of the ribosomal proteins are only just emerging, in large part by the efforts of Wool and co-workers, and by the cloning and sequencing of DNAs encoding the proteins.

14.3.4. Initiation Factors

There are three initiation factors in *E. coli:* IF1 (8.1 kDa); IF2α (97.3 kDa) and IF3 (20.7 kDa) (reviewed in References 10 and 11). A smaller form of IF2, IF2β (79.7 kDa), also is present in cells and is expressed by independent translational initiation on the same mRNA that expresses IF2α. However, different functional roles for IF2α and IF2β have not been identified. The initiation factors are moderately abundant proteins, with cellular levels of about 0.2 molecules per ribosome (ca. 5000 molecules per cell). They have been purified to near-homogeneity, their genes have been cloned (see Table 1), and amino acid sequences are derived. The initiation factors were isolated on the basis of their stimulation of fMet-tRNA$_f$ binding to 30S or 70S ribosomes in the presence of AUG, synthetic oligonucleotides or natural mRNAs. *In vitro* assays for protein synthesis are very inefficient, such that it is not possible to conclude that all essential initiation factors have been identified. A genetic approach to this problem has proven difficult and inconclusive. Conditional mutations in *infC* and *infB* indicate that IF3 and IF2 are essential for bacterial growth.[53, 54] This view is reinforced for IF2, whose cellular level was artificially reduced by recombinant DNA techniques, leading to a reduction in polysome size and a reduced rate of protein synthesis.[55] There are no mutants of *infA* reported, so it is not known if IF1 is essential for protein synthesis *in vivo*.

The functional roles of the initiation factors have been studied by *in vitro* assays (reviewed in References 11, 40, and 56). IF2 binds to GTP and stimulates the binding of fMet-tRNA$_f$ to 30S ribosomes; it also possesses a

70S ribosome-dependent GTPase activity. It binds to 30S ribosomes with rather low avidity, but in the presence of IF1 and IF3 the association constant is $2 \times 10^8 \ M^{-1}$, tight enough to saturate free 30S subunits at physiological concentrations.[57] The primary structure of IF2 contains a region homologous to the GTP binding domains of EF-Tu and EF-G. This region, called the G domain, has been modeled in three dimensions based on the X-ray crystallographic model of EF-Tu.[58] The domains responsible for fMet-tRNA binding and the factor's interaction with ribosomes have not yet been identified. The factor's major function may be to assure that the initiator fMet-tRNA$_f$ rather than some other aa-tRNA is placed on the 30S ribosome, thereby establishing that phasing occurs at the AUG initiator codon rather than nearby through an interaction with noninitiator tRNAs.

IF3 binds to 30S ribosomal subunits with an association constant of $3 \times 10^7 \ M^{-1}$. Its binding reduces the rate of association with 50S subunits, leading to dissociated ribosomes. IF3 is not required for binding fMet-tRNA$_f$ to 30S ribosomes with the triplet, AUG, although it stimulates fMet-tRNA$_f$ binding with natural mRNAs. mRNAs with the potential for a strong interaction between the Shine/Dalgarno region and the rRNA may be less dependent on IF3, yet evidence suggests that IF3 does not directly influence the stability of the RNA-RNA interaction.[59] The role of IF3 may be to favor kinetically a correct interaction in the ribosome decoding site between the fMet-tRNA$_f$ and its codon, and to disfavor other tRNA-mRNA interactions by increasing their rates of dissociation from the ribosome.[60] If correct, both IF2 and IF3 function to assure that fMet-tRNA$_f$ is placed on the 30S subunit, by affecting kinetic parameters of the pathway.

The role of IF1 is less clear. It increases the avidity of IF2 for 30S ribosomal subunits, and stimulates fMet-tRNA$_f$ binding to 30S or 70S ribosomes in the presence of IF2. It also stimulates both the rate of ribosome dissociation and subunit association, thereby increasing the rate that equilibrium is reached. Nevertheless, a specific or unique role of IF1 has not been identified.

Comparison of bacterial initiation factors with their mammalian counterparts is difficult since the functional roles of neither is well defined. The closest parallel is between IF2 and eIF-2. Both factors bind GTP, although a G-domain similar to EF-Tu has not been detected in eIF-2. Instead, GTP binding to eIF-2 appears to be shared by both the β- and γ-subunits. Both factors interact with the initiator tRNA, although the binding is weak for IF2 whereas for eIF-2 a stable complex of eIF-2-GTP-Met-tRNA$_i^{met}$ is isolated. The protein structures are entirely different; IF2 shares no obvious regions of homology with either the α- or β-subunits of eIF-2 (the γ-subunit has not yet been cloned and sequenced). Bacterial IF2 also appears to exchange GDP for GTP in a rapid reaction, and no exchange factor appears to be required. IF3 shares some functional aspects with eIF-3: both factors bind stably to the small ribosomal subunit in the absence of other components of the initiation process, and both act as anti-association factors. However, IF3 is a small basic protein whereas eIF-3 is large and complex, comprising at least seven polypeptide subunits and a mass of over 500,000 Da. IF3 and the eukaryotic factors involved in mRNA binding (eIF-4A, eIF-4B, and eIF-4F) also share no structural or functional similarities.

14.3.5. Elongation Factors

The elongation cycle is catalyzed by three elongation factors: EF-Tu (43.2 kDa); EF-Ts (30.3 kDa); and EF-G (77.4 kDa). EF-Tu is the most abundant protein in *E. coli*, present in greater than five copies per ribosome, whereas EF-Ts and EF-G levels are stoichiometric with ribosomes. The proteins have been purified and sequenced and their genes have been cloned and characterized (see Table 1). EF-Tu is unusual in that two genes, *tufA* and *tufB*, code for the factor; the encoded proteins differ only at the C-terminal residue and appear to function identically. The eukaryotic equivalents of EF-Tu and EF-Ts are eEF-1α and eEF-1βγ, respectively, and the two sets of factors are thought to function very similarly. In fact, EF-Tu can replace eEF-1α for *in vitro* protein synthesis on mammalian ribosomes and eEF-1α can function in the prokaryotic system.[61] eEF-2 is functionally comparable to EF-G and shares considerable sequence identity within an N-terminal region of 163 residues (the G-domain) as well as elsewhere in the primary sequence.

A high resolution X-ray crystallographic structure has been determined for EF-Tu.[62, 63] The protein comprises three domains, one of which is responsible for binding GTP or GDP (the G-domain). The G-domain structure is very similar to that of the *ras* protein, P21, and serves as a model of the prototype G protein. The secondary structural elements of both the EF-G and IF-2 G-domains can be modeled into a three-dimensional array based on the EF-Tu structure. Considerable effort is being made to understand the nature of the conformational change induced in EF-Tu by GTP hydrolysis. Such knowledge will be useful in understanding how all three translational G-proteins are ejected more rapidly from ribosomes in the GDP conformation. It may also serve to help explain how other G-proteins function in signal transduction pathways. The EF-Tu structure enables researchers to study important parameters for recognizing GTP and for its hydrolysis. Site-directed mutagenesis of DNA encoding the G-domain was employed to alter Asp[138] to Asn, leading to a change in specificity where XTP rather than GTP is the preferred binding substrate for the mutant EF-Tu protein.[64]

14.3.6. Termination Factors

E. coli contains two termination factors, RF-1 (35.9 kDa) and RF-2 (38.4 kDa) which are present at low concentrations in the cell (ca. one to two copies per ribosome). A third factor, RF-3, has been identified which stimulates the termination reaction and hydrolyzes GTP. The genes for RF-1 and RF-2 have been cloned and sequenced. Most remarkable is the finding that the RF-2 gene is split into two reading frames; its expression therefore depends on an efficient frame-shift event.[65] In mammalian cells, termination is promoted by a single factor, RF, which recognizes all three termination codons. The bacterial and mammalian factors appear to function comparably, although molecular details of how they promote hydrolysis of the peptidyl-tRNA bond are lacking in both systems. Finally, another factor, termed ribosome release factor (RRF) has been identified in bacteria[30] and its gene has been cloned.[66] RRF functions to promote dissociation of components from the 70S ribosome following termination. No comparable factor has been identified in eukaryotic cells.

14.4. TRANSLATIONAL CONTROLS

Evaluation of how bacteria modulate the rate of protein synthesis provides interesting insights and contrasts to the mechanism of translational control in eukaryotes. Three aspects in particular are used by bacteria to influence the efficiency of translation of mRNAs: the choice of initiator codon and surrounding sequence; the presence of mRNA secondary structure in the region of the ribosome binding site; and the action of *trans*-acting protein factors or antisense RNAs.

14.4.1. Initiator Codon and Context

Most cistrons utilize AUG as the initiator codon. However, about 10% of mRNAs employ other codons (usually GUG) that may be essential in regulating translation. mRNAs beginning with non-AUG codons are less efficiently translated *in vitro*.[67] The initiator codon AUU is employed to initiate the *infC* gene and is an essential element in feed-back regulation by IF3.[68] Thus GUG and other codons may be used to dampen the efficiency of protein synthesis, but the effect appears to depend on other aspects of the mRNA as well. The context of the initiator codon has been discussed above, with the Shine/Dalgarno sequence as well as other regions known to interact with the 30S ribosomal subunit. The rate of initiation is likely determined by the sum of these various interactions. When one or more is optimal, still others may be less important. By employing a non-AUG initiation codon and thereby diminishing the strength of interaction at this point, elements such as the Shine/Dalgarno region become more important and their contributions may become critical.

14.4.2. mRNA Secondary Structure

The most important element regulating protein synthesis in prokaryotes is mRNA secondary structures that inhibit the rate of initiation. The influence of secondary structure was demonstrated in early studies of protein synthesis with the RNA bacteriophages; both the maturation A protein and the replicase cistrons are masked and translation begins only at the coat cistron.[69] The generally accepted view is that secondary structures that include in particular the initiator codon and Shine/Dalgarno region, but any part of the ribosome binding site, reduce initiation efficiency. Such sensitivity to secondary structure near the initiator codon is peculiar to prokaryotes and is not seen in eukaryotes. Eukaryotes employ RNA helicases (eIF-4A and eIF-4B) and/or the 40S scanning mechanism to melt out such secondary structures, whereas bacteria have no such functions. Access to the ribosome binding site is therefore of great importance in initiation and can be readily modulated by events, such as translation itself, that melt out the structures.

14.4.3. *Trans*-acting Protein Factors and Antisense RNAs

A common mechanism for inhibiting translation involves the binding of a repressor protein at or near the ribosome binding site. The bound repressor may block access to ribosomes, may stabilize secondary structure which masks the ribosome binding site, or may allow ribosome binding but inhibit a subsequent step in the initiation pathway. Inhibition by *trans*-acting repressor proteins is seen with ribosomal protein operons, where initation of the first cistron is inhibited by one of the ribosomal proteins encoded by the mRNA.[44] Other well studied examples of autogenous regulation include the *thrS* operon inhibited by threonyl-tRNA synthetase,[70] and translation of the bacteriophage T4 gene 32 mRNA inhibited by its product.[71] These examples are formally similar to the regulation of ferritin synthesis in mammalian cells, where the ferritin repressor binds to a specific sequence in the 5'-untranslated region and inhibits initiation[72] (and Part II, Chapter 10 of this book). However, the actual mechanism whereby the repressor inhibits ferritin mRNA translation is not yet known.

Instead of a protein repressor, RNA complementary to a region of the mRNA may inhibit bacterial protein synthesis. Antisense RNAs have been identified that inhibit gene expression by binding to mRNAs. Examples of genes so regulated are the IS10 transposase gene: *ompF* (an outer membrane protein); *ant* (phage P22 antirepressor protein); and *crp* (catabolite repressor protein) (reviewed in Reference 73). The IS10 transposase, *ompF* and *ant* antisense RNAs are thought to block 30S ribosome binding to the initiator region of the mRNAs due to complementary binding there. This is not the only mechanism of inhibition of gene expression by antisense RNAs, however, as exemplified by antisense RNA binding to nascent *crp* mRNA leading to premature termination of transcription. The possibility also exists that complementary RNAs might inhibit translation by induction of mRNA cleavage by RNase III. Whether or not eukaryotic antisense RNAs, such as the tcRNAs affecting the translation of myosin heavy chain mRNA,[74] function in an analogous manner is yet to be established.

ACKNOWLEDGMENTS

Research was supported by NIH Grants GM22135 and GM40082.

REFERENCES

1. Littlefield JW, Keller EB, Gross J, and Zamecnik PC: Studies on cytoplasmic ribonucleoprotein particles from the liver of the rat. *J. Biol. Chem.* 217:111 (1955).
2. Hoagland MB: An enzymic mechanism for amino acid activation in animal tissues. *Biochim. Biophys. Acta* 16:288 (1955).
3. Hoagland MB, Zamecnik PC, Stephenson ML: Intermediate reactions in protein biosynthesis. *Biochim. Biophys. Acta* 24:215 (1957).
4. Fessender JM and Moldave K: Evidence for two protein factors in the transfer of amino acids from sRNA to ribonucleoprotein particles. *Biochem. Biophys. Res. Commun.* 6:232 (1961).
5. Bishop JO and Schweet RS: Role of glutathione in transfer of amino acids from amino acyl-ribonucleic acid to ribosomes. *Biochim. Biophys. Acta* 49:235 (1961).
6. Jacob F and Monod J: Genetic regulatory mechanisms in the synthesis of proteins. *J. Mol. Biol.* 3:318 (1961).
7. Nirenberg MW and Matthaei JH: The dependence of cell-free protein synthesis in *E. coli* upon naturally occurring or synthetic polyribonucleotides. *Proc. Natl. Acad. Sci. U.S.A..* 47:1588 (1961).
8. Capecchi M and Gussin GN: Suppression *in vitro:* identification of a serine-sRNA as a "nonsense" suppressor. *Science* 149:417 (1965).
9. Lucas-Lenard J and Lipmann F: Separation of three microbial amino acid polymerization factors. *Proc. Natl. Acad. Sci. U.S.A..* 55:1562 (1966).
10. Hershey JWB: Protein synthesis, in Neidhardt FD (Ed): *Escherichia coli and Salmonella typhimurium: Cellular and Molecular Biology,* Washington, D.C., American Society for Microbiology, 613, 1987.
11. Maitra U, Stringer EA, and Chadhuri A: Initiation factors in protein biosynthesis. *Annu. Rev. Biochem.* 51:869 (1982).

12. Grunberg-Manago M, Buckingham RH, Cooperman BS, and Hershey JWB: Structure and function of the translational machinery. *Symp. Soc. Gen. Microbiol.* 28:27 (1978).

13. Hartz D, McPheeters DS, Traut R, and Gold L: Extension inhibition analysis of translation initiation complexes. *Methods Enzymol.* 164:419 (1988).

14. Zucker FH and Hershey JWB: Binding of *Escherichia coli* protein synthesis initiation factor IF1 to 30S ribosomal subunits measured by fluorescence polarization. *Biochemistry* 25:3682 (1986).

15. van Duin J, Overbeek GP, and Backendorf C: Functional recognition of phage RNA by 30-S ribosomal subunits in the absence of initiator tRNA. *Eur. J. Biochem.* 110:593 (1980).

16. Jay G and Kaempfer R: Initiation of protein synthesis. Binding of mRNA. *J. Biol. Chem.* 250:5742 (1975).

17. Gualerzi C, Risuleo G, and Pon CL: Initial rate kinetic analysis of the mechanism of initiation complex formation and the role of initiation factor IF-3. *Biochemistry* 16:1648 (1977).

18. Ellis S and Conway TW: Initial velocity kinetic analysis of 30S initiation complex formation in an *in vitro* translation system derived from *E. coli. J. Biol. Chem.* 259:7606 (1984).

19. Gold L, Pribnow D, Schneider T, Shinedling S, Singer B, and Stormo G: Translational initiation in prokaryotes. *Annu. Rev. Microbiol.* 35:365 (1981).

20. Jacob WF, Santer M and Dahlberg AE: A single base change in the Shine-Dalgarno region of 16S rRNA of *Escherichia coli* affects translation of many protiens. *Proc. Natl. Acad. Sci. U.S.A..* 84:4757 (1987).

21. Hui A and deBoer H: Specialized ribosome system: preferential translation of a single mRNA species by subpopulation of mutated ribosomes in *Escherichia coli. Proc. Natl. Acad. Sci. U.S.A* 84:4762 (1987).

22. Kozak M: Comparison of initiation of protein synthesis in procaryotes, eucaryotes, and organelles. *Microbiol. Rev.* 43:1 (1983).

23. Suryanarayana T and Subramanian AR: An essential function of ribosomal protein S1 in mRNA translation. *Biochemistry* 22:2715 (1983).

24. Steitz JA, Sprague KU, Yuan RC, Laughren M, Moore PB, and Wahba AJ: RNA-RNA and protein-RNA interactions during the initiation of protein synthesis, in Vogel HJ (Ed): *Nucleic Acid-Protein Recognition,* New York, Academic Press, 491, 1977.

25. van der Hofstad GAJM, Buitenhek A, Bosch L, and Voorma HO: Initiation factor IF-3 and the binary complex between initiation factor IF-2 and fMet-tRNA are mutually exclusive on the 30S ribosomal subunit. *Eur. J. Biochem.* 89:213 (1978).

26. Bermek E: Mechanisms in polypeptide chain elongation on ribosomes. *Prog. Nucleic Acid Res. Mol. Biol.* 21:63 (1978).

27. Clark BFC: The elongation step of protein biosynthesis. *Trends Biochem. Sci.* 5:207 (1980).

28. Moazed D and Noller HF: Intermediate state in the movement of tRNA in the ribosome. *Nature* 342:142 (1989).

29. Inoue-Yokosawa N, Ishikawa C, and Kaziro Y: The role of guanosine triphosphate in translocation reaction catalyzed by elongation factor G. *J. Biol. Chem.* 249:4321 (1974).

30. Ryoji M, Karpen JW, and Kaji A: Further characterization of ribosome releasing factor and evidence that it prevents ribosomes from reading through a termination codon. *J. Biol. Chem.* 256:5798 (1981).

31. King TC and Schlessinger D: Processing of RNA transcripts, in Neidhardt FC (Ed): *Escherichia coli* and *Salmonella typhimurium: Molecular and Cellular Biology*, Washington, D.C., American Society for Microbiology, p. 703 (1987).

32. Gouy M, and Gautier C: Codon usage in bacteria: correlation with gene expressivity. *Nucleic Acid Res.* 10:7055 (1982).

33. Bennetzen JL and Hall BD: Codon selection in yeast. *J. Biol. Chem.* 257:3026 (1982).

34. DeBoer HA and Kaestelein RA: Biased codon usage: an exploration of its role in optimizing translation, in Reznikoff W and Gold L (Eds): *Maximizing Gene Expression* Boston, Butterworths, 225, 1986.

35. Sprinzl M and Gauss DH: Compilation of tRNA sequences. *Nucleic Acids Res.* 12:r1 (1984).

36. Fournier MJ and Ozeki H: Structure and organization of the transfer ribonucleic acid genes of *Escherichia coli* K-12. *Microbiol. Rev.* 49:379 (1985).

37. Rich A and RajBhandary UL: Transfer RNA: molecular structure, sequence, and properties. *Annu Rev. Biochem.* 45:805 (1976).

38. Goddard JP: The structures and functions of transfer RNA. *Prog. Biophys. Mol. Biol.* 32:233 (1977).

39. Ikemura T: Correlation between the abundance of *Escherichia coli* transfer RNAs and the occurrance of the respective codons in its protein genes: a proposal for a synonymous codon choice that is optimal for the *E. coli* translational system. *J. Mol. Biol.* 151:389 (1981).

40. Grunberg-Manago M: Regulation of the expression of aminoacyl-tRNA synthetases and translation factors, in Neidhardt FC (Ed): *Escherichia coli* and *Salmonella typhimurium: Cellular and Molecular Biology*, Washington, D.C., American Society for Microbiology, p. 1386 (1987).

41. Rould MA, Perona JJ, Söll D, and Steitz TA: Structure of *E. coli* glutaminyl-tRNA synthetase complexed with tRNA[Gln] and ATP at 2.8 Å resolution. *Science* 246:1135 (1989).

42. Yang DHC, Garcia JV, Johnson YD, and Wahib S: Multi-enzyme complexes of mammalian aminoacyl-tRNA synthetases. *Curr. Topics Cell. Regul.* 26:325 (1985).

43. Elson NA, Adams SL, Merrick WC, Safer B, and Anderson WF: Comparison of fMet-tRNA[f] and Met-tRNA from *Escherichia coli* and rabbit liver in initiation of hemoglobin synthesis. *J. Biol. Chem.* 250:3074 (1975).

44. Jinks-Robertson S and Nomura M: in Neidhardt FC (Ed): *Escherichia coli* and *Salmonella typhimurium: Molecular and Cellular Biology*, Washington, D.C., American Society for Microbiology, p. 1358 (1987).

45. Noller HF, Nomura M: Ribosomes. in Neidhardt FC (Ed): *Escherichia coli* and *Salmonella typhimurium: Cellular and Molecular Biology*, Washington D.C., American Society for Microbiology, p.104 (1987).

46. Moore PB: The ribosome returns. *Nature* 331:223 (1988).

47. Dahlberg AE: The functional role of ribosomal RNA in protein synthesis. *Cell* 57:525 (1989).

48. Brimacombe R, Atmada J, Stiege W, and Schüler D: A detailed model of the three-dimensional structure of *E. coli* 16S rRNA *in situ* in the 30S subunit. *J. Mol. Biol.* 199:115 (1988).

49. Stern S, Powers T, Changchien L-M, and Noller HF: RNA-protein interactions in 30S ribosomal subunits: folding and function of 16S rRNA. *Science* 244:783 (1989).

50. Moazed D and Noller HF: Interaction of tRNA with 23S rRNA in the ribosomal A, P, and E sites. *Cell* 57:585 (1989).

51. Hardesty B and Kramer G (Eds): *Structure, Function and Genetics of Ribosomes.* Berlin, Springer-Verlag, 1986.

52. Hill W (Ed): *Ribosomes.* Washington, D.C., American Society for Microbiology, 1990.

53. Springer M, Graffe M, and Hennecke H: Specialized transducing phage for the initiation factor IF3 gene in *E. coli. Proc. Natl. Acad. Sci. U.S.A..* 74:3970 (1977).

54. Shiba J, Ito K, Nakamura Y, Dondon J, and Grunberg-Manago M: Altered translation initiation factor 2 in the cold-sensitive *ssyG* mutant affects protein export in *E. coli. EMBO J.* 5:3001 (1986).

55. Cole JR, Olsson CL, Hershey JWB, Grunberg-Manago M, and Nomura M: Feedback regulation of rRNA synthesis in *Escherichia coli:* requirement for initiation factor IF2. *J. Mol. Biol.* 198:383 (1987).

56. Gualerzi C, Pon CL, Pinolik RT, Canonaco MA, Paci M, and Wintermeyer W: Role of the initation factors in *E. coli* translational initiation, in Hardesty B and Kramer G (Eds): *Structure, Function and Genetics of Ribosomes,* Berlin, Springer-Verlag. (1986).

57. Weiel J, and Hershey JWB: The binding of fluorescein-labeled protein synthesis initiation factor 2 to *Escherichia coli* 30S ribosomal subunits determined by fluorescence polarization. *J. Biol. Chem.* 257:1215 (1982).

58. Cenatiempo Y, Deville F, Dondon J, Grunberg-Manago M, Sacerdot C, Hershey JWB, Hansen HF, Clark BFC, Kjeldgaard M, LaCour TFM, Mortensen KK, Nyborg J, and Petersen HU: The protein synthesis initiation factor 2 G-domain. Study of a functionally active C-terminal 65-kilodalton fragment of IF2 from *Escherichia coli. Biochemistry* 26:5070 (1987).

59. Canonaco MA, Gualerzi CO, and Pon CL: Alternative occupancy of a dual ribosomal binding site by mRNA affected by translation initiation factors. *Eur. J. Biochem.* 182:501 (1989).

60. Gualerzi CO, Calogero RA, Canonaco MA, Brombach M, and Pon CL: Selection of mRNA by ribosomes during prokaryotic translational initiation, in Tuite MF, Picard M, and Bolotin-Fukuhara (Eds): *Genetics of Translation, New Approaches,* NATO ASI Series 4, Vol. 14, Springer-Verlag, p. 317 (1988).

61. Grasmuk H, Nolan KD, Drews J: Interchangeability of elongation factor-Tu and elongation factor-1 in aminoacyl-tRNA binding to 70S and 80S ribosomes. *FEBS Lett.* 82:237 (1977).

62. la Cour FM, Nyborg J, Thirup S, and Clark BFC: Structural details of the binding of guanosine diphosphate to elongation factor Tu from *E. coli* as studied by X-ray crystallography. *EMBO J.* 4:2385 (1985).

63. Jurnak F: Structure of the GDP domain of EF-Tu and location of the amino acids homologous to *ras* oncogene proteins. *Science* 230:32 (1985).

64. Hwang Y-W and Miller DL: A mutation that alters the nucleotide specificity of elongation factor Tu, a GTP regulatory protein. *J. Biol. Chem.* 262:13081 (1987).

65. Craigen WJ, Cook RG, Tate WP, and Caskey CT: Bacterial peptide chain release factors: conserved primary structure and possible frameshift regulation of release factor 2. *Proc. Natl. Acad. Sci. U.S.A.*. 82:3616 (1985).

66. Ichikawa S, Ryoji M, Siegfried Z, and Kaji A: Localization of the ribosome-releasing factor gene in the *E. coli* chromosome. *J. Bacteriol.* 171:3689 (1989).

67. Brown JC and Doty P: Kinetic preference for initiation of protein synthesis at AUG codons. *Biochim. Biophys. Acta* 228:746 (1971).

68. Butler JS, Springer M, Dondon J, Graffe M, and Grunberg-Manago M: *E. coli* protein synthesis initiation factor IF3 controls its own gene expression at the translational level *in vivo*. *J. Mol. Biol.* 192:767 (1986).

69. Iserentant D and Fiers W: Secondary structure of mRNA and efficiency of translation initiation. *Gene* 9:1 (1980).

70. Springer M, Graffe M, Dondon J, and Grunberg-Manago M: Translational control in *E. coli:* the case of threonyl-tRNA synthetase, in Tuite MF, Picard M, and Bolotin-Fukuhara M (Eds): *Genetics of Translation, New Approaches* NATO ASI Series 4, Vol. 14, Berlin, Springer-Verlag, p. 463 (1988).

71. Gold L: Initiation complex formation: a simple assay for translational regulation, in Tuite MF, Picard M, and Bolotin-Fukuhara M (Eds): *Genetics of Translation, New Approaches,* NATO ASI Series 4, Vol. 14, Berlin, Springer-Verlag, p. 443 (1988).

72. Rouault TA, Hentze MW, Haile DJ, Harford JB, and Klausner RD: The iron-responsive element binding protein: a method for the affinity purification of a regulatory RNA-binding protein. *Proc. Natl. Acad. Sci. U.S.A.*. 86:5768 (1989).

73. Simons RW and Kleckner N: Biological regulation by antisense RNA in prokaryotes. *Annu. Rev. Genet.* 22:567 (1988).

74. Heywood SM: tcRNA as a naturally occurring antisense RNA in eukaryotes. *Nucleic Acids Res.* 14:6771 (1986).

Chapter

15

Inhibitors of Eukaryotic Protein Synthesis

Juan P. G. Ballesta
Centro de Biologia Molecular
C.S.I.C. and U.A.M. Canto Blanco
Madrid, Spain

15.1. INTRODUCTION

In addition to their possible practical interest, inhibitors are useful tools for the investigation of the underlying processes they inhibit. In the case of bacterial protein synthesis the use of several well known antibiotics, i.e., puromycin, chloramphenicol, streptomycin, etc., have been of great importance in the study of basic steps of the process as well as in determining important information concerning ribosomal structure. Similarly, in eukaryotic systems the analysis of an inhibitor mode of action can provide important clues for understanding the peculiarities of their protein synthetic apparatus.

A large number of compounds are known to inhibit the process of protein synthesis in eukaryotic cells at different levels; some of them can be considered as antibiotics (although this term is far from being clearly defined) while others are either complex macromolecules, including plant and microbial toxins, or very simple chemicals.

Several extensive reviews on inhibitors of protein synthesis have been published previously.[1-4] The well established facts already included in these publications will not be commented upon here in detail and the reader will be referred to those publications for more extensive information. A detailed consideration will be given to new data reported in the last few years.

15.2. MODE OF ACTION

15.2.1. Inhibitors of Initiation

Kasugamycin—Kasugamycin is an unusual aminoglycoside antibiotic that has been almost exclusively studied on bacterial cells (which, in fact, are rather resistant to the antibiotic) although the drug has practical interest as an antifungal agent.[5] In bacterial extracts, kasugamycin blocks the interaction of fMet-tRNA with the mRNA-programmed small ribosomal subunit and in addition can disrupt the previously formed 30S initiation complex releasing the initiator tRNA molecule. However, once the 50S subunit is bound, the 70S complex becomes resistant to the antibiotic.[6, 7] Kasugamycin is supposed to act in a similar manner in eukaryotic systems but detailed studies are lacking.

Low levels of resistance to kasugamycin have been associated with the lack of dimethylation of A_{1518} and A_{1519} in the 16S ribosomal RNA (rRNA) of different prokaryotic species (see Reference 7 for a review). Similar modified residues are found in equivalent positions in the small subunit rRNA from nearly all organisms studied.[8] The N^6,N^6-dimethyadenosines have been located by immunoelectron microscopy at the end of the platform of the wheat germ 40S ribosomal subunit.[9] These results suggest the possible location for the kasugamycin binding site at the mRNA binding domain of the ribosomal subunit[10, 11] which is compatible with the mode of action of the drug.

Edeine—Edeine A_1 and B_1 are peptide antibiotics considered as inhibitors of initation of protein synthesis in bacterial and eukaryotic systems *in vitro* although *in vivo* they preferentially seem to block DNA synthesis.[12] Edeine A binds with high affinity to the ribosomes (K_A approximately 10^{10} M^{-1}) but its mode of action is not completely understood since contradictory data have been reported. It is generally accepted that edeine, interacting with the small ribosomal subunit affects the initation complex in such a way that it inhibits indirectly the coupling of the large subunit. Preformed 80S initiation complexes seem to become resistant to the action of the drug (see References 2 and 3) and it has been suggested that edeine might affect the mRNA interaction with the small ribosomal subunit.[13] In agreement with this notion, an increased affinity of initiation complexes for mRNA has been shown to confer resistance to edeine in a *Saccharomyces cerevisiae* mutant.[14] Interestingly, this effect seems to be related to mRNA associated polypeptides rather than to the ribosomal particles.[15]

Using chemical footprinting techniques the drug has been shown to protect residues G_{693}, A_{794}, C_{795}, and C_{926} in the *E. coli* 16S rRNA P-site[16] which may be somehow involved in the binding of the drug and therefore related to those functions inhibited by edeine.

Pactamycin—Like the two previous antibiotics, pactamycin is active in prokaryotic and eukaryotic cells although unlike those drugs, its mode of action has been studied in more detail in the latter case.[1, 2] At low concentrations pactamycin interacts specifically with the small ribosomal subunit and inhibits preferentially polypeptide chain initiation. At higher concentrations the drug blocks elongation as well, probably due to interaction with a lower affinity site in the large ribosomal subunit. Depending on the conditions, and on the system used, pactamycin can either block the coupling of the large subunit to the 40S initiation complex or induce the accumulation of abortive 80S complexes which in some conditions can catalyze the formation of the first peptide bond.[1, 2]

Affinity labeling of the pactamycin binding site on prokaryotic and eukaryotic ribosomes has been carried out[17–19] and the results show that ribosomal protein S10 is the most specifically labeled protein in the yeast ribosome. This protein is the counterpart of the mammalian ribosomal protein S6 which has been implicated in the interaction of the tRNA with the ribosome and seems to have a role in the selection of the mRNA (see Part II, Chapter 9 of this book). It is therefore conceivable that pactamycin interacting with this protein, and perhaps other components of the binding site, affects the initiation of protein synthesis.

Some proteins from the large ribosomal subunit are also labeled by pactamycin[17, 18] suggesting a subunit interface location of the drug binding site, an area where other antibiotics are thought to interact.[20, 21]

Chemical footprinting analysis of pactamycin-treated bacterial ribosomes indicates that residues G_{693} and C_{795} are protected by the drug in the 16S rRNA (J. Woodcock and D. Moazed; J. Egebjerg, referred in Reference 54). These nucleotides are in a region where ribosomal proteins S6 + S18 and S11 interact with the 16S rRNA[22] and is considered as part of the ribosomal decoding domain.[11] Since no systematic correlation between eukaryotic and prokaryotic ribosomal proteins has been performed it is difficult to establish whether any of these proteins correspond to yeast ribosomal protein S10 which, as commented above, has been targeted by pactamycin affinity labeling.

Similar nucleotides are protected by pactamycin and edeine suggesting the existence of overlapping binding sites in spite of the relatively different biological effects of both drugs.

Miscellaneous Inhibitors of Initiation—A number of heterologous compounds including triphenylmethane dyes (aurintricarboxylic acid, pyrochatechol violet, fuchsine, etc), adrenochrome, polydextran sulfate, polyvinyl sulfate have been shown to block protein synthesis initiation.[1] Their mode of action seems, however, to be rather unspecific since they bind to multiple sites in the ribosome and inhibit more than one step of the synthetic process.

Better defined is the activity of different methylated derivatives of guanosine and more specifically that of $m^7G^{5'}p$ (7-methyl-guanosine-monophosphate). This nucleotide and its analogs have been shown to inhibit eukaryotic translation *in vitro* although with a variable efficiency depending on the system used. It is has been assumed that they inhibit translation by competing with capped mRNAs for binding to the initiation factors. However, the fact that translation of some uncapped mRNAs is sensitive to cap analogs indicates that these can inhibit some other steps as well.[3, 23]

In addition, NaF and KF inhibit eukaryotic protein synthesis by blocking the joining of the large subunit to the 40S initiation complex.[1] The mechanism by which the fluoride ions are able to block this specific step of protein synthesis is totally unknown but it might be related to their well known phosphatase inhibitory activity, thereby interfering with some phosphorylation-dephosphorylation process. In this sense, it is interesting to note that phosphorylation of the exchangeable acidic proteins of the large ribosomal subunit is required for binding of these polypeptides to the ribosome[24] and also, that ribosomes from resting cells have half the amount of acidic proteins present compared with those from exponentially growing cells.[24a] The fluoride ions could interfere with phosphorylation-dephosphorylation of the acidic proteins rending the large subunit unable to bind to the 40S initiation complex.

15.2.2. Inhibitors of Elongation

15.2.2.1. Aminoacyl-tRNA Binding

Tetracyclines—In spite of their extensive clinical use as antibacterial drugs, tetracyclines are equally active on eukaryotic cells which are resistant *in vivo* due to the fact that they lack the permeation mechanism that concentrates the drug inside the bacterial cells. These drugs are generally accepted as specific inhibitors of the enzymatic and nonenzymatic binding of aminoacyl-tRNA to the ribosomal A-site[1, 2] although it seems that, in bacterial systems, the binding to the P-site can be inhibited as well at high aminoacyl-tRNA to ribosome ratio.[25] Tetracycline, however, does not interfere directly with the codon-anticodon interaction nor with the 3' end of the aminoacyl-tRNA at the peptidyl transferase center and therefore its binding site must be located somewhere along the L-shaped structure of the tRNA molecule between these two extreme sites. Affinity labeling experiments carried out in bacterial ribosomes indicate that ribosomal protein S7 is the target for the drug and the position of this protein on the ribosome structure is compatible with this notion.[26]

Tetracycline has been shown to protect A_{892} in the bacterial 16S rRNA from chemical modification.[16] Although this nucleotide is not within the rRNA interaction site of ribosomal protein S7 it is not far away.[27]

Fusidic Acid—The steroid antibiotic fusidic acid inhibits cell-free protein synthesizing systems from bacteria as well as eukaryotes. Its mode of action has been worked out in great detail in bacteria. The drug binds to the elongation factor (EF)-G forming a stable structure with the ribosome which inhibits the binding of the incoming aminoacyl-tRNA-EF-Tu complex.[1, 2] The data available indicate that fusidic acid acts in a similar way in eukaryotic systems, stabilizing the EF-2-ribosome complex.[28]

15.2.2.2. Peptide Bond Formation

Puromycin—The universal inhibitor of peptide bond formation, active in all the systems so far studied, is puromycin, an analog of the aminoacyl-adenosine at the 3' end of the aminoacyl-tRNA. Puromycin interacts at the A-site of the peptidyl transferase center and is able to accept the peptidyl moiety of the donor tRNA molecule in the transfer reaction. In this way, puromycin interrupts the elongation of the peptide chain which is then released into the cytoplasm covalently bound to the drug.[1, 2]

It has been recently shown that O-demethyl-puyromycin, in which the methyl group blocking the OH group of the *p*-methoxy-L-phenylalanine residue has been removed, is considerably less active than puromycin.[29] On the other hand, substitution of the ribose moiety by a carbocyclic ring has little effect on its biological activity.[30] These results confirm previous data[31] indicating that the base and the amino acid are the most important parts of the molecule for biological activity.

In recent years, the puromycin binding site has been studied using affinity labeling techniques. Ribosomal proteins L23, S7, and S14 have been found labeled in the *E. coli* ribosome[32] indicating a subunit interface position for the binding site. Similar studies have been performed in rat liver[33] and in *Drosophila melanogaster.*[34] Ribosomal proteins L10 and S3 were labeled the first case and ribosomal proteins L2, L17, S8, and S22 in the second case.

Sparsomycin Group—Under this title a number of inhibitors of peptide bond formation in prokaryotic and eukaryotic ribosomes having relatively similar structures are considered. It includes sparsomycin, amicetin, gougerotin, blasticidin S, actinobolin, and anthelmycin. All of them, except actinobolin, carry a modified uracyl moiety in the molecule and they are mutually exclusive in respect to their binding sites on the ribosome. They interact at the peptidyl transferase A site inhibiting the binding of the acceptor substrate and, at the same time stimulate the binding of the donor substrate at the P site, albeit to a different extent depending on the antibiotic.[1]

Sparsomycin is the best known drug of this group and has recently been the subject of detailed studies (for reviews see References 35 and 36). A structure-activity relationship analysis has been performed taking advantage of the total synthesis of the drug[37-40] which has allowed the preparation of a large number of sparsomycin analogs carrying single structural alterations.[41-44]

The data indicate that the drug probably interacts with the ribosome by hydrogen bonding through the O and N atoms in the uracyl group. It has also been found that a decrease in the hydrophilicity at the S-Me end of the molecule results in a dramatic increase of its inhibitory activity[41, 42, 45] resulting in derivatives that are up to tenfold more active than sparsomycin and which are of great therapeutic interest as antitumor drugs.[46] The stimulatory effect of the hydrophobic substituents is related to an increase in the affinity of the drug for the ribosomes, probably due to appearance of an additional interaction site at a hydrophobic region of the peptidyl transferase center. This region must correspond to the postulated recognition site for the hydrophobic side chains of amino acids. Indeed, based on structural comparisons, it has been proposed that, like puromycin, sparsomycin might be an analog of aminoacyl-adenine of the 3' end of the acceptor tRNA[45, 47] and according to this model, the hydrophobic substituents of sparsomycin are, in fact, in an equivalent position to the lateral amino acid chain in aminoacyl-tRNAs as in puromycin (see for details References 35 and 36).

The negative effect of the O-demyethylation, which increases the hydrophilicity of the puromycin aminoacyl residue, on the drug's inhibitory activity[29] supports the proposed model and stresses the role of the hydrophobic region in the interaction of these compounds with the ribosome.

The competitive character of the inhibition of peptide bond formation by sparsomycin using puromycin as an acceptor molecule[45, 48] is in agreement with the existence of closely related interaction sites for both antibiotics and therefore supports the proposed model for the mode of action of sparsomycin. However, it is also clear that not all the sparsomycin effects, like the stimulatory effect of the binding at the P-site as well as its structural distortion[49] can be explained simply by competitive inhibition (see Reference 35 for a lengthy discussion of this subject).

Structural considerations have also led to the proposal of an interesting model of covalent interaction of sparsomycin with either a ribosomal component or the growing peptide through the sulfoxide function present in the drug molecule.[47] However, experiments carried out using radiolabeled analogs have clearly shown the noncovalent character of the drug interaction with the ribosomal particle.[49a]

Anisomycin Group—Anisomycin is the best known representative of a group of compounds including the trichothecene antibiotics (the best known being trichodermin), the Amaryllidaceae alkaloids headed by narciclasine, the *Cephalotaxus* alkaloids represented by harringtonine, bruceantine and tenuazonic acid. They block peptide bond formation exclusively in eukaryotic cells by interacting with the 60S ribosomal peptidyl transferase at closely related binding sites.[1]

A competitive inhibition of the puromycin reaction by anisomycin as well as an effect on the binding of substrates to the A- and P-sites have been reported, although it is not totally clear which of the different steps in peptide bond formation, substrate interaction, and/or transfer reaction are inhibited by these drugs. They interact with the ribosome in a mutually exclusive way and in addition, mutants resistant to any one of them show cross-resistance to some of the others. Interestingly, these mutants are hypersensitive to sparsomycin.[50] A similar phenotype has been found in resistant mutants of the anisomycin sensitive archaebacteria *Methanobacterium formicicum*[51] and in the ribosomes from *Myrothecium verrucaria* the producer of verrucarin, a trichothecene antibiotic.[52]

There are data, however, indicating that the ribosomal interaction site is not the same for all these compounds, and for instance, it has been shown that the *Cepahlotaxus* alkaloids, bruceantin and some members of the trichotecene family like verrucarin A, are unable to inhibit polysomes, probably because their binding sites are not accessible in these structures. Moreover, only lycorine and pseudolycorine bind tightly to trichodermin resistant ribosomes.[1, 2]

Taking advantage of the anisomycin sensitivity of *Halobacterium* species having single-copy rRNA genes it has been possible to obtain anisomycin-resistant mutants that have an altered rRNA. Thus, G_{2466}, C_{2471}, and A_{2472} have been found mutated in the 23S ribosomal RNA of different *Halobacterium halobium* and *Halobacterium cutirubrum* anisomycin-resistant mutants.[53] These bases are located in a highly conserved region of domain V, which is supposed to be part of the peptidyl transferase center, where other mutations which give resistance to a number of other bacterial peptide bond formation inhibitors have been mapped (see Reference 54 for a review). Based on these results a unitarian model for the mode of action of different peptidyl transferase inhibitors, including chloramphenicol, puromycin, and anisomycin, has been proposed which considers that part of their molecules can mimic the peptide bond.[55]

An attempt to map directly the trichodermin binding site by affinity labeling has been carried out using chemically reactive derivatives of the drug in *Drosophila melanogaster* ribosomes. Labeling of ribosomal proteins L1, L3, and L25 in the large subunit as well as some small subunit proteins, including protein S8 which is labeled also by puromycin, has been reported.[56]

Structure-activity relationship of bruceantin have been studied using chemically modified derivatives, indicating an important role of the lactone ring and of a double bond in ribosome interaction.[57]

15.2.3. Translocation

Two functionally homogeneous groups of translocation inhibitors have been described.[1] One included the well known antibiotic cycloheximide together with the less studied pederine and the second is formed by cryptopleurine, emetine, tutbulosine, and the *Tylophora* alkaloids tylocrebrin and tylophorine.

Cycloheximide and Pederine—Cycloheximide has pleiotropic effects *in vivo* as well as in *in vitro* systems. In the cells it affects not only protein but also DNA and RNA synthesis. These effects are probably related to the inhibition of protein synthesis since, in yeast, cycloheximide resistance has been associated with the presence of resistant ribosomes and is linked to a single mutation in L29, a large ribosomal subunit protein.[58] However a direct effect on RNA polymerase I *in vitro* has been reported.[1] In model systems the drug affects enzymic and nonenzymic translocation but also inhibits polysome formation suggesting a possible effect on initiation.

Pederine, like cycloheximide, has also multiple effects on model systems but, similarly, it very efficiently blocks the translocation step in model systems.

Both drugs interact with the 60S ribosomal subunit, but their binding sites must be different since they do not show cross-resistance.[59]

Cryptopleurine Group—All the compounds included in this group inhibit the enzymic translocation model reaction by interacting with the small ribosomal subunit. At high concentrations they also inhibit peptide bond formation.[1] Mutants of yeast and Chinese hamster ovary cells resistant to cryptopleurine show cross-resistance to the other members of the group confirming the close relation of their binding sites.[60, 61]

The interaction of cryptopleurine with the ribosome has been directly studied using radioactive drugs.[62] One high affinity binding site ($K_d = 3.9 \times 10^{-8} M$) was found in the 40S subunit which is responsible for the effect at the translocation step. Several low affinity sites were detected in the large subunit as well that might be the cause of the inhibition of peptide bond formation reported at high concentrations. All the drugs in this group compete with cryptopleurine for binding to ribosomes, although tylophorine and tylocrebrine are better competitors than emetine and tubulosine, suggesting the existence of different although closely related binding sites for the last two compounds.[1]

Aminoglycoside Antibiotics—Some aminoglycoside antibiotics have been shown to inhibit and induce mistranslation in eukaryotic systems but at higher concentrations than in bacterial systems (for an extensive review see Reference 3) although *Tetrahymena* sp. seem to be especially sensitive to them.[63] Recent *in vivo* experiments have confirmed the accumulation and biological effects of these drugs in mammalian cells.[64, 65]

A mechanism to explain the pleiotropic effects of these drugs on bacterial cells based on their mistranslation activity has been proposed[66] that in principle could be extended to eukaryotic cells as well. Similarly, an attempt to explain the mode of action of aminoglycosides according to the allosteric three-site model for aminoacyl-tRNA binding to bacterial ribosomes has been carried out,[67] but at this moment it is difficult to evaluate its applicability to eukaryotic systems.

Resistance to paromomycin and hygromycin associated to point mutations in the 17S rRNA have been reported.[68] The changes take place near the 3' end of the rRNA, in a highly conserved region where alterations in prokaryotic aminoglycoside-resistant mutants have also been found.[54]

15.2.4. Termination

Only one drug, lycorine, an alkaloid from the Amaryllidaceae, has been reported to produce effects in a HeLa *in vivo* system which are compatible with an inhibition of polypeptide chain termination.[69] As indicated previously, lycorine is also an inhibitor of peptide bond formation.

15.2.5. Plant and Microbial Toxins

A large number of proteins and glycoproteins from different microorganisms and plants have been shown to inhibit protein synthesis in eukaryotic cells by blocking the chain elongation step (see Reference 4 for a review). From a functional point of view they can be divided into two groups. Those that inactivate the elongation factor EF-2 and those that inactivate the ribosome.

Members of the first group are diphtheria toxin produced by *Corynebacterium diphtheriae* and PA toxin produced by *Pseudomonas aeruginosa* which enzymatically inactivate the factor, thereby producing ADP-ribosyl-EF-2. The modified factor is able to form a complex with GTP but it has a very low affinity for the ribosome as well as a low rate of GTP hydrolysis and is therefore very inefficient in promoting translocation.

All the plant and fungal toxins are in the ribosome inactivating group. Some bacterial toxins, like the Shiga toxin from *Shigella dysenteriae*[70] can also be included in this group. These toxins can be composed of either only one single polypeptide, the fungal α-sarcin being the best studied example, or two or more subunits like the plant toxins ricin and abrin. In the latter case one subunit carries the protein synthesis inactivating activity and the rest is required for cell penetration. The toxin-inactivated ribosomes are defective in the enzymatic binding of aminoacyl-tRNA and depending on the toxin, and possibly on the system used, the elongation factor EF-2-dependent reactions can be affected as well. Thus ricin-treated ribosomes seem unable to perform translocation while α-sarcin only affects the aminoacyl-tRNA binding. The bacterial Shiga toxin effects are similar to those of α-sarcin.[71]

In all cases studied the ribosome inactivation caused by these toxins has been associated with an alteration on the ribosomal RNA. The enzymic cleavage of about 320 nucleotides of the large subunit rRNA from yeast ribosomes by α-sarcin was the first report showing a toxin-associated RNAase activity.[72] Afterwards, it was confirmed that α-sarcin indeed had an unspecific RNAase activity when tested on naked RNA[73] but is highly specific, cleaving exclusively the phosphodiester bond between G_{4325} and A_{4326} of the 28S rRNA when acting on ribosomes[74] (also Part I, Chapter 1 of this book).

The fact that rRNA from ricin-treated ribosomes was intact when tested by SDS gel electrophoresis[75] ruled out the existence of an endonuclease activity associated to the toxin. Recently, Endo et al.[76] have shown that ricin cleaves the N-glycosidic bond of A_{4324} in rat liver 28S rRNA releasing the base and producing a labile polynucleotide carrying a phosphodiester bond which is very sensitive to mild chemical treatments. Other plant toxins have been shown to have similar N-glycosidase activity.[77-79]

Since the sites of action of the different toxins on the large ribosomal rRNA are very closely related, their similar effects on the ribosomal activity are not surprising. However, there are enough differences in the resulting modification (for example, α-sarcin cleaves physically to rRNA while ricin leaves the phosphodiester chain intact) to explain the peculiarities of each specific toxin. In fact, it has been shown that the conformation effect caused by α-sarcin and ricin on the inactivated ribosome is not the same.[80]

15.3. RESISTANCE TO ANTIBIOTICS

Mutants resistant to some antibiotics have been artificially induced in eukaryotic organisms, usually yeast and CHO cells. In most cases the resistance has been related to the alteration of a ribosomal protein. In yeast, the resistance is associated with proteins L29, L3, and rp59 for cycloheximide, tricodermin, and cryptopleurine[58, 81, 82] and the loci are located in chromosomes VII, XV, and III, respectively.[83]

A phenotypic resistance to several antibiotics associated with ribosomal protein phosphorylation upon starvation in *Tetrahymena thermophila* has been described.[84] The resistance seems to be due to a conformational change resulting from the phosphorylation of a single small subunit protein but it affects the large subunit as well since resistance to anisomycin and cycloheximide is induced.

In some highly resistant strains carrying dominant cycloheximide resistance mutations, however, no ribosomal protein alterations have been detected and the possibility of an unspecified ribosome enzymatic modification has been considered.[85]

The high copy number of rRNA genes in higher cells precludes, in principle, the isolation by mutagenesis of mutations directly affecting ribosomal RNA. *Tetrahymena,* having a single active rRNA gene is probably an exception and resistance to some aminoglycosides associated to rRNA mutations has been reported in this organism.[68] Taking advantage of the presence of only one rRNA gene in some archaebacteria sensitive to eukaryotic antibiotics it has been possible to obtain resistant mutants having rRNA alterations, as commented for anisomycin.[53] Interesting new results using this approach can be forecasted.

Naturally occurring antibiotic resistance, mostly in antibiotic producing organisms, is frequently associated with inducible alterations in the rRNA, generally methylations. This fact has allowed the characterization of modified nucleotides probably implicated in the ribosomal binding sites for a number of bacterial inhibitors.[54] Analogous situations possibly occur in the case of eukaryotic organisms. In fact, the ribosomes of *Myrothecium verrucaria* the producer of verrucarin have been shown to be resistant to the antibiotic.[52] In addition, the existence of an inducible, although biochemically undefined, cycloheximide resistance mechanism in *Candida maltosa*,[86] subsequently cloned in *Saccharomyces cerevisiae*,[87] has been reported. This is, however, a rather unexplored but very promising field.

15.4. INHIBITORS AS RESEARCH TOOLS

In addition to being used as tools for the study of the protein synthetic machinery, inhibitors are also extensively used as probes for the dependence of biological processes on active protein synthesis. This is a useful approach which is, however, limited by the specificity of the selected inhibitor and it should be kept in mind that totally specific inhibitors are scarce. Most of them are able to affect, either directly or indirectly, other biological processes in addition to protein synthesis, especially at high concentrations or when acting for a prolonged period of time, and these conditions are difficult to control in some *in vivo* experiments. Some inhibitors are however better than others, and from this point of view it is unfortunate that the most extensively used inhibitor so far has been cycloheximide which reportedly affects, as commented above, DNA and RNA synthesis as well. Lately, however, the more specific anisomycin has tended to be used in studies of this type.[88–90] The most specific inhibitors of protein synthesis in eukaryotic cells are probably the plant and microbial toxins which would be the inhibitors of choice when feasible.

In any case, the strong interdependence of all the metabolic pathways make the interpretation of some *in vivo* experiments difficult. In this sense, it is interesting to note that the well characterized RNA synthesis inhibitor actinomycin has been shown to inhibit very specifically, although indirectly, the initiation of protein synthesis.[91]

15.5. FUTURE DEVELOPMENTS

Basic research in the field of eukaryotic protein synthesis inhibitors will have probably to be focused in the near future on two main topics. On one hand, the structural aspects of the inhibitor-ribosome interaction, mainly the identification and characterization of the binding sites, will have to follow the path opened by the investigation of the prokaryotic counterparts. Direct identification of the binding site components by affinity labeling, protection methods and analysis of naturally and artificially induced resistant strains will be required. These studies are especially valuable in the case of broad spectrum inhibitors that also act in bacterial systems, since a correlation of the results in each system will be extremely useful.

On the other hand, a re-evaluation of the data available on the modes of action of many inhibitors is required in light of new developments in the understanding of the process of protein synthesis. Thus, the confirmation of the three-site model for the binding of tRNA to the ribosome will require additional experimental data to complement our present view of the mode of action of many inhibitors that interfere with that process. Such a re-evaluation has been already started in the case of inhibitors of bacterial systems.[67]

An interesting application of ribosomal antibiotic resistance to the phylogenetic study of archaebacteria has provided useful information for the systematization of this heterogeneous kingdom.[92] The eukaryotic organisms that traditionally have been supposed to have very homogeneous biological properties are becoming more and more heterogeneous as new uncommon species are being studied. An analogous application of the different antibiotic sensitivity of their ribosomes, i.e., lower eukaryotes seem to be resistant to tenuazonic acid,[1] might be of interest as an additional tool for phylogenetic analysis.

15.6. TECHNIQUES USED IN THE PRESENT ANTIBIOTIC RESEARCH

rRNA methylation by enzymes inducing antibiotic resistance and identification of the methylated residues.[93] Affinity labeling of antibiotic binding sties.[32, 94] Protection of rRNA chemical modification by antibiotics.[22] Determination of binding constants by equilibrium dialysis.[95] Radioactive labeling of antibiotics by mild iodination.[96]

ACKNOWLEDGMENTS

I thank Dr. D. J. Holmes and Dr. E. Cundliffe for reading and commenting on the manuscript and Fundacion Ramon Areces for support through an Institutional grant to the Centro de Biologia Molecular.

REFERENCES

1. Vazquez, D: Inhibitors of protein synthesis. *Mol. Biol. Biochem. Biophys.* 30:1 (1979).
2. Gale EF, Cundliffe E, Reynolds PE, Richmond MH, and Waring MJ: *The Molecular Basis of Antibiotic Action,* London, John Wiley & Sons, 402 (1981).
3. Vazquez D, Zaera, E, Dölz H, and Jimenez A: in Perez-Bercoff R (Ed): *Protein Biosynthesis in Eukaryoties.* New York, Plenum Press, 311 (1982).
4. Jimenez A and Vazquez D: Plant and fungal protein and glycoprotein toxins inhibiting eukaryotic protein synthesis. *Annu. Rev. Microbiol.* 39:649 (1985).
5. Umezawa H, Okami Y, Hashimoto T, Suhara Y, and Takeuchi T: A new antibiotic, kasugamycin. *J. Antibiot.* 18:101 (1965).
6. Poldermans B, Goosen N, and van Knippenberg PH: Studies on the function of two adjacent N^6, N^6-dimethyladenosines near the 3′ end of 16S ribosomal RNA of *Escherichia coli*. I. Effect of kasugamycin on initiation of protein synthesis. *J. Biol. Chem.* 254:9085 (1979).
7. van Knippenberg PH: Structural and functional aspects of the N^6,N^6 dimethyladenosines in 16S ribosomal RNA, in Hardesty B and Kramer G (Eds): *Structure, Function and Genetics of Ribosomes,* New York, Springer-Verlag, 412 (1986).
8. Raué HA, Klootwijk J, and Musters W: Evolutionary conservation of structure and function of high molecular weight ribosomal RNA. *Prog. Biophys. Mol. Biol.* 51:77 (1988).
9. Montesano L and Glitz DG: Wheat germ cytoplasmic ribosomes. Structure of ribosomal subunits and location of N^6,N^6-Dimethyladenosine by immunoelectron mycroscopy. *J. Biol. Chem.* 263:4932 (1988).
10. Wittmann HG: in Hardesty B and Kramer G (Eds): *Structure, Function and Genetics of Ribosomes,* New York, Springer-Verlag, 1 (1986).
11. Oakes M, Henderson E, Scheinman A, Clark AAM, and Lake JA: in Hardesty B and Kramer G (Eds): *Structure, Function and Genetics of Ribosomes.* New York, Springer-Verlag, 48 (1986).
12. Kurylo-Borowska Z and Szer W: Inhibition of bacterial DNA synthesis by edeine. Effect on *Escherichia coli* mutants lacking DNA polymerase I. *Biochim. Biophys. Acta* 287:236 (1972).
13. Kozak M: Migration of 40S ribosomal subunits on messenger RNA when initiation is perturbed by lowering magnesium or adding drugs. *J. Biol. Chem.* 254:4731 (1979).
14. Herrera F, Moreno N, and Martinez JA: Increased affinity for mRNA causes resistance to edeine in a mutant of *Saccharomyces cerevisiae. Eur. J. Biochem.* 145:339 (1984).
15. Herrera F, Triana L, and Bosch I: Importance of polysomal mRNA-associated polypeptides for protein synthesis initation in yeast. *Eur. J. Biochem.* 175:87 (1988).
16. Moazed D and Noller HF: Interaction of antibiotics with functional sites in 16S rbosomal RNA. *Nature* 327:389 (1987).
17. Tejedor F, Amils R, and Ballesta JPG: Photoaffinity labeling of the pactamycin binding site on eukaryotic ribosomes. *Biochemistry* 24:3667 (1985).

18. Tejedor F, Amils R, and Ballesta JPG: Pactamycin binding site on archaebacterial and eukaryotic ribosomes. *Biochemistry* 26:652 (1987).
19. Synetos D, Amils R, and Ballesta JPG: Photolabeling of protein components in pactamycin binding site of rat liver ribosomes. *Biochim. Biophys. Acta* 868:249 (1986).
20. Tejedor F and Ballesta JPG: Ribosome structure: binding site of macrolides studied by phototaffinity labeling. *Biochemistry* 24:467 (1985).
21. Arevalo MA, Tejedor F, Polo F, and Ballesta JPG: Protein components of the arythromycin binding site in bacterial ribosomes. *J. Biol. Chem.* 263:58 (1988).
22. Stern S, Moazed D, and Noller HF: Structural analysis of RNA using chemical and enzymatic probing monitored by primer extension. *Methods Enzymol.* 164:481 (1988).
23. Fresno M and Vazquez D: Inhibitor effects of "cap" analogues on globin mRNA and encephalomyocarditis RNA transcription in reticulocyte cell-free sytems. *Eur. J. Biochem.* 103:125 (1980).
24. Juan-Vidales F, Saenz-Robles MT, and Ballesta JPG: Acidic proteins of the large ribosomal subunit in *Saccharomyces cerevisiae:* effect of phosphorylation. *Biochemistry* 23:390 (1984).
24a. Saenz-Robles MT, Remacha M, Vilella MD, Zinker S, and Ballesta JPG: The acidic ribosomal proteins as regulators of the eukaryotic ribosomal activity. *Biochem. Biophys. Acta* 1050:51 (1990).
25. Geigenmüller U and Nierhaus KH: Tetracycline can inhibit tRNA binding to the ribosomal P site as well as to the A site. *Eur. J. Biochem.* 161:723 (1986).
26. Goldman RA, Hasan T, Hall CC, Strycharz A, and Cooperman BS: Photoincorporation of tetracycline into *Escherichia coli* ribosomes. Identification of the major proteins photolabeled by native tetracycline and tetracycline photoproducts and implications for the inhibitory action of tetracycline on protein synthesis. *Biochemistry* 22:359 (1983).
27. Powers T, Changchien L-M, Craven GR, Noller HF: Probing the assembly of the 3' major domain of 16S ribosomal RNA. Quaternary interactions involving ribosomal proteins S7, S9 and S19. *J. Mol. Biol.* 200:309 (1988).
28. Carrasco L and Vazquez D: Ribosomal sites involved in binding of aminoacyl-tRNA and EF 2. Mode of action of fusidic acid. *FEBS Lett.* 32:152 (1973).
29. Vara J, Perez-Gonzalez A, and Jimenez A: Biosynthesis of puromycin by *Streptomyces alboniger.* Characterization of puromycin N-acetyl transferase. *Biochemisty* 24:8074 (1985).
30. Vince R, Daluge S, and Brownell J: Carbocyclic puromycin: synthesis and inhibition of protein biosynthesis. *J. Med. Chem.* 29:2400 (1986).
31. Nathans D and Neidle A: Structural requirements for puromycin inhibition of protein synthesis. *Nature* 197:1076 (1963).
32. Cooperman BS, Hall CC, Kerlavage AR, Weitzmann CJ, Smith J, Hassan T, and Freidlander JD: in Hardesty B and Kramer G (Eds): *Structure, Function and Genetics of Ribosomes.* New York, Springer-Verlag, 362 (1986).
33. Reboud AM, Dubost S, Buisson M, and Reboud JP: Photoincorporation of puromycin into rat liver ribosomes and subunits. *Biochemistry* 20:5281 (1981).

34. Gilly M and Pellegrini M: Puromycin photoaffinity labels small- and large-subunit proteins at the A site of the *Drosophila* ribosome. *Biochemistry* 24:5781 (1985).

35. Ottenheijm HCJ, van den Broek LAGM, Ballesta JPG, and Zylicz Z: Chemical and biological aspects of sparsomycin, an antibiotic from *Streptomyces*. *Prog. Med. Chem.* 23:219 (1986).

36. Ballesta JPG and Lazaro E: Peptidyl transferase inhibitors. Structure-activity relationships analysis by chemical modification. Proc. of the ASM Meeting on Ribosomes, Glacier Park (Montana), August 1989.

37. Helquist P and Shekhani MS: Total synthesis of (Rc)-Sparsomycin. *J. Am. Chem. Soc.* 101:157 (1979).

38. Ottenheijm HC, Liskamp RMJ, van Nispen SPJH, Boots HA, and Tijhuis MW: Total synthesis of the antibiotic sparsomycin: a modified uracil amino acid monoxodithioacetal. *J. Org. Chem.* 46:3273 (1981).

39. Liskamp RMJ, Zeegers HJM, and Ottenheljm HCJ: Synthesis and ring-opening reactions of functionalized sultines. A new approach to Sparsomycin. *J. Org. Chem.* 46:5408 (1981).

40. Hwang DR, Helquist P, and Shikhani MS: Total synthesis of (+) − sparsomycin. Approaches using cysteine and serine inversion. *J. Org. Chem.* 50:1264 (1985).

41. van den Broek LAGM, Liskamp RMJ, Colstee JH, Lieleveld P, Remacha M, Vazquez D, Ballesta JPG, and Ottenheijm HCJ: Structure-activity relationships of sparsomycin and its analogues. Inhibition of peptide bond formation in cell-free systems and of L1210 and bacterial growth. *J. Med. Chem.* 30:325 (1987).

42. van den Broek LAGM, Lazaro E, Zylicz Z, Fennis PJ, Missler FAN, Lieleveld P, Garzotto M, Wagener DJT, Ballesta JPG, and Ottenheijm HCJ: Lipophilic analogues of sparsomycin as strong inhibitors of protein synthesis and tumor growth. *J. Med. Chem.* 32:2002 (1989).

43. van den Broek LAGM, Fennis PJ, Arevalo MA, Lazaro E, Ballesta JPG: Lelieveld P, and Ottenheijm HCJ: The role of the hydroxymethyl function on the biological activity of the antitumor antibiotic sparsomycin. *Eur. J. Med. Chem.* 24:503 (1989).

44. Lazaro E, van den Broek LAGM, Ottenheijm HCJ, Lieleveld P, Ballesta JPG: Structure-activity relationships of sparsomycin: Modification at the hydroxyl group. *Biochemie* 69:849 (1987).

45. Ash RJ, Flynn GA, Liskamp RMJ, and Ottenheijm CHJ: Sulfoxide configuration in sparsomycin determines time-dependent and competitive inhibition of peptidyl transferase. *Biochem. Biophys. Res. Commun.* 125:784 (1984).

46. Zylicz Z: Sparsomycin and its Analogues. A Preclinical Study on Novel Anticancer Drugs. PhD. dissertation. University of Nijmegen, The Netherlands (1988).

47. Flynn GA and Ash RJ: Necessity of sulfoxide moiety for the biochemical properties of an analog of sparsomycin. *Biochem. Biophys. Res. Commun.* 114:1 (1983).

48. Coutsogeorgopoulos C, Miller JT, and Hann DM: Inhibitors of protein synthesis V. Irreversible interaciton of antibiotics with an initiation complex. *Nucleic Acids Res.* 2:1053 (1975).

49. Perez-Gosalbez M, Leon Rivera G, and Ballesta JPG: Affinity labeling of peptidyl transferase center using the 3' terminal pentanucleotide from amino acyl-tRNA. *Biochem. Biophys. Res. Commun.* 113:941 (1983).

49a. Lazaro E, San Felix A, van den broek LAGM, Ottenheijm HCJ, and Ballesta JPG: Interaction of the antibiotic sparsomycin with the ribosome. *Antimicrob. Agents Chemother.* 35:10 (1991).

50. Jimenez A and Vazquez D: Quantitative binding of antibiotics to ribosomes from a yeast mutant altered on the peptidyl-transferase center. *Eur. J. Biochem.* 54:483 (1975).

51. Hummel H and Böck A: Mutations in *Methanobacterium formicicum* confering resistance to anti-80S ribosome-targeted antibitotics. *Mol. Gen. Genet.* 198:529 (1985).

52. Hobden AN and Cundliffe E: Ribosomal resistance to the 12,13-epoxytrichotecene antibiotics in the producing organism *Myrothecium verrucaria. Biochem. J.* 190:765 (1980).

53. Hummel H and Böck A: 23 S ribosomal RNA mutations in halobacteria confering resistance to the anti-80S ribosome targeted antiobiotic anisomycin. *Nucleic Acids Res.* 15:2431 (1987).

54. Cundliffe E: Recognition Sites for Antibiotics within Ribosomal RNA. Proc. of the ASM Meeting on Ribosomes, Glacier Park, Montana, August 1989.

55. Vester B and Garret RA: The importance of highly conserved nucleotides in the binding region of chloramphenicol at the peptidyl transferase centre of *Escherichia coli* 23S ribosomal RNA. *EMBO J.* 7:3577 (1988).

56. Gilly M, Benson NR, and Pellegrini M: Affinity labeling the ribosome with eukaryotic-specific antibiotics: (bromoacetyl) trichodermin. *Biochemistry* 24:5787 (1985).

57. Considine C, Willingham W, Chaney SG, Wyrick S, Hall I, and Lee K-H: Structure-activity relationships for binding and inactivation of rabbit reticulocyte ribosomes by quassinoid antineoplastic agents. *Eur. J. Biochem.* 132:157 (1983).

58. Fried HM and Warner JR: Molecular cloning and analysis of yeast gene for cycloheximide resistance and ribosomal protein L29. *Nucleic Acids Res.* 10:3133 (1982).

59. Jimenez A, Carrasco L, and Vazquez D: Enzymic and non-enzymic translocation by yeast polysomes. Site of action of a number of inhibitors. *Biochemistry* 16:4727 (1977).

60. Grant PG, Schindler D, and Davies JE: Mapping of trichodermin resistance in *Saccharomyces cerevisiae:* a genetic locus for a component of the 60S ribosomal subunit. *Genetics* 83:667 (1976).

61. Gupta RS and Siminovitch L: Mutants of CHO cells resistant to protein synthesis inhibitors, cryptopleurine and tylocrebrine:genetic and biochemical evidence for common site of action of emetine, cryptopleurine, tylocrebrine and tubulosine. *Biochemisty* 16:3209 (1977).

62. Dölz H, Vazquez D, and Jimenez A: Quantitation of specific interaction of [14a-3H] Cryptopleurine with 80S and 40S ribosomal species from the yeast *Saccharomyces cerevisiae. Biochemistry* 21:3181 (1982).

63. Eustice DC and Wilhelm JM: Mechanisms of action of aminoglycoside antibiotics in eukaryotic protein synthesis. *Antimicrob. Agents Chemother.* 26:53 (1984).

64. Buss WC, Piatt MK, and Kauten R: Inhibition of mammalian microsomal protein synthesis by aminoglycoside antibiotics. *J. Antimicrob. Chemother.* 14:231 (1984).
65. Buchanan JH, Stevens A, and Sidhu J: Aminoglycoside treatment of human fibroblasts: intracellular accumulation, molecular changes and loss of ribosomal accuracy. *Eur. J. Cell Biol.* 43:141 (1987).
66. Davis BD: Mechanism of bactericidal action of aminoglycosides. *Microbiol. Rev.* 51:341 (1987).
67. Hausner T-P, Geigenmüller U, and Nierhaus KH: The allosteric three-site model for the elongation ribosomal elongation cycle. New insights into the inhibition mechanisms of aminoclycosides, thiostrepton and viomycin. *J. Biol. Chem.* 263:13103 (1988).
68. Spangler EA and Blackburn EH: The nucleotide sequence of the 17S ribosomal RNA gene of *Tetrahymena thermophila* and identification of point mutations resulting in resistance to antibiotics paromomycin and hygromycin. *J. Biol. Chem.* 260:6334 (1985).
69. Vrijsen R, van den Berghe D, Vlietinck AJ, and Boeye A: Lycorine: a eukaryotic termination inhibitor? *J. Biol. Chem.* 261:505 (1986).
70. Reisbig R, Olsnes S, and Eiklid K: The cytotoxic activity of Shigella toxin. Evidence for catalytic inactivation of the 60S ribosomal subunits. *J. Biol. Chem.* 256:8739 (1981).
71. Obrig TG, Moran TP, and Brown JE: The mode of action of Shiga toxin on peptide elongation of eukaryotic protein synthesis. *Biochem. J.* 244:287 (1987).
72. Schindler D and Davies JE: Specific cleavage of ribosomal RNA caused by α-sarcin. *Nucleic Acids Res.* 4:1097 (1977).
73. Endo Y, Hubert PW, and Wool IG: The ribonuclease activity of the cytotoxin a-sarcin. *J. Biol. Chem.* 258:2662 (1983).
74. Endo Y and Wool IG: The site of action of α-sarcin on eukaryotic ribosomes. *J. Biol. Chem.* 257:9054 (1982).
75. Mitchell SJ, Hedblom M, Cawley D, and Houston LL: Ricin does not act as an endonuclease on L cell polysomal RNA. *Biochem. Biophys. Res. Commun.* 68:763 (1976).
76. Endo Y, Mitsui K, Motizuki M, and Tsurugi K: The mechanism of action of ricin and related toxic lectins on eukaryotic ribosomes. *J. Biol. Chem.* 262:5908 (1987).
77. Endo Y and Tsurugi K: RNA N-glycosidase activity of ricin A chain. Mechanism of action of the toxic lectin ricin on eukaryotic ribosomes. *J. Biol. Chem.* 262:8128 (1987).
78. Endo Y, Tsurugi K, and Franz H: The site of action of the A-chain of mistletoe lectin I on eukaryotic ribosomes. *FEBS Lett.* 231:378 (1988).
79. Stirpe F, Bailey S, Miller SP, and Bodley JW: Modification of ribosomal RNA by ribosome-inactivating proteins from plants. *Nucleic Acids Res.* 1349 (188).
80. Terao K, Uchiumi T, Endo Y, and Ogata K: Ricin and a-sarcin alter the conformation of 60S ribosomal subunits at neighboring but different sites. *Eur. J. Biochem.* 174:429 (1988).
81. Fried HM and Warner JR: Cloning of yeast gene for trichodermin resistance and ribosomal protein L3. *Proc. Natl. Acad. Sci. USA* 78:238 (1981).

82. Larkin JC and Woolford JL Jr: Molecular cloning and analysis of the CRY1 gene: a yeast ribosomal protein gene. *Nucleic Acids Res.* 11:403 (1983).

83. Mortimer RK, Schild D, Contopoulou CR, and Kans JA: Genetic map of *Saccharomyces cerevisiae. Yeast* 5:321 (1989).

84. Hallberg RL, Wilson PG, and Sutton C: Regulation of ribosome phosphorylation and antibiotic sensitivity in *Tetrahymena thermophila:* a correlation. *Cell* 26:47 (1981).

85. Fleming GH, Boynton JE, and Gillham NW: The cytoplasmic ribosomes of *Chlamydomonas reinhardtii:* characterization of antibiotic sensitivity and cycloheximide-resistant mutants. *Mol. Gen. Genet.* 210:419 (1987).

86. Takagi M, Kawai S, Tukata Y, and Tanaka N, Sunairi M, Miyazaki M, and Yana K: Induction of cycloheximide resistance in *Candida maltosa* by modifying the ribosomes. *J. Gen. Appl. Microbiol.* 31:267 (1985).

87. Takagi M, Kawai S, Shibuya I, Miyazaki M, and Yano K: Cloning in *Saccharomyces cerevisiae* of a cycloheximide resistance gene from the *Candida maltosa* genome which modifies ribosomes. *J. Bacteriol.* 168:417 (1986).

88. Milton AS and Sawhney VK: The effects of the protein synthesis inhibitor anisomycin on the febrile responses to intracerebroventricular injections of bacterial pyrogen, arachidonic acid and prostaglandin E2. *Naunym Schmied. Arch Pharmacol.* 336:332 (1987).

89. Inouye ST, Takahashi JS, Wollnik F, and Turek FW: Inhibitor of protein synthesis phase shifts a circadian pacemaker in mammalian SCN. *Am J. Physiol.* 255:R1055 (1988).

90. Frey U, Krug M, Reymann KG, and Matthies H: Anisomycin, an inhibitor of protein synthesism blocks late phases of LTP phenomena in hippocampal CA1 region *in vitro. Brain Res.* 452:57 (1988).

91. Kostura M and Craig N: Treatment of chinese hamster ovary cells with the transcriptional inhibitor actinomycin D inhibits binding of mRNA to ribosomes. *Biochemistry* 25:6384 (1986).

92. Amils R and Sanz JL: In Hardesty B and Kramer G (eds): *Structure, Function and Genetics of Ribosomes,* New York, Springer-Verlag, 605 (1986).

93. Beauclerk AAD and Cundliffe E: Sites of action of two ribosomal methylases responsible for resistance to aminoglycosides. *J. Mol. Biol.* 193:661 (1987).

94. Cooperman BS: Affinity labeling of ribosomes. *Methods Enzymol.* 164:341 (1988).

95. Teraoka H and Nierhaus KH: Measurement of binding of antibiotics to ribosomal particles by means of equilibrium dialysis. *Methods Enzymol.* 59:862 (1979).

96. Tejedor F and Ballesta JPG: Iodination of biological samples without loss of functional activity. *Anal. Biochem.* 127:143 (1982).

Chapter

16

Nomenclature of Initiation, Elongation and Termination Factors for Translation in Eukaryotes
Recommendations 1988*

Brian Safer
National Insitutes of Health
Bethesda, Maryland

In the 10 years since a system for naming eukaryotic initiation, elongation, and termination factors was agreed by several research groups,[1] knowledge of their structures and functions has advanced substantially, and the original proposals fall short of meeting present needs. A new system is therefore recommended, which conserves the original one as far as possible.

* These are recommendations of the Nomenclature Committee of the International Union of Biochemistry (NC-IUB), whose members are J. F. G. Vliegenthart (Chairman), C. R. Cantor, M. A. Chester, C. Liébecq (representing the IUB Committee of Editors of Biochemical Journals), W. Saenger, N. Sharon and P. Venetianer. UC-IUB thanks a panel whose members are J. Hershey (USA), W. C. Merrick (USA), B. Safer (USA, convenor) and H. O. Voorma (The Netherlands) for drafting these recommendations. NC-IUB also thanks members of the IUPAC-IUB Joint Commission on Biochemical Nomenclature (JCBN) and former members of NC-IUB and JCBN, namely, H. Bielka, J. R. Bull, A. Cornish-Bowden, H. B. F. Dixon, P. Karlson, K. L. Loening, G. P. Moss, J. Reedijk, E. J. Van Lenten and E. C. Webb, for consultation. Comments may be sent to any member of NC-IUB, or to its secretary, M. A. Chester, BioCarb AB, S-22370 Lund, Sweden, or to the convenor of the panel, B. Safer, Laboratory of Molecular Hematology, National Heart, Lung and Blood Institute, Building 10, Room 7D18, National Institutes of Health, 9000 Rockville Pike, Bethesda, Maryland, USA 20892.

16.1. GENERAL PRINCIPLES OF SYMBOLISM

The three types of eukaryotic translation factors are designated eIF, eEF, and eRF for initiation factors, elongation factors, and termination (release) factors, respectively. The initial 'e' stands for eukaryotic.

16.2. INITIATION FACTORS

16.2.1. Main Groups of Initiation Factors

Eukaryotic initiation factors are grouped according to the major functions served, with a number assigned to each group analogous to the established nomenclature for prokaryotic initiation factors. Symbols are obtained by prefixing the appropriate number of eIF, e.g., eukaryotic initiation factor 4 is symbolized as eIF-4. Descriptive names implying a specific functional role are avoided, because it is likely that in some cases such names will prove misleading or incorrect. The five groups are as follows:

eIF-1: factors designated as eIF-1, eIF-1A etc. are involved with pleiotropic stimulation of initiation complex assembly.

eIF-2: the designation of initiation factors as eIF-2, eIF-2A, etc. indicates that the primary function is to facilitate the binding of Met-tRNA$_i$ to ribosomes.

eIF-3: factors designated as eIF-3, eIF-3A, etc. have as their primary function the formation of native ribosomal subunits.

eIF-4: the designation of an initiation factor as eIF-4, eIF-4A, etc. indicates that the primary function is to facilitate the binding of mRNA to ribosomes.

eIF-5: initiation factors designated as eIF-5, eIF-5A etc. have as their primary function the facilitation of ribosomal subunit joining and the positioning of Met-tRNA$_i$ for synthesis of the first peptide bond.

16.2.2. New Initiation Factors

As new initiation factors are identified they will receive the next available letter designation within the correct group. To avoid confusion, new names should not be the same as obsolete names from the previous classification.[1]

Thus new factors related to eIF-4 should not be designated as eIF-4C, eIF-4D, eIF-4E or eIF-4F (which have been used previously), but rather as eIF-4G, eIF-4H, etc.

16.2.3. Subunits of Initiation Factors

Subunits of initiation factors are designated α, β, γ —- starting with the smallest subunit. Individual subunits that can be purified but that do not function independently of the factor complex do not receive a separate name. Thus the α subunit of eIF-4, previously called eIF-4E (see [24] in Table 1) is designated eIF-4α. Once prefixes have been assigned, subsequent revision of the subunit molecular masses will not alter the designations. For example, the designation of eIF-3α will remain unchanged even if future work were to show it as larger than eIF-3β.

16.3. ELONGATION FACTORS

There are two groups of elongation factors, eEF-1 and eEF-2, corresponding to the prokaryotic EF-Tu · Ts complex and EF-G, respectively. The first of these consists of three subunits designated eEF-1α, eEF-1β and eEF-1γ, of which eEF -1α corresponds to prokaryotic EF-Tu and the eEF-1βγ complex to prokaryotic EF-Ts.

16.4. TERMINATION FACTOR

One eukaryotic termination factor is known, designated eRF.

16.5. LIST OF KNOWN MAMMALIAN FACTORS

Eukaryotic initiation, elongation and termination factors known at the time of these recommendations are listed in Table 16-1. Whenever possible, corresponding factors isolated from other eukaryotes should use an equivalent nomenclature.

16.6. FUTURE REVISION

It is hoped that this approach will provide a flexible framework to accommodate new activites as they are found and to correlate factors studied in different laboratories. It is recognized that significant revision may be required as new information becomes available.

TABLE 16–1.
Initiation, Elongation and Termination Factors for Translation in Eukaryotes

Factor	Activities	Complete Molecule Molecular mass (kDa)	pI	Subunits Molecular mass (kDa)	pI	Previous Nomenclature
eIF-1	Stimulates formation and stabilizes 43S and 48S preinitiation complexes	15		15		eIF-1 (1)[a], IF-E1 (2)
eIF-1A	Stimulates formation and stabilizes 43S and 48S preinitiation complexes		5.6	17.6	5.6	IF-E7 (2), IF-M2Bβ (3), eIF-4C (4)[b]
eIF-2	GTP-dependent Met-tRNA$_1$ binding to 43S ribosomal subunits	130	6.4	α 36.1 β 38.4 γ 55.3	5.1 5.4 8.9	IF-MP (5), IF-1 (6,7), IF-E2 (2,8,9), EIF-3 (9)
eIF-2A	AUG-dependent Met-tRNA$_1$ binding to 40S ribosomal subunits	65	9.0	65	9.0	eIF-2A (1)[a], IF-M1 (10)
eIF-2B	GTP:GDP exchange	272		α 26 β 39 γ 58 δ 67 ε 82		eRF (11), GEF (12), ESP (13), RF (14), SP (15), sRF (16), anti-HRI (17), SF (18), Co-eIF 2 (19)
eIF-3	Formation of native 43S ribosomal subunits	550	6.7	α 35 β 36 γ 40 δ 44 ε 47 ζ 66 η 115 θ 170	5 5.85 6.45 6.25 5.7 5.3 6.7	IF-3 (1)[a], IF-E3 (2), IF-M5 (20), eIF-3 (21)
eIF-3A	Formation of native 60S ribosomal subunits	25		25		eIF-6 (23)[b]

Factor	Function					Other designations
eIF-4	RNA-dependent ATPase; mRNA binding to 43S preinitiation complex; recognition of mRNA 5' cap structure	270		α 25.1[c] β 44.4[c] γ 220[c]	6.2 5.75 6.6	CBP-II (24), eIF-4F (25), eIF-4E (26)[b], CBP-I (24)
eIF-4A	RNA-dependent ATPase; mRNA binding to 43S preinitiation complex	44.4	5.75	44.4	5.75	eIF-4A (1)[a], IF-E4 (2), IF-M4 (20), IF$_{EMC}$ (27)
eIF-4B	Stimulates mRNA binding to 43S preinitiation complex	160		80	5.8	eIF-4B (1)[a], IF-E6 (2), IF-M3 (28)
eIF-4F[d]	RNA-dependent ATPase; mRNA binding to 43S preinitiation complex; recognition of mRNA 5' cap structure			α 25.1[d] β 44.4 γ 220	6.2 5.75 6.6	CBP-II (24), eIF-4F (25)
eIF-5	GTP-dependent release of bound factors upon 60S ribosomal subunit joining to the 43S preinitiation complex	150	6.4	150	6.4	eIF-5 (1)[a], IF-E5 (2), IF-M2A (29), IF-II (30), IF-L2 (31), IF-3 (32)
eIF-5A	Stimulates ribosomal subunit joining and enhances 80S-bound Met-tRNA$_i$ reactivity with puromycin	15	6.1	16.7	6.1	eIF-4D (1)[b], IF-M2Bα (33)
eEF-1	Binding of aminoacyl-tRNA	(150)$_n$ $\alpha\beta\gamma$ complex	6.4	α 51 β 48 γ 30	9.2 5.9 6.8	EF-1 (34)
eEF-1α	Binding of aminoacyl-tRNA	51		51	9.2	EF-1 (34), EF-1α (35)
eEF-1$\beta\gamma$	Recycling of eEF-1α	80		β 48 γ 30	5.9 6.8	EF-1 (34), EF-1$\beta\gamma$ (36)
eEF-2	Translocation	95	6.8	95	6.8	EF-2 (37)
eRF	Chain termination and polypeptide release	110		55		RF (38)

[a]Designations maintained from the previously adopted nomenclature.[1]

[b]Designations changed from the previous nomenclature.[1]

[c]The designation of eIF-4 as the three non-identical subunits shown represents the view of nearly all experts consulted but there was a minority opinion that it is premature and possibly incorrect. The designation of eIF-4 (see below) may also be used.

[d]The designation of a single factor responsible for mRNA binding as eIF-4 cannot be made at this time. For the purpose of accurate communication, the widely used designation eIF-4F is retained. It is recognized, however, that subunit stoichiometry reported in the literature has been variable, possibly reflecting complex-dependent as well as complex-independent activities of the individual subunits.

REFERENCES

1. Anderson WF, Bosch L, Cohn WE, Lodish H, Merrick WC, Weissbach H, Wittmann HG, and Wool G: *FEBS Lett.* 76:1 (1977).
2. Schreier MH, Erni B, and Staehelin, T: *J. Mol. Biol.* 116:727 (1977).
3. Merrick WC, Kemper WM, and Anderson WF: *J. Biol. Chem.* 250:5556–5562 (1975).
4. Thomas, A, Goumans H, Voorma HO, and Benne, R: *Eur. J. Biochem.* 107:39 (1980).
5. Safer B, Anderson WF, and Merrick WC: *J. Biol. Chem.* 250:9067 (1975).
6. Gupta NK, Woodley CL, Chen YC, and Bose KK: *J. Biol. Chem.* 248:4500 (1973).
7. Dettman GL and Stanley WM Jr: *Biochim. Biophys. Acta* 299:142 (1973).
8. Schreier MH and Staehelin T: *Nature (London)* 242:35 (1973.
9. Ranu RS and Wool IG: *J. Biol. Chem.* 251:1926 (1976).
10. Merrick WC and Anderson WF: *J. Biol. Chem.* 250:1197 (1975).
11. Salimans M, Goumans, H, Amesz H, Benne R, and Voorma HO: *Eur. J. Biochem.* 145:91 (1984).
12. Siekierka J, Manne V, and Ochoa A: *Proc. Natl. Acad. Sci. U.S.A.* 81:352 (1984).
13. de Haro C and Ochoa S: *Proc. Natl. Acad. Sci. U.S.A.* 76:2163 (1979).
14. Matts RL, Levin DH, and London IM: *Proc. Natl. Acad. Sci. U.S.A.* 80:2559 (1983).
15. Siekierka J, Manne V, Mauser L, and Ochoa S: *Proc. Natl. Acad. Sci. U.S.A.* 80:1232 (1983).
16. Ralston RO, Das A, Dasgupt A, Roy R, Palmieri S, and Gupta NK: *Proc. Natl. Acad. Sci. U.S.A.* 75:4858 (1978).
17. Amesz H, Goumans H, Haubrich-Morree T, Voorma HO, and Benne R: *Eur. J. Biochem.* 98:513 (1979).
18. Ranu RS and London IM: *Proc. Natl. Acad. Sci. U.S.A.* 76:1079 (1979).
19. Bagchi M, Bannerjee A, Roy R, Chakrabarty I, and Gupta NK: *Nucleic Acids Res.* 10:6501 (1982).
20. Safer B, Adams SL, Kemper WM, Berry KW, Lloyd M, and Merrick WC: *Proc. Natl. Acad. Sci. U.S.A.* 73:2584 (1976.
21. Thompson HA, Sadnik I, Scheinbuks J, and Moldave K: *Biochemistry* 16:2221 (1977).
22. Heywood S, Kennedy DS, and Bester AJ: *Proc. Natl. Acad. Sci. U.S.A.* 17:2478 (1974).
23. Russell DW and Spremulli LL: *J. Biol. Chem.* 254:8796 (1979).
24. Tahara S, Morgan MA, and Shatkin AJ: *J. Biol. Chem.* 256:7691 (1981).
25. Grifo JA, Tahara SM, Morgan MA, Shatkin AJ, and Merrick WC: *J. Biol. Chem.* 258:5804 (1983).
26. Altman M, Edery I, Sonenberg N, and Trachsel H: *Biochemistry* 24:6085 (1985).
27. Wigle DT and Smith AE: *Nature New Biol.* 242:136 (1973).
28. Prichard PM and Anderson WF: *Methods Enzymol.* 30:136 (1974).
29. Merrick WC, Kemper WM, and Anderson WF: *J. Biol. Chem.* 250:5556 (1975).
30. Cashion LM and Stanley WM Jr: *Proc. Natl. Acad. Sci. U.S.A.* 71:436 (1974).

31. Levin PH, Kyner D, and Acs G: *J. Biol. Chem.* 248:6416 (1973).
32. Gupta NK, Woodley CL, Chen YC, and Bose KK: *J. Biol. Chem.* 248:4500 (1973).
33. Kemper WM, Berry KW, and Merrick WC: *J. Biol. Chem.* 251:5551 (1976).
34. Miller DL and Weisbach H: in Weissbach H and Pestka S (eds): *Molecular Mechanisms of Protein Synthesis* New York, Academic Press, 323 (1977).
35. Iwasaki K and Kaziro Y: *Methods Enzymol.* 60:657–676 (1979).
36. Lauer SJ, Burks E, Irving JD, and Ravel JM: *J. Biol. Chem.* 259:1644 (1984).
37. Bielka H and Stahl J: in Arnstein HRV (ed): *Amino acids and protein synthesis II* Baltimore, University Park Press, 79 (1978).
38. Tate WP, Beaudet AL, and Caskey CT: *Proc. Natl. Acad. Sci. U.S.A.* 70:2350 (1973).

INDEX

A

aaRS, see Aminoacyl-tRNA synthetases
Abrin, 383
Acceptor tRNA, 363, 380
Acidic proteins (A proteins), 11
Actin, 88
Actinobolin, 379
Actinomycin, 80, 385
Adenosine diphosphate (ADP), 163
Adenosine diphosphate (ADP)-ribosyl-elon-
    gation factor 2, 336, 383
Adenosine monophosphate (AMP), 59
    cyclic, see cAMP
Adenosine triphosphatase (ATPase), 97,
    122, 124, 126, 154
Adenosine triphosphate (ATP), 43, 129,
    131, 213
    affinity for, 44, 122
    antagonistic effect of, 218
    binding of, 202, 211
    covalent modification and, 237
    decrease in cellular, 297
    depletion of, 132
    hydrolysis of, 112, 117, 122, 124–126,
        283
    inhibition and, 195
    initiation factor interactions with, 124
    as initiation inhibitor, 193
    Met-tRNA binding and, 210, 211, 216–
        218
    peptide chain termination and, 181
    requirements for, 109, 112, 117, 122
Adenovirus, 209, 214, 274, 277, see also
    specific types
Adipocytes, 300
ADP, see Adenosine diphosphate
Adrenochrome, 377
Affinity labeling, 386
Alanine, 50
Alanylation, 55
Alarmone biosynthesis, 49–50
Albumin, 83
Alfalfa mosaic virus (AMV), 114, 115,
    120, 305
Alkylation, 103
Amicetin, 379
Amino acids, 293, 301, see also specific
    types

aminoacyl-tRNA synthetases and, 41
    availability of, 247
    biosynthesis of, 48–49, 247
    deficiency of, 295
    elongation and, 327
    regulation of biosynthesis of, 48–49
    in ribosomes, 4–9
    transfer of, 59
Amino acid starvation, 260, 294–296, 310
Aminoacylation, 45, 47
    glutamate, 48
    incomplete, 59
    in vivo, 50
    of tRNA, 49, 51–60, 237
        aminoacyl-tRNA synthetases and, 52–
            58
        incomplete, 59
        as kinetic process, 51–52
        metabolic processes and, 60
Aminoacylation plateau theory, 59
Aminoacyl-tRNA, 35, 149, 179
    binding of, 325
        aminoglycosides and, 382
        elongation inhibition and, 378–379
        to ribosomes, 153–155
        toxins and, 383
    concentration of, 332
    elongation and, 332–334
    ribosome interaction with, 154
    synthetases of, see Aminoacyl-tRNA
        synthetases
Aminoacyl-tRNA ligases, see Aminoacyl-
    tRNA synthetases
Aminoacyl-tRNA synthetases, 40–51
    alanine binding to, 50
    in alarmone biosynthesis, 49–50
    amino acid deficiency signal and, 295
    amino acids and, 41, 48–49
    autoantibodies against, 50
    in autoregulation, 47–48
    bacterial, 50
    biosynthesis of, 50–51
    bonding of to uncharged tRNA, 260
    in chlorophyll biosynthesis, 48
    compartmentation of, 339
    core, 45–46
    covalent modification and, 237–238
    defined, 40

cAMP-stimulated kinases, 302
Cap-binding factor, 300
Cap-binding protein (CBP), 82, 121, 275–277
Capping, 112–115, 243, 275
Carboxyl groups of tyrosine, 43
Cardioviruses, 274, 277, see also specific types
Casein kinases, 153, 198, 202, 234, 236, 237
Catalysis, 15, 26, 46, 47
CBP, see Cap-binding protein
cDNA, 5, 9, 78, 81, 232
CDP-ethanolamine, 158
Cell differentiation, 302
Cell-free translational systems, 111
Cell growth, 299–303
Cell lysates, 132, see also specific types
Cell proliferation, 299, 301
Cell wall peptides, 60
cGMP-dependent protein kinase, 236
Charging, 58–60, 112, 295
Chloramphenicol, 375
Chlorophyll biosynthesis, 48
Chloroplastic mRNA, 74
Cis-acting RNA, 262–263
Cis-trans model for translation control, 261–269
Competitive concentrations of mRNA, 127
Consensus RNA-binding, 123
Consensus sequence, 118
Covalent modification, 20, 231–238
Coxsackievirus, 274
CPAI, 245–246
Cross-linking, 80, 100, 124, 156, 364
  efficiency of, 276
  of initiation factor 4, 276
  of mRNP, 83
  photochemical, 121, 125
  of polypeptides, 88
  of RNA and proteins, 80, 81, 83
  of 4-thiouridine, 49
  of tRNA to aminoacyl-tRNA synthetases, 53–54
Cross-pathway control system, 51
Cryptopleurines, 381, 382, see also specific types
CSK, see Cytoskeletal framework
Cyclic AMP, see cAMP
Cyclic DNA, see cDNA
Cyclins, 307, 309
Cycloheximide, 87, 236, 274, 342, 381–382, 384, 385

Cytokines, 293, 299, 302, see also specific types
Cytoplasmic calcium, 298
Cytoplasmic kinases, 301
Cytoplasmic mRNA, 243
Cytoplasmic mRNP, 79–81, 84
Cytoplasmic structure of tRNA, 36–38
Cytosine arabinoside, 85
Cytoskeletal fraction, 128
Cytoskeletal framework (CSK), 71, 86, 88
Cytoskeleton, 86–88
Cytosol, 262
Cytotoxins, 154, see also specific types

D

DAI, 209–215, 302
  discovery of, 193
  interferon and latent, 211–213
  viruses and, 213–215
Daudi cells, 302
Deacylase, 204
Deacylation, 59
Deadenylation, 310
O-Demethylation, 380
2-Deoxy-glucose, 296
Dephosphorylation, 197, 205, 208, 211, 221–222, 231, 337
Dimethyl sulfoxide, 84
Diphthamide, 162, 336
Diphtheria, 336, 383
Discriminatory factor, 127
Disulfide bonds, 299
DNA
  automated synthesizers of, 54
  blocking of synthesis of, 85
  c (cyclic), 5, 9, 78, 81, 232
  cessation of synthesis of, 76
  cyclic, 5, 9, 78, 81, 232
  genomic, 9
  inhibition of synthesis of, 75, 84
  recombinant, 365
  repair of, 50
  synthesis of, 75, 84, 382, 385
  transcription of, 243
Double-stranded, see ds
Downstream ORF, 280, 283
dsRNA, 193, 302
dsRNA-activated inhibitor, see DAI
dsRNA-activated protein kinase, 209–211
Dyes, 377, see also specific types